VOLUME NINETY EIGHT

# CURRENT TOPICS IN DEVELOPMENTAL BIOLOGY

# Transcriptional Switches During Development

VOLUME NINETY EIGHT

# CURRENT TOPICS IN DEVELOPMENTAL BIOLOGY

# Transcriptional Switches During Development

*Edited by*

SERGE PLAZA AND FRANÇOIS PAYRE
*Centre de Biologie du Développement*
*Université Paul Sabatier*
*CNRS UMR 5547*
*Toulouse, France*

ELSEVIER

AMSTERDAM • BOSTON • HEIDELBERG • LONDON
NEW YORK • OXFORD • PARIS • SAN DIEGO
SAN FRANCISCO • SINGAPORE • SYDNEY • TOKYO
Academic Press is an imprint of Elsevier

Academic Press is an imprint of Elsevier
525 B Street, Suite 1900, San Diego, CA 92101-4495, USA
225 Wyman Street, Waltham, MA 02451, USA
32, Jamestown Road, London NW1 7BY, UK
Linacre House, Jordan Hill, Oxford OX2 8DP, UK

First edition 2012



ISBN: 978-0-12-386499-4
ISSN: 0070-2153

For information on all Academic Press publications
visit our website at elsevierdirect.com

Printed and bound in USA

12 13 14 9 8 7 6 5 4 3 2 1

# Contents

# Contributors

**Stein Aerts**
Laboratory of Computational Biology, Center for Human Genetics, Katholieke Universiteit (K.U.) Leuven, Leuven, Belgium

**Scott Barolo**
Department of Cell & Developmental Biology, University of Michigan Medical School, Ann Arbor, Michigan, USA

**Laetitia Bataillé**[1]
GReD INSERM UMR1103, CNRS UMR6293, University of Clermont-Ferrand, Clermont-Ferrand, France

**Philip N. Benfey**
Department of Biology and Duke Center for Systems Biology, Duke University, Durham, North Carolina, USA

**Ken M. Cadigan**
Department of Molecular, Cellular and Developmental Biology, University of Michigan, Ann Arbor, Michigan, USA

**Lionel Christiaen**
Department of Biology, Center for Developmental Genetics, New York University, New York, New York, USA

**Albert J. Courey**
Department of Chemistry & Biochemistry and Molecular Biology Institute, University of California, Los Angeles, California, USA

**Cyrille de Joussineau**
GReD INSERM UMR1103, CNRS UMR6293, University of Clermont-Ferrand, Clermont-Ferrand, France

**Carlos Estella**[2]
Department of Biochemistry and Molecular Biophysics, Columbia University, New York, New York, USA

[1] Present address: Developmental Biology Centre, CNRS UMR5547/Université Paul Sabatier, Toulouse, France
[2] Current address: Fundación Instituto Valenciano de Infertilidad (FIVI), Valencia University, and Instituto Universitario IVI/INCLIVA, Valencia, Spain

**Nicole C. Evans**
Department of Cell & Developmental Biology, University of Michigan Medical School, Ann Arbor, Michigan, USA

**Bassem A. Hassan**
Laboratory of Neurogenetics, VIB Center for the Biology of Disease, VIB; Center for Human Genetics, KU Leuven School of Medicine, Leuven, Belgium, and Janelia Farm Research Campus, Howard Hughes Medical Institute, Ashburn, Virginia, USA

**Krzysztof Jagla**
GReD INSERM UMR1103, CNRS UMR6293, University of Clermont-Ferrand, Clermont-Ferrand, France

**Teresa Jagla**
GReD INSERM UMR1103, CNRS UMR6293, University of Clermont-Ferrand, Clermont-Ferrand, France

**Richard S. Mann**
Department of Biochemistry and Molecular Biophysics, Columbia University, New York, New York, USA

**Cédric Maurange**
Aix-Marseille Univ, IBDML, and CNRS, UMR7288, Marseille, France

**Miguel A. Moreno-Risueno**
Department of Biology and Duke Center for Systems Biology, Duke University, Durham, North Carolina, USA

**Xiao-jiang Quan**
Laboratory of Neurogenetics, VIB Center for the Biology of Disease, VIB, and Center for Human Genetics, KU Leuven School of Medicine, Leuven, Belgium

**Ariane Ramaekers**
Laboratory of Neurogenetics, VIB Center for the Biology of Disease, VIB, and Center for Human Genetics, KU Leuven School of Medicine, Leuven, Belgium

**Christos Samakovlis**
Department of Developmental Biology, Wenner-Gren Institute, Stockholm University, Stockholm, Sweden

**Christina I. Swanson**[3]
Department of Cell & Developmental Biology, University of Michigan Medical School, Ann Arbor, Michigan, USA

[3] Present address: Department of Biology, University of North Carolina, Chapel Hill, North Carolina, USA

**Wiam Turki-Judeh**
Department of Chemistry & Biochemistry and Molecular Biology Institute, University of California, Los Angeles, California, USA

**Jaimie M. Van Norman**
Department of Biology and Duke Center for Systems Biology, Duke University, Durham, North Carolina, USA

**Roumen Voutev**
Department of Biochemistry and Molecular Biophysics, Columbia University, New York, New York, USA

**Shenqiu Wang**[4]
Department of Developmental Biology, Wenner-Gren Institute, Stockholm University, Stockholm, Sweden

**Wei Wang**
Department of Biology, Center for Developmental Genetics, New York University, New York, New York, USA

[4] Present address: Cancer Genetics Branch, National Human Genome Research Institute, National Institutes of Health, Bethesda, Maryland, USA

# Preface

Fifty years ago, Francois Jacob and Jacques Monod published a seminal study in the *Journal of Molecular Biology* on the role of "regulator genes" in controlling the expression of "structural genes" in bacteria. This turned out to accurately define the transcription factors that act to determine the subset of genes to be expressed in each cell, in interaction with the external milieu. If these findings were rapidly recognized as a milestone, the contributions of this paper to collective scientific thought are even more impressive with regard to recent functional genomics. The authors already proposed in 1961 "*that the genome contains not only a series of blueprints, but a coordinated program of protein synthesis and the means of controlling its execution.*" This represents an extraordinary vision of the way biological systems are hardwired by genomic information, a central concept in our current view of physiology, development, and evolution.

Genetic analyses in model systems later provided an unbiased portal into the molecular control of embryonic development, showing the importance of transcription factors and signaling pathways that regulate their outputs. For example, Eric Wieschaus and Christiane Nüsslein-Volhard showed that most mutations affecting the embryonic pattern in *Drosophila* alter regulator genes. The discovery of so-called homeotic genes by Ed Lewis and their products (which contain the homeodomain DNA-binding motif) by Walter Gehring and colleagues further demonstrate that transcription factors govern both body plan establishment and the final differentiation of tissues and organs. Three decades later, major advances have been made in our comprehension of how various protein motifs mediate specific DNA recognition, exemplified by recent successes in designing artificial proteins to target a given genomic region. Deciphering the genome of hundreds of species has revealed that transcription factors are an abundant class of molecules (representing 5–10% of the total number of proteins), displaying unexpected levels of evolutionary conservation. It bears reminding that, even in genetically amenable systems, the function of most transcription factors has yet to be discovered.

While the general concepts of transcriptional regulation are now well established, a major challenge resides in understanding the logic and physical elements implementing these regulatory interactions within a given cell in a developing organism. A central problem to be addressed is how transcriptional programs are set up and then modified throughout the successive steps of embryonic development. This volume provides a broad range of insights into the molecular determinants of such transcriptional switches during development.

Far from attempting to be exhaustive, the following chapters summarize recent efforts to tackle this question in various model systems, each with their advantages and limitations. These contributions also show that different scales of analysis are required together to solve the problem of gene expression control during development. We believe that the combination of both genome-wide approaches (experimental or predictive) and functional analyses scrutinizing the intimate details of each element is the most promising way to unravel the control of terminal differentiation and its evolutionary diversification.

A general idea is that a single transcription factor can influence gene expression either positively or negatively, depending on the cell state and in interactions with neighboring tissues. Chapter 1 by Ken M. Cadigan presents an up-to-date view of the mechanisms regulating TCF activity upon receiving Wnt signal. While the growing diversity of TCF partners can partially explain how the same signal triggers different transcriptional responses in various tissues, recent work also reveals the existence of profound variations in how Wnt signaling can regulate TCF-dependent gene expression. In Chapter 2, Shenqiu Wang and Christos Samakovlis focus on Grainy head transcription factors, which play an evolutionarily conserved role in epithelial differentiation and wound repair. Grainy head proteins can function as activators or repressors, a switch that is effected by different protein isoforms. Recent studies show that Grainy head activity is further controlled by posttranslational modifications in response to extracellular signaling, for example, for its function during tissue regeneration upon epithelial injury.

Coactivators and corepressors, while essential components of transcription complexes, are often viewed as auxiliary factors that only refine the information provided by transcription factors, since they themselves lack sequence-specific DNA-binding domains. Wiam Turki-Judeh and Albert J. Courey in Chapter 3 review compelling evidence that Groucho represents a corepressor that instead exhibits instructive properties during development. In the same vein, Groucho is subjected to various posttranslational modifications, thus switching the activity of transcriptions factors in response to signal transduction pathways.

Transcription factors regulate gene expression by binding to discrete regions called enhancers or cis-regulatory modules (CRMs). These CRMs are composed of clustered binding sites for various transcription factors collectively required for gene expression. A debated question in the field is whether CRM activity relies only on the number and type of different binding sites (billboard model) or whether sites need to be organized in a specific arrangement, or grammar (enhanceosome model). In Chapter 4, Scott Barolo and colleagues report on recent works focusing on a CRM taken from the Drosophila dPax2 gene, which reveals unexpected functional properties that challenge the current view of enhancer structure and organization. It also highlights the importance of number versus affinity of binding sites for a given transcription factor. A considerable challenge still ahead is to identify the register of genomic regions that harbor CRMs. Transcription factors recognize short DNA motifs that occur with high frequency in the genome. It is with difficulty that we are learning to discriminate, as transcription factors do, which among the huge number of total motifs correspond to functional CRMs. Stein Aerts in Chapter 5 surveys approaches aimed at locating functional CRMs using bioinformatics. Other genome-wide approaches, such as ChIP-Seq, often reveal myriad *in vivo* binding sites for a transcription factor, and this chapter shows how computational studies can complement experimental approaches to recognize functional cis-regulatory elements. Novel experimental systems are also likely to provide a way to identify a larger number of animal CRMs. In Chapter 6, Wei Wang and Lionel Christiaen review the potential that lies in the use of simple chordate species of the *Ciona* genus for high-throughput discovery of CRMs. Similar approaches can then be used for the functional dissection of CRMs and the identification of cis-elements mediating their activity.

The expression of transcription factors is itself controlled at the transcriptional level, leading to complex gene networks comprising cascades of regulatory interactions. Reviewing leg morphogenesis in flies, Richard S. Mann and colleagues in Chapter 7 discuss how the analysis of key CRMs can bring specific insights into how these networks define a spatial pattern of gene expression. Recent studies also shed light on the often-overlooked importance of temporal control of transcriptional programs. Cédric Maurange in Chapter 8 reviews a series of works showing that neural stem cells are sequentially specified by a series of transcription factors in *Drosophila*. This chapter also explains how this temporal sequence implements regulatory loops that govern waves of differentiation competence. Chapter 9 by Philip N. Benfey and colleagues shows that several aspects of plant development also rely on both spatial and temporal regulatory interactions to switch the fate of differentiating cells and build a coherent pattern. They also discuss novel data on the regulatory cues leading to oscillating gene expression, which is required for positioning lateral roots during postembryonic development.

One particularly puzzling aspect of studying the transcriptional control of development is that most of the downstream targets of transcription factors that have been identified so far are, themselves, transcription factors. It is possible that our current view has been biased by direct genetic approaches, relying on the observation of gross phenotypical abnormalities to infer regulatory interaction. One can speculate that, by definition, the inactivation of a single "structural gene" would often lead to a subtle phenotype when compared to that observed in the absence of a "regulator gene" controlling the expression of hundreds of genes. Chapters 10 and 11 address this question and show how alternative multidisciplinary strategies can be used to identify the cellular effectors directly responsible for changing cell behavior in response to regulatory cascades. Bassem A. Hassan and colleagues focus on the transcriptional control of retinal differentiation in flies. They review recent work that succeeded in defining cellular effectors downstream of the transcription factor Atonal, linking upstream regulatory cascades to terminal differentiation. Krzysztof Jagla and colleagues provide a detailed view of our current understanding of muscle differentiation in the fly embryo and how upstream regulatory cascades are parsed into control of discrete subsets of cellular effectors, which contribute to diversification of muscle types.

This volume on transcriptional switches during development brings to specialists an updated snapshot of each of the fields covered by the different chapters. It may also illustrate for younger colleagues and students the general view that decoding the regulatory genome is one of the most exciting and promising directions to explore experimentally. Besides the intellectual challenge of unraveling the complexity and intimate logic of this code, it represents a key step toward understanding how biological information is used for normal development and how it can be altered in various pathological states.

Serge Plaza and François Payre
Centre de Biologie du Développement,
Toulouse, France

CHAPTER ONE

# TCFs and Wnt/β-catenin Signaling: More than One Way to Throw the Switch

Ken M. Cadigan

## Contents



Department of Molecular, Cellular and Developmental Biology, University of Michigan, Ann Arbor, Michigan, USA

*Current Topics in Developmental Biology*, Volume 98
ISSN 0070-2153, DOI: 10.1016/B978-0-12-386499-4.00001-X

## Abstract

Wnts are conserved, secreted signaling proteins that can influence cell behavior by stabilizing β-catenin. Accumulated β-catenin enters the nucleus, where it physically associates with T-cell factor (TCF) family members to regulate target gene expression in many developmental and adult tissues. Recruitment of β-catenin to Wnt response element (WRE) chromatin converts TCFs from transcriptional repressors to activators. This review will outline the complex interplay between factors contributing to TCF repression and coactivators working with β-catenin to regulate Wnt targets. In addition, three variations of the standard transcriptional switch model will be discussed. One is the Wnt/β-catenin symmetry pathway in *Caenorhabditis elegans*, where Wnt-mediated nuclear efflux of TCF is crucial for activation of targets. Another occurs in vertebrates, where distinct TCF family members are associated with repression and activation, and recent evidence suggests that Wnt signaling facilitates a "TCF exchange" on WRE chromatin. Finally, a "reverse switch" mechanism for target genes that are directly repressed by Wnt/β-catenin signaling occurs in *Drosophila* cells. The diversity of TCF regulatory mechanisms may help to explain how a small group of transcription factors can function in so many different contexts to regulate target gene expression.

## 1. Introduction to Wnts and Wnt/β-catenin Signaling

Wnts are cysteine-rich secreted proteins that mediate short-range paracrine signaling. Wnts can influence cell behavior by activating several signaling pathways, determined by a combination of particular Wnts and cell surface receptors (van Amerongen and Nusse, 2009). This review will focus on canonical Wnt (Wnt/β-catenin) signaling. In this pathway, Wnt stimulation leads to elevated levels of β-catenin in the nucleus, where it serves as a potent transcriptional coregulator (Cadigan and Peifer, 2009; MacDonald *et al.*, 2009). Wnt/β-catenin signaling is required throughout metazoans during development and in adult-tissue homeostasis and regeneration (Archbold *et al.*, 2011; Grigoryan *et al.*, 2008; Haegebarth and Clevers, 2009; Logan and Nusse, 2004; Nusse, 2008; Petersen and Reddien, 2009). In addition, aberrant Wnt/β-catenin signaling has been linked to several human diseases, most notably cancer (Clevers, 2006; Polakis, 2007) and bone disorders (Baron *et al.*, 2006).

The broad outlines of the Wnt/β-catenin pathway are shown in Fig. 1.1 and have been reviewed elsewhere (Cadigan and Peifer, 2009; MacDonald *et al.*, 2009). In the absence of signaling, β-catenin is bound and phosphorylated by a "destruction complex" containing the adenomatous polyposis coli (APC) and Axin proteins, as well as glycogen synthase kinase 3 (GSK3)

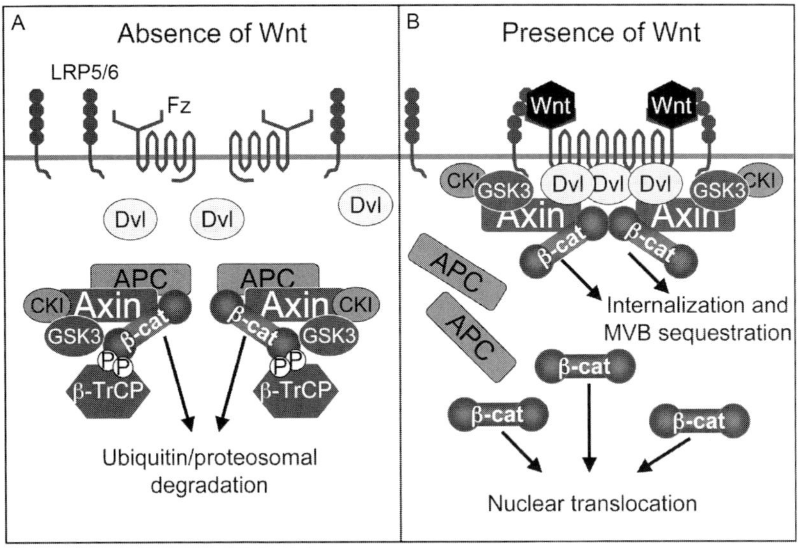


**Figure 1.1** Model of Wnt/β-catenin signaling in the absence (A) or presence (B) of Wnt stimulation. See text for further description. (For color version of this figure, the reader is referred to the Web version of this chapter.)

and casein kinase I (CKI). Phosphorylated β-catenin is bound by the F box protein Slimb/β-TrCP and polyubiquitinated, leading to proteosomal degradation. In addition, the complex acts to prevent nuclear localization of β-catenin (Brocardo and Henderson, 2008; Tolwinski and Wieschaus, 2004). Upon Wnt binding to Frizzled (Fz) and low-density lipoprotein-related proteins 5 and 6 (LRP5/6), GSK3, Axin, and other destruction complex components are recruited to the receptor complex (Cadigan and Peifer, 2009; MacDonald *et al.*, 2009), where they are internalized and sequestered in multivesicular bodies (MVBs) (Taelman *et al.*, 2010). Newly synthesized β-catenin escapes degradation and is free to enter the nucleus (Fig. 1.1B).

T-cell factor (TCF) proteins are major mediators of nuclear β-catenin, recruiting it to Wnt target gene chromatin. TCFs are defined by an N-terminal β-catenin-binding domain, along with a high mobility group (HMG) domain followed by a short stretch of basic residues (sometimes called the basic tail) that bind DNA specifically (Archbold *et al.*, 2011). In addition, almost all invertebrate TCFs and some vertebrate isoforms contain an additional DNA-binding domain called the C-clamp, which binds sequences adjacent to the HMG-binding site (Archbold *et al.*, 2011; Atcha *et al.*, 2007; Chang *et al.*, 2008). The *cis*-regulatory elements bound by TCFs that respond to Wnt signaling will be referred to hereafter as Wnt response elements (WREs). In the absence of β-catenin, TCFs often mediate transcription repression of target genes, in conjunction with corepressors.

When β-catenin binds to TCFs on WREs, it displaces corepressors and recruits coactivators, leading to increased transcription of target genes (Archbold *et al.*, 2011; Cadigan and Peifer, 2009).

This chapter will examine how β-catenin mediates the switch of TCFs from repression to activation. While models for the TCF switch often invoke a single TCF molecule mediating both sides of the switch in concert with various coregulators, this is likely to be an oversimplification. In *Caenorhabditis elegans*, Wnt signaling-dependent removal of TCF from the nucleus is essential for activation of many targets. In vertebrates, which contain multiple TCFs, the switch likely involves replacement of repressive TCFs with ones predisposed to transcriptional activation. In addition, some Wnt targets are directly repressed by the pathway, and the role of TCF in this mode of regulation will be described.

## 2. Historical Perspective on TCF's Dual Role in Regulating Wnt Targets

Shortly after the discovery of TCFs as important nuclear regulators of Wnt signaling (Behrens *et al.*, 1996; Molenaar *et al.*, 1996), several lines of evidence supported their dual role in regulating transcription. Mutation of the TCF-binding sites in the *siamois* WRE revealed that they are required for activation of WRE activity in dorsal blastomeres of *Xenopus* embryos, but they are also required for repression of the WRE in ventral blastomeres (Brannon *et al.*, 1997; Fan *et al.*, 1998). Complementing these results, genetic analysis of the *pangolin* (*pan*) gene in *Drosophila*, which encodes the fly TCF, revealed that it repressed Wingless (Wg, a fly Wnt) targets in the absence of signaling (Cavallo *et al.* 1998). *pan* mutants have a defect in embryonic epidermal patterning that is qualitatively similar as *wg* mutants, but less severe (Brunner *et al.*, 1997; van de Wetering *et al.*, 1997). Interestingly, *wg*, *pan* double mutants appear identical to *pan* single mutants (Cavallo *et al.* 1998). In other words, loss of *pan* suppresses the *wg* mutant phenotype, which can be explained by the fly TCF acting as a switch. Loss of Wg signaling leaves TCF in the repressive state, causing extremely low expression of Wg targets. But loss of *pan*, even in a *wg* mutant background, results in intermediate expression of these genes due to loss of TCF repression (known as derepression). In this way, *pan* mutants are conceptually similar to the *siamois* WRE reporter when the TCF sites were mutated: activation by Wnt signaling is compromised, but derepression results in an intermediate level of expression of the WRE or Wnt targets.

Studies of the *POP-1* gene, which encodes the *C. elegans* TCF, initially suggested that it acted solely through repression of Wnt targets. At the four-cell stage of nematode embryogenesis, the EMS blastomere is polarized by Wnt signaling, undergoing an asymmetric cell division producing MS (mesodermal progenitor) and E (endodermal progenitor) cells (Bowerman, 1995; Fig. 1.2A). Loss of Wnt signaling results in both daughter cells becoming MS cells (Rocheleau *et al.*, 1997; Thorpe *et al.*, 1997; Fig. 1.2B), while loss of *POP-1* (the worm TCF) has the opposite phenotype (Lin *et al.*, 1995; Fig. 1.2C). Wnt signaling inhibits nuclear localization of POP-1 in E cells (Rocheleau *et al.*, 1997; Thorpe *et al.*, 1997). Nuclear POP-1 represses transcription of *end-1* and *end-3*, which are required for the endoderm cell fate (Maduro *et al.*, 2002). However, further analysis of an *end-1* WRE reporter revealed that POP-1 is also required for transcriptional activation (Shetty *et al.*, 2005). Mutation of a single TCF-binding site in the *end-1* reporter resulted in derepression in the MS cell, but there was also a reduction in expression levels in the E cell (Shetty *et al.*, 2005; Fig. 1.2D). These results are reminiscent of the *siamois* WRE (Brannon *et al.*, 1997). Taken together, the data summarized in this section indicate that TCFs in a wide array of organisms both repress and activate Wnt target gene expression.

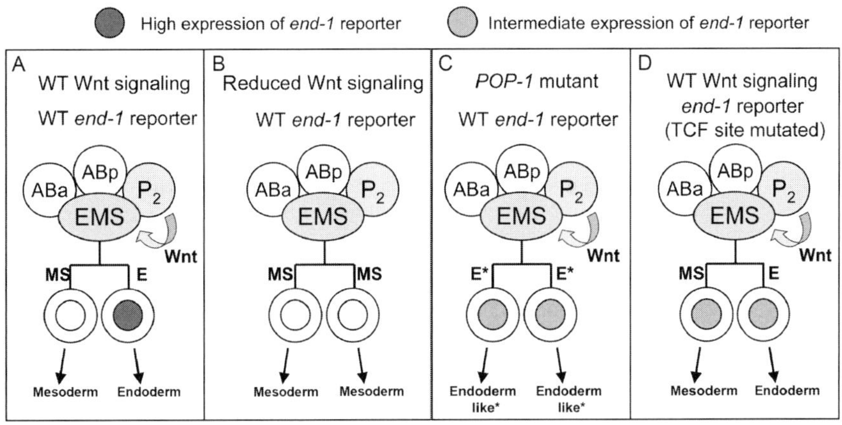


**Figure 1.2** The dual role of TCF in regulating Wnt target gene expression is exemplified by *end-1* regulation in the endodermal precursor cells of *C. elegans*. Cartoons are of four-cell stage embryos. (A) In wild-type (WT) embryos, Wnt signaling from the $P_2$ cell promotes an asymmetric division of the EMS cell. The daughter further away from $P_2$ develops into mesoderm (M), while the adjacent one develops into endoderm (E). A reporter containing the *end-1* WRE is normally expressed at high levels in E cells (Shetty *et al.*, 2005). (B) In mutants where Wnt signaling is reduced, endoderm specific genes are not expressed and both daughters adopt the M cell fate (Huang *et al.*, 2007; Shetty *et al.*, 2005). (C) In *POP-1* mutants, the M cell expresses endoderm genes and both daughters adopt an endoderm-like fate. However, expression of endoderm genes is lower than found in WT E cells (denoted as light green vs. dark green nuclei; Shetty *et al.*, 2005). (D) When a single TCF site is mutated in the *end-1* reporter in an otherwise WT embryo, the expression of the reporter is similar to panel C, despite M and E cell fates being correctly specified (Shetty *et al.*, 2005). These data support a model where POP-1 directly represses *end-1* in the M cell but is also required for full activation in the E cell. (For interpretation of the references to color in this figure legend, the reader is referred to the Web version of this chapter.)

## 3. Gro/TLEs Proteins and TCF Transcriptional Repression

How do TCFs repress Wnt targets in the absence of β-catenin? One important group of corepressors linked to TCF are members of the transducin-like enhancer of split (TLE) family. *Drosophila* has one TLE, known as Groucho (Gro), which serves as a corepressor for several transcription factors (Chen and Courey, 2000). Reduction of *gro* gene activity partially rescues the *wg* mutant phenotype, similar to *pan* mutations (Cavallo *et al.*, 1998). There are five *TLE* genes in mammals, known as *Groucho related genes* (*Grgs*) in mice (Jennings and Ish-Horowicz 2008). Grgs were identified as binding partners for TCF, and Grg4 strongly suppressed the ability of β-catenin to activate several Wnt readouts (Roose *et al.*, 1998). Thus Gro/TLEs appear to act as a classic corepressor for TCFs (Fig. 1.3B).

Gro/TLE proteins can be divided into five domains (Fig. 1.3A), the first two of which are involved in Wnt signaling. The conserved N-terminal glutamine-rich Q domain undergoes homotetramerization, known to be crucial for repression (Chen *et al.*, 1998; Song *et al.*, 2004) and is sufficient for binding to TCFs (Brantjes *et al.*, 2001; Daniels and Weis, 2005). The glycine/proline (GP) domain of Gro/TLEs promote interactions with histone deacetylases (HDACs; Brantjes *et al.*, 2001; Chen *et al.*, 1999). The deacetylation of histones is strongly correlated with transcriptional repression (Narlikar *et al.*, 2002). In addition to homooligomerization, the Q domain of Gro/TLEs binds to the central domain of TCFs, including the HMG domain (Arce *et al.*, 2009; Brantjes *et al.*, 2001; Daniels and Weis, 2005; Roose *et al.*, 1998; Fig. 1.3A).

The N-terminal portion of TCFs are necessary and sufficient for binding to β-catenin (Behrens *et al.*, 1996; Graham *et al.*, 2000; Molenaar *et al.*, 1996; Poy *et al.*, 2001; van de Wetering *et al.*, 1997). Therefore, the location of the Gro/TLE binding domain on TCFs does not immediately indicate that it would compete with β-catenin for binding. However, there is evidence that the central portion of TCF which partially overlaps the Gro/TLE binding domain also contributes to β-catenin binding (Daniels and Weis, 2005; Fig. 1.3A). Consistent with this, binding of the Q domain and β-catenin to lymphoid enhancer-binding factor 1 (LEF1, one of four human TCFs) are mutally exclusive (Daniels and Weis, 2005). This relationship is also apparent in studies of a *c-myc* WRE, where chromatin immunoprecipitation (ChIP) demonstrated that β-catenin and TLE1 cannot reside on the chromatin simultaneously (Sierra *et al.*, 2006). These data support a model where β-catenin and Gro/TLE compete for binding to TCFs, promoting transcriptional activation and repression, respectively (Fig. 1.3B and C).

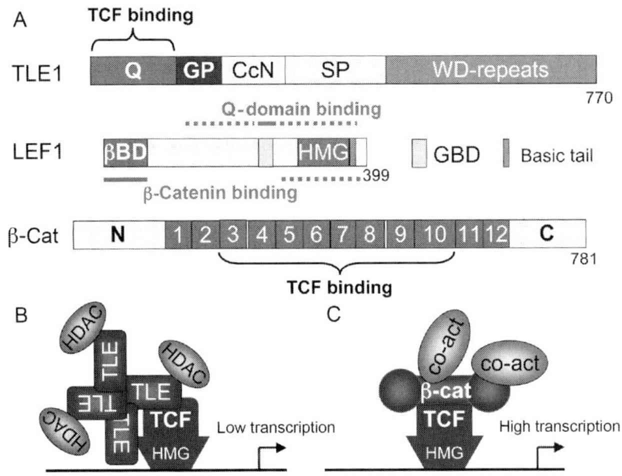


**Figure 1.3** Relationships between TCFs, β-catenin, and Gro/TLEs. (A) Diagram of the primary sequences of TLE1, LEF1, and β-catenin, with the regions implicated in direct protein–protein interactions indicated. Solid lines above and below LEF1 denote the primary regions that interact with Gro/TLE and β-catenin, respectively (Arce *et al.*, 2009). Dashed lines indicate secondary regions that contribute to binding. As found in other Gro/TLEs, TLE1 contains an N-terminal glutamine-rich domain (Q) that binds TCFs, a glycine–proline (GP) rich domain that binds HDACs, as well as a CcN, SP, and WD domains that may not be crucial for interactions with TCFs, but function with other factors (Jennings and Ish-Horowicz, 2008). Like other TCFs, LEF1 contains an N-terminal β-catenin-binding domain and an HMG domain followed by a basic tail. The HMG-basic tail region contains a secondary β-catenin-binding domain (Daniels and Weis, 2005) that partially overlaps with the region that binds to the Q domain of Gro/TLEs. β-Catenin contains 12 centrally located Arm repeats, flanked by an N-terminal domain that is phosphorylated and ubiquitinated by the destruction complex, and a C-terminal domain that recruits coactivators (some coactivators also bind to the Arm repeats). (B) Cartoon of a TCF–TLE–HDAC complex repressing a Wnt target gene. The Q domain of TLE interacts with TCF and other TLEs, while the GP domain recruits HDACs, which deacetylate the chromatin and promote transcriptional repression. (C) Cartoon of a TCF–β-catenin-coactivator complex activating a Wnt target. β-Catenin displaces TLE, causing derepression. In addition, β-catenin recruits a variety of coactivators that promote gene activation. (For color version of this figure, the reader is referred to the Web version of this chapter.)

Vertebrates also contain a shortened Gro/TLE family member containing only the Q and GP domains, known as amino-terminal enhancer of split (AES) in humans and Grg5 in mice (Beagle and Johnson, 2010). Grg5 can bind to TCFs but unlike the other Gro/TLEs, it potentiates β-catenin activation of TCF-dependent transcription (Brantjes *et al.*, 2001; Roose *et al.*, 1998). While the Q domain of Grg5/AES is competent for homo-tetramerization, the GP domain cannot bind to HDACs, suggesting a model where the shorter protein acts as a dominant negative to inhibit Gro/TLE activity (Brantjes *et al.*, 2001).

## 4. The Balance Between TCF Repression and Activation Is Not the Same in All Wnt Targets

While the TCF switch mechanism is a productive way to view Wnt/β-catenin transcriptional regulation, it should be noted that *cis*-regulatory elements are complex and contain considerable information besides TCF-binding sites that control expression. For example, the previously mentioned results where derepression of the *siamois* or *end-1* WRE reporters occurs when TCF sites are mutated (Brannon *et al.*, 1997; Fan *et al.*, 1998; Shetty *et al.*, 2005) implies the existence of other local activators that are now free to act in the absence of TCF repression (Barolo and Posakony, 2002).

The available data indicates that some WREs (e.g., *even-skipped* and *Ultrabithorax* in fly mesoderm) are similar to *siamois* and *end-1*, that is, TCF site mutants have reduced expression in cells where Wnt signaling normally activates the WRE, combined with derepression in adjacent cells (Knirr and Frasch, 2001; Riese *et al.*, 1997). But mutation of TCF sites in other WREs reveals little evidence for a repressive role for TCF, that is, loss of Wnt-induced expression is the only defect observed. These include WREs from *sloppy paired 1* (Lee and Frasch, 2000), *naked cuticle*, and *notum/wingful* (Chang *et al.*, 2008) in flies, *Brachyury* in mice (Yamaguchi *et al.* 1999), and *CEH-22* in *C. elegans* (Lam *et al.*, 2006). These WREs do not contain sequences that promote significant activation in the absence of TCF binding.

Conversely, there are WREs where the repressive role of TCF is the dominant determinant of expression levels. In a *decapentapleigic* (*dpp*) WRE that is normally expressed in a narrow region (where Wg is expressed) of the fly embryonic visceral mesoderm, TCF site mutations result in high levels of expression in the entire tissue (Yang *et al.*, 2000). For such WREs, the major role of the Wnt/β-catenin pathway is to relieve TCF-mediated repression, allowing other non-Wnt pathway factors to activate expression. While the balance of TCF repression versus TCF/β-catenin activation can vary widely among individual WREs, it is likely that both repression and activation by TCFs are crucial for regulating the full repertoire of Wnt targets in a particular tissue or organ.

## 5. Factors Contributing to TCF Repression

β-catenin stability and nuclear localization are major regulators of the transcription switch, but the control of these processes will not be considered here. Rather, this section addresses factors repressing TCF targets that fall into three broad groups: (1) those bound directly to TCFs, (2) those bound to WREs independently of TCFs, and (3) those that prevent β-catenin from interacting with TCFs.

## 5.1. Factors binding to TCFs

Several proteins interact with TCFs and appear to work analogously to Gro/TLEs. Corepressor of Pan (Coop) binds to the HMG domain of TCF/Pan in a manner that does not interfere with DNA binding (Song *et al.*, 2010). Like Gro/TLE, it competes with Arm/β-catenin for binding to TCF/Pan and inhibits Wg signaling in cell culture and fly tissues (Song *et al.*, 2010). Consistent with a classic corepressor, it inhibits Wg target genes in the absence of Wg signaling (Song *et al.*, 2010). In mammals, several myeloid translocation gene (MTG) family members bind to TCFs and prevent β-catenin binding and Wnt signaling (Moore *et al.*, 2008). MTGs are known to bind to HDACs and other corepressors (Linggi *et al.*, 2005) suggesting a similar mechanism as proposed for Gro/TLEs (Fig. 1.4A).

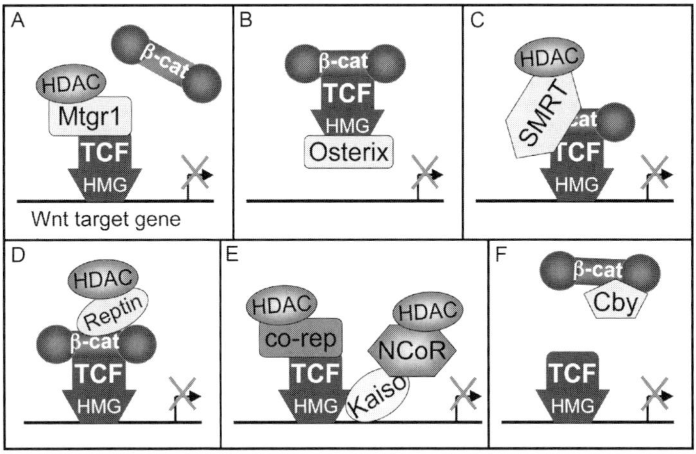


**Figure 1.4** Summary of several mechanisms by which nuclear proteins inhibit activation of Wnt targets. (A) Like Gro/TLEs, factors such as Mtgr-1 can bind to TCFs and recruit HDACs to inhibit Wnt target gene expression. These corepressors are displaced by β-catenin binding to TCF (not shown but see Fig. 1.3C). (B) Other corepressors, such as SMRT, are recruited to WRE chromatin through interactions with β-catenin and TCFs. These factors may dampen the ability of β-catenin to activate target genes. (C) A similar role is proposed for Reptin, which binds to β-catenin and HDACs containing complexes. (D) In contrast, Osterix is thought to act by inhibiting TCF binding to DNA. (E) Other factors, like Kaiso, may act in parallel to TCF to inhibit gene expression. (F) Still others, like Cby, bind to β-catenin and prevent it from binding TCFs. See text for more details. (For color version of this figure, the reader is referred to the Web version of this chapter.)

In addition, other proteins, like Hydrogen peroxide-inducible clone (Hic-5) can also bind TCFs and exclude β-catenin from target gene chromatin (Ghogomu *et al.*, 2006; Li *et al.*, 2011a).

Several other TCF-interacting proteins are also implicated in regulating Wnt targets, but through different mechanisms. Osterix, which inhibits TCF reporter expression in osteoblasts, prevents TCF from binding to DNA (Zhang *et al.*, 2008; Fig. 1.4B). *S*ilencing *M*ediator for *R*etinoid and *T*hyroid hormone receptor (SMRT) and *N*uclear receptor *Co-R*epressor (NCoR) bind directly to both TCFs and β-catenin and antagonize Wnt readouts (Song and Gelmann, 2008). Other proteins, like Reptin (also known as TIP49B) and 12-O-Tetradecanoylphorbol-13-acetate-induced sequence 7 (TIS7), bind β-catenin and inhibit the TCF-β-catenin complex from activating transcription (Bauer *et al.*, 2000; Rottbauer *et al.*, 2002; Vietor *et al.*, 2005). Reptin and TIS7 exist in HDAC-containing complexes (Olson *et al.*, 2006; Vietor and Huber, 2007), and the DNA-dependent ATPase activity of Reptin is also linked to its ability to inhibit Wnt activation of transcription (Rottbauer *et al.*, 2002). These factors act to dampen activation of Wnt targets after the TCF-β-catenin complex has formed on WREs (Fig. 1.4C and D).

## 5.2. WRE-bound factors acting in parallel to TCFs

In addition to corepressors being recruited to WREs by binding to TCFs, other factors may act in parallel with TCF to maintain low target gene expression in the absence of signaling. A DNA-binding protein called Kaiso contains a POZ/BTB domain at its N-terminus and three Zn-fingers at its C-terminus. The Zn-finger domain can recognize a specific DNA sequence (Daniel *et al.*, 2002). Mutation of a Kaiso-binding site in the *siamois* WRE led to derepression, and Kaiso depletion caused elevated expression of *siamois* and other Wnt targets, likely due to increased β-catenin bound to the *siamois* WRE. Kaiso also associated with TCF3 in co-IP assays (Park *et al.*, 2005). The POZ/BTB domain of xKaiso, which is known to bind to NCoR (Yoon *et al.*, 2003), is required for antagonizing Wnt/β-catenin signaling (Park *et al.*, 2005), suggesting the model illustrated in Fig. 1.4E.

Kaiso was originally identified as a binding partner for p120-catenin, an Arm domain protein involved in cell adhesion (Daniel, 2007). p120-Catenin antagonizes Kaiso repression of Wnt targets by promoting Kaiso export from the nucleus (Park *et al.*, 2005, 2006). Like β-catenin, p120-catenin's stability is enhanced by Wnt signaling via inhibition of the destruction complex (Hong *et al.*, 2010; Park *et al.*, 2006). These data support a model where Kaiso and TCF act in concert to repress Wnt targets, and p120-catenin and β-catenin are required to inhibit this repression.

The proposed role for Kaiso described above is controversial because another report found that the sequence-specific DNA-binding activity of Kaiso was not required during *Xenopus* development (Ruzov *et al.*, 2009b).

An interaction between Kaiso and TCFs was observed, but this blocked TCF from binding to DNA, which could explain the ability of Kaiso to inhibit activation of Wnt targets when overexpressed (Ruzov *et al.*, 2009a). No effect on *siamois* expression was observed in Kaiso morphants (Ruzov *et al.*, 2009b). More strikingly, another report found that morpholino-depletion of Kaiso caused a significant decrease in *siamois* WRE reporter expression (Iioka *et al.*, 2009). In another discrepancy with the earlier reports, mutation of the Kaiso-binding site in the *siamois* WRE reduced its activity (Iioka *et al.*, 2009).

Another factor where different mechanisms have been proposed for influencing Wnt target gene expression is C-terminal-binding protein (CtBP). TCFs can associate with CtBP *in vitro* (Brannon *et al.*, 1999; Valenta *et al.*, 2003), and the presence of classic CtBP-binding sites in TCFs has been correlated with increased repressive ability (Cuilliere-Dartigues *et al.*, 2006; Tang *et al.*, 2008). CtBP has also been proposed to act with the hypermethylated in cancer 1 (HIC1) protein to recruit TCF4 and β-catenin away from target gene chromatin and into nuclear bodies (Valenta *et al.*, 2006). Another report found no interacton between CtBP and TCFs, instead showing that a CtBP–APC complex could sequester β-catenin away from TCF (Hamada and Bienz, 2004). CtBP is required for repression of some Wg targets in the absence of signaling, and CtBP localized to WRE chromatin in fly cells (Fang *et al.*, 2006). However, this binding was TCF independent and simultaneous depletion of CtBP and TCF resulted in synergistic derepression. These results suggest that CtBP and TCF act in parallel to repress WRE activity (Fang *et al.*, 2006). Unlike the situation with Kaiso, where the controversy exists between different reports using similar approaches in the same organism, in the case of CtBP, the different mechanisms proposed are not mutually exclusive and may occur to different extents in different cells or species.

HDACs are important in repressing TCF targets (Billin *et al.* 2000) and their involvement in TCF repression is a recurring theme (see Fig. 1.4). But there are also examples where ATP-dependent chromatin remodelers act to repress Wnt targets. In *Drosophila*, the ATPase ISWI and its binding partner ACF1 are required for inhibition of targets in the absence of signaling (Liu *et al.*, 2008b). Recruitment of these factors to WRE chromatin is not dependent on TCF and was decreased by activation of Wg signaling (Liu *et al.*, 2008b). In another fly study, components of the Brahma complex, a SWI/SNF chromatin-remodeling complex, were also found to repress some Wg targets (Collins and Treisman, 2000), though it is not clear whether this effect is direct.

## 5.3. TCF-β-catenin buffers

In addition to acting as corepressors, Gro/TLE, Mtgr-1, and Coop also compete with β-catenin for TCF binding (Daniels and Weis, 2005; Moore *et al.*, 2008; Song *et al.*, 2010). One can think of these factors as TCF-β-catenin buffers that act to set a threshold of nuclear β-catenin necessary to form a complex with TCFs on target gene chromatin. In this section, we review other proteins that play a similar role, by binding to either TCF or β-catenin.

In *Drosophila*, Creb-binding protein (CBP, also known as *nejire* in flies), a histone acetyltransferase (HAT), binds and acetylates TCF, reducing its ability to bind Arm (Waltzer and Bienz, 1998). Consistent with this, reduction of *nejire* gene activity results in elevated Wg signaling (Waltzer and Bienz, 1998). siRNA depletion of CBP and its close relative p300 also caused an elevation of TCF reporter activity in human cells (Li *et al.*, 2007). As will be discussed in Section 6, CBP/p300 also plays an essential role as a β-catenin coactivator.

Two small proteins called Chibby (Cby) and Inhibitor of β-catenin and TCF (ICAT) bind to the C-terminal region of β-catenin and inhibit its binding to TCFs (Daniels and Weis, 2002; Graham *et al.*, 2002; Tago *et al.*, 2000; Takemaru *et al.*, 2003; Fig. 1.4F). Cby has also been reported to act with 14-3-3 chaperones to promote β-catenin efflux from the nucleus (Li *et al.*, 2008, 2010a). Another factor, the p15Ink4b-related protein P15RS, has also been reported to inhibit TCF-β-catenin association through binding to both proteins (Wu *et al.*, 2010).

Another TCF-β-catenin buffer has been more closely tied to components of the β-catenin destruction complex. The SRY-box 9 (Sox9) protein inhibits Wnt/β-catenin signaling in developing cartilage and intestine (Akiyama *et al.*, 2004; Bastide *et al.*, 2007). Sox9 binds directly to β-catenin and prevent access to TCF (Akiyama *et al.*, 2004). In addition, Sox9 can also promote β-catenin degradation by promoting the nuclear localization of GSK3, β-TrCP, and other destruction components to the nucleus (Topol *et al.*, 2009). The ability of Sox9 to enter the nucleus is tightly correlated with its ability to promote β-catenin degradation (Topol *et al.*, 2009).

Whether acting at the level of WRE chromatin or in the nucleoplasm, the numerous mechanisms described in this section all serve to tightly control the expression of Wnt targets by TCF-β-catenin. But it is likely that these mechanisms operate in a cell-specific manner, though more genetic studies are needed to establish this point. The significance of different cells using diverse strategies to ensure that subthreshold levels of nuclear β-catenin do not throw the TCF transcriptional switch to activation is not clear and remains an unexplored aspect of Wnt signaling.

## 6. Factors Contributing to TCF-β-catenin Transcriptional Activation

β-catenin is known to bind to a dizzying array of coactivators that have been demonstrated to promote Wnt transcritptional activation in some context (Cadigan and Peifer, 2009; Mosimann *et al.*, 2009; Parker *et al.*, 2007; Willert and Jones, 2006). This section will describe some well studied examples, with an emphasis on more recent findings.

## 6.1. Factors facilitating β-catenin binding to TCF

As discussed in Section 3, the N-terminus of TCFs is necessary and sufficient to bind to β-catenin (Behrens *et al.*, 1996; Graham *et al.*, 2000; Molenaar *et al.*, 1996; Poy *et al.*, 2001; van de Wetering *et al.*, 1997). But whether this interaction is sufficient under physiological conditions is not clear. Indeed, two studies have provided evidence for additional factors being required to reinforce the TCF-β-catenin complex on WRE chromatin. Transducin β-like protein 1 (TBL1) and TBL1-related protein (TBLR1) are required for β-catenin/Armadillo recruitment to WREs and activation of Wnt targets by the pathway in several cell types (Li and Wang, 2008). TBL1 can bind both TCFs and β-catenin, suggesting that it strengthens their physical association. A similar function has also been proposed for the Jerky/Earthbound (Ebd1) protein during *Drosophila* development (Benchabane *et al.*, 2011). Interestingly, Ebd1 is expressed in a tissue-restricted manner and is only required for Wg signaling in the pupal eye and indirect flight muscles (Benchabane *et al.*, 2011). There is no sequence similarity between Ebd1 and TBL1/TBLR1, and there may be additional proteins that strengthen TCF-β-catenin association.

Two recent reports demonstrate that under some conditions the N-terminal β-catenin-binding domain of TCF is dispensable for recruiting β-catenin to WREs. Expression of an N-terminally truncated form of TCF blocks Wg signaling in flight muscles, but this phenotype is suppressed by overexpression of Arm (Benchabane *et al.*, 2011). In human osteoblasts, a naturally occurring LEF1 isoform lacking the N-terminus can still associate with β-catenin and activate Wnt targets (Hoeppner *et al.*, 2011). It seems likely that factors such as Ebd1 or TBL1/TBLR1 help promote formation of the TCF-β-catenin complex under these conditions.

## 6.2. The N-terminal transactivation domain of β-catenin: Lgs/Bcl9 and Pygo

The ability of β-catenin to activate transcription independently of TCFs has been demonstrated by fusing the protein to the DNA-binding domain of Gal4 (Hoffmans *et al.*, 2005; Hsu *et al.*, 1998; Natarajan *et al.*, 2001). The transactivation ability of β-catenin can be subdivided into smaller domains, for example, one residing in the N-terminal half of β-catenin, localized to the first four Arm repeats (Stadeli and Basler, 2005). This section focuses on this N-terminal transactivation domain.

The best characterized coactivator that interacts with the N-terminal domain of Arm/β-catenin is known as Legless (Lgs) in flies and Bcl9 and Bcl9-2 in vertebrates. Lgs/Bcl9 interacts with Pygopus (Pygo) proteins, which are required for Wnt signaling in flies and vertebrates (Jessen *et al.*, 2008). In flies, Lgs and Pygo are major contributors to Wg signaling (Belenkaya *et al.*, 2002; Kramps *et al.*, 2002; Parker *et al.*, 2002; Thompson *et al.*, 2002), but in mice, double knockouts of *bcl9*, *bcl9-2* or *pygo1*, *pygo2* exhibit a milder reduction in Wnt signaling that varies from target to target (Brack *et al.*, 2009; Deka *et al.*, 2010; Schwab *et al.*, 2007).

Both Lgs and Pygo proteins contain short conserved regions important for their function. The homology domain 2 (HD2) of Lgs/Bcl9 binds to the first Arm repeat of Arm/β-catenin (Hoffmans and Basler, 2004; Sampietro *et al.*, 2006) while HD1 binds to the PHD domain of Pygo (Kramps *et al.*, 2002), a zinc-finger like domain found in many transcriptional regulators (Aasland *et al.*, 1995). Pygo also contains an N-terminal homology domain (NHD) which has transactivation activity (Stadeli and Basler, 2005). These data favor a model where an Lgs–Pygo complex is recruited to TCF–Arm complexes on WRE chromatin to activate Wg targets (Kramps *et al.*, 2002; Fig. 1.5A).

Furthering this "chain of adaptors" model, the NHD of Pygo interacts with several factors and complexes involved in transcriptional activation. In flies, Pygo interacts with Med12 and Med13 (Carrera *et al.*, 2008), subunits of the mediator complex, which is thought to form a bridge between transcription factors and the RNA polymerase II complex (Malik and Roeder, 2010). Med12 and Med13 are required for activation of Wnt targets in several organisms (Carrera *et al.*, 2008; Rocha *et al.*, 2010; Yoda *et al.*, 2005). TAF4, a TFIID subunit, physically interacts with Pygo and is required for activation of a Wg target (Wright and Tjian, 2009). The NHD of human Pygo2 has also been shown in interact with the GCN5 HAT complex (Andrews *et al.*, 2009), as well as the CBP HAT and the MLL2 histone methyltransferase complex (Chen *et al.*, 2010). Depletion of human Pygo2 reduces the recruitment of GCN5 and MLL2 components to WRE chromatin, as well as H3-K9ac and H3-K4me3, the respective chromatin marks these complexes catalyze (Chen *et al.*, 2010).

The PHD domain of Pygo can bind to mono-, di-, and trimethlylated forms of H3-K4 (Fiedler *et al.*, 2008; Gu *et al.*, 2009; Kessler *et al.*, 2009). This binding is allosterically enhanced by the HD1 domain of Lgs/Bcl9 (Fiedler *et al.*, 2008; Miller *et al.*, 2010). While the physiological importance of PHD domain binding to H3-K4me is disputed in *Drosophila* (Kessler *et al.*, 2009), the interaction appears to be required for Pygo2 activity in mammary progenitor cells (Gu *et al.*, 2009). This suggests a model where Pygo both recruits MLL2 to WREs and then "reads" the chromatin mark created by MLL2 to achieve transcriptional activation.

Pygo and Lgs/Bcl9 have also been implicated in promoting high levels of nuclear Arm/β-catenin (Brack *et al.*, 2009; Townsley *et al.*, 2004). Pygo has also been found to localize with TCF in the absence of signaling (de la Roche and Bienz, 2007), where it may act to antagonize Gro activity

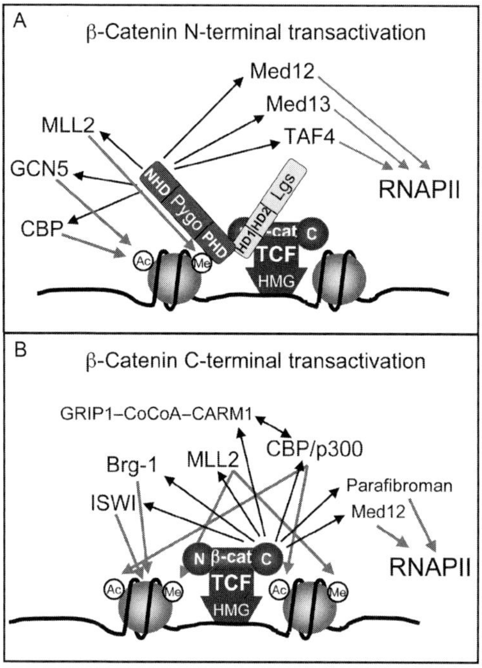


**Figure 1.5** Summary of some of the coactivators that interact with the N-terminal and C-terminal transactivation domains of β-catenin. (A) The first Arm repeat of β-catenin binds to Lgs through its HD2 domain, while the PHD domain of Pygo binds to Lgs through its HD1 domain. The NHD domain of Pygo can recruit several proteins, which either interact with RNA polymerase II (RNAPII) or modify chromatin. The PHD domain of Pygo can also bind to H3 histones methylated on K4. (B) The C-terminal transactivation domain binds to a variety of coactivators, some of which are shown that are thought to act on chromatin or with RNAPII. See text for further explanation. (For color version of this figure, the reader is referred to the Web version of this chapter.)

(Mieszczanek *et al.*, 2008). These alternative functions for Pygo may act in parallel with its aforementioned ability to recruit multiple transcriptional coactivators to WREs (Fig. 1.5A).

In flies, it is thought that the only function for Lgs is as an adaptor for Arm and Pygo (Kramps *et al.*, 2002). However, Bcl9 and Bcl9-2 also contain conserved domains in their C-terminal region not found in invertebrates. This region of Bcl9 and Bcl9-2 contains transactivation activity (Adachi *et al.*, 2004; Brembeck *et al.*, 2004; Sustmann *et al.*, 2008). This activity of Bcl9 is cell-type specific and independent of Pygo, though the mechanism is not yet determined (Sustmann *et al.*, 2008).

While the Lgs/Bcl9-Pygo complex is the best characterized mediator of Arm/β-catenin N-terminal transactivation activity, other factors have also been implicated. The DNA-dependent helicase Pontin (also known at

TIP49) can potentiate TCF/β-catenin activation of targets (Bauer *et al.*, 2000; Feng *et al.*, 2003; Rottbauer *et al.*, 2002). CtBP, discussed previously as a repressor of Wnt targets, is also required for activation of some Wg targets in flies (Bhambhani *et al.*, 2011; Fang *et al.*, 2006). While oligomeric CtBP is required for repression, monomeric CtBP retains the ability to activate transcription, acting at some point downstream of Pygo (Bhambhani *et al.*, 2011).

## 6.3. The C-terminal transactivation domain of β-catenin: A plethora of interacting partners

The C-terminus of β-catenin downstream of the Arm repeats also possesses the ability to activate transcription (Stadeli and Basler, 2005; van de Wetering *et al.*, 1997; Vleminckx *et al.*, 1999). However, since the last two Arm repeats are required for binding of several coactivators, the C-terminal transactivation domain is thought of as beginning with Arm repeat 11 (Mosimann *et al.*, 2009; Sierra *et al.*, 2006). This section focuses of some of the coactivators that interact with this region.

Activation of transcription is highly correlated with acetylation of the N-termini of histone subunits H3 and H4 (Grewal and Moazed, 2003). CBP and the closely related protein p300 are major HATs that catalyze this histone modification (Narlikar *et al.*, 2002). Both HATs bind to the C-terminal half of β-catenin and activate Wnt targets in several contexts (Hecht *et al.*, 2000; Li *et al.*, 2007; Ma *et al.*, 2005; Sun *et al.*, 2000; Takemaru and Moon, 2000). CBP/p300 is recruited to WREs in a Wnt signaling-dependent manner (Kioussi *et al.*, 2002; Parker *et al.*, 2008; Sierra *et al.*, 2006) and is correlated with H3/H4 acetylation (Kioussi *et al.*, 2002; Parker *et al.*, 2008; Sierra *et al.*, 2006). Although CBP and p300 are often thought of as interchangeable in their ability to activate Wnt targets, there is also evidence that they possess distinct, nonredundant functions (Teo and Kahn, 2010).

Several other proteins also act with β-catenin and p300/CBP to activate Wnt targets. Coactivators originally associated with nuclear receptor gene regulation, including glucocorticoid receptor-interacting protein 1 (GRIP1), coactivator-associated arginine methyl transferase (CARM1), and coiled-coil coactivator (CoCoA), bind β-catenin and p300 and enhance TCF-dependent transcription (Koh *et al.*, 2002; Li *et al.*, 2004; Ou *et al.*, 2011; Yang *et al.*, 2006a,b).

Many other factors that interact with the C-terminal transactivation domain of β-catenin remain to be characterized (Sierra *et al.*, 2006), but some have been explored in detail. These include the MLL2 histone methyltransferase, which is recruited to WRE chromatin in a Wnt-dependent fashion (Sierra *et al.*, 2006). Brahma (Brm) and Brahma related gene 1 (Brg-1), the fly and vertebrate ATPase subunits of the SWI/SNF chromatin-remodeling complex, promote Wnt signaling when coexpressed with β-catenin (Barker *et al.*, 2001) and are required for Wnt target gene activation, along with other complex components (Barker *et al.*, 2001; Eroglu *et al.*, 2006; Mahmoudi *et al.*, 2010; Major *et al.*, 2008). TERT, the protein component of the telomerase complex, associates with Brg-1 and is required for maximal activation of Wnt targets (Park *et al.*, 2009). ISWI, another chromatin-remodeling ATPase, also interacts with the C-terminus of β-catenin (Sierra *et al.*, 2006; Song *et al.*, 2009) and is required for Wg signaling in flies (Song *et al.*, 2009). In this activating role, ISWI functions in a complex containing NURF301 (Song *et al.*, 2009), while ISWI also acts with ACF1 (the ACF complex) to repress several Wg targets (Liu *et al.*, 2008b).

## 6.4. Interactions between the β-catenin transactivation domains

The N- and C-terminal halves of β-catenin are sufficient for activating transcription when fused to a DNA-binding domain (Fang *et al.*, 2006; Hsu *et al.*, 1998; Natarajan *et al.*, 2001; Stadeli and Basler, 2005). However, the two transactivation domains clearly interact with each other, as illustrated by studies of Pygo in *Drosophila* cells. Consistent with the protein interaction data, Pygo is required for the N-terminal transactivation domain but not the C-terminal (Stadeli and Basler, 2005). However, when the entire Arm sequence is fused to Gal4, its ability to activate transcription is absolutely dependent on Pygo (Stadeli and Basler, 2005). In addition, Hyrax/Parafibroman, a component of the polymerase-associate factor (PAF1) complex that binds to the C-terminal transactivation domain of β-catenin/Arm, activates Wnt signaling in a Pygo-dependent manner (Mosimann *et al.*, 2006). In the context of the whole protein, both transactivation domains are required for β-catenin/Arm to facilitate transcriptional activation.

The molecular mechanism(s) by which the two transactivation domains function is not well understood, but there are several reports of physical associations between coactivators recruited to either domain. The Med12 subunit of the Mediator complex interacts with Pygo, but it also has been shown to bind to the C-terminus of β-catenin (Kim *et al.*, 2006). The same is true of MLL2 (Chen *et al.*, 2010; Sierra *et al.*, 2006). Pygo has also been shown to physically and functionally interact with CBP (Andrews *et al.*, 2009). These studies suggest a model where coactivators are recruited to β-catenin bound to chromatin through multiple interactions, perhaps in a sequential manner (Mosimann *et al.*, 2009).

Several of the coregulators described above participate in both repression and activation of Wnt targets, for example, Kaiso, CtBP, ISWI, and Brahma/Brg-1. These findings emphasize the complexities of transcriptional regulation, where the same factor can promote different transcriptional outputs, likely due to the chromatin state of the target gene and/or the presence of additional coregulators on the WRE chromatin.

## 6.5. The role of chromatin modifications in activating Wnt targets

Athough the HATs, HDACs, and histone methyltransferases mentioned above are often assumed to act on Wnt targets through histones, some also modify nonhistone proteins. For example, acetylation of TCF/Pan and POP-1 regulate their function (Gay *et al.*, 2003; Waltzer and Bienz, 1998). In addition, CBP and an associated HAT, PCAF, can acetylate β-catenin, increasing its stability or its affinity for TCFs (Ge *et al.*, 2009; Levy *et al.*, 2004; Wolf *et al.*, 2002). The four and a half LIM-only protein 2 (FHL2) binds to β-catenin and p300 and potentiates β-catenin acetylation and transcriptional activity of Wnt targets (Labalette *et al.*, 2004, 2010). While these nonhistone mechanisms contribute to Wnt gene regulation, activation of Wnt targets is also correlated with histone modification of target gene chromatin, which will be summarized below.

A Wnt signaling-dependent increase in H3 and H4 acetylation has been described at many WREs (Chen *et al.* 2010; Kioussi *et al.*, 2002; Lyu *et al.*, 2011; Parker *et al.*, 2008; Sierra *et al.*, 2006), though this increase is not always observed (Blythe *et al.*, 2010; Wohrle *et al.*, 2007). In *Drosophila*, the effect is very dramatic, and a widespread (up to 30kb) increase in histone acetylation was observed at Wnt target genes, centered on the WREs (Parker *et al.*, 2008). This was not a by-product of transcription, since elevated histone acetylation by Wg signaling was not blocked by a-amanitin, a RNA polymerase II inhibitor (Parker *et al.*, 2008). These histone modifications were completely dependent on CBP, though CBP was only found at the WRE. These data support a model where localized CBP somehow promotes widespread histone acetylation, which may be required to overcome widespread repressors such as ACF1 on Wg targets (Liu *et al.*, 2008b; Parker *et al.*, 2008).

Other chromatin marks that are found at Wnt targets in response to pathway activation include H3K4me3 (Blythe *et al.*, 2010; Chen *et al.*, 2010; Parker *et al.*, 2008; Sierra *et al.*, 2006), consistent with the positive role of MLL2 discussed earlier. Monomethylation of H4K20 by the histone methylase SET8 has also been linked to Wnt target gene activation (Li *et al.*, 2011b). SET8 interacts with the HMG domain of TCFs in a manner that is facilitated by β-catenin (Li *et al.*, 2011b). The arginine methyltransferase CARM1, required for TCF-β-catenin transcriptional activation (Koh *et al.*, 2002), mediates H3R17 methylation on Wnt target genes (Ou *et al.*, 2011).

Another member of the protein arginine methyltransferase family (PMRT) has also been linked to Wnt target gene activation in *Xenopus* development. PRMT2 was found to be associated with β-catenin biochemically and is necessary and sufficient for *siamois* expression in frog embryos

(Blythe *et al.*, 2010). Interestingly, β-catenin dependent asymmetric dimethylation of H3K8 and H3K8me was found to occur on the *siamois* WRE before the midblastula transition, the developmental stage when *siamois* transcription begins. This is consistent with data using an inducible β-catenin that indicates an early requirement in *Xenopus* development to "lock in" target gene for later expression (Blythe *et al.*, 2010).

Another histone methyltransferase, DOT1L (also called DOT1), specific for H3K79, appears to be involved in Wnt target gene transcription downstream of initiation events. DOT1L and one of its binding partners, MLLT10/AF10, are required for Wnt signaling in human and zebrafish cells (Mahmoudi *et al.*, 2010). DOT1L-MLLT10/AF10 is implicated in transcriptional elongation (Steger *et al.*, 2008). Consistent with this, elevated levels of H3K79me2 and H3K79me3 are found along the transcribed regions of Wnt targets, and the presence of the elongating form of RNA polymerase II is DOT1L MLLT10/AF10 dependent (Mahmoudi *et al.*, 2010). Microarray analysis of Wnt-induced genes in HEK293T cells revealed that MLLT10/AF10 and DOT1 are largely dedicated to Wnt-induced genes, in contrast to Brg-1 and p300, which regulate many genes not induced by Wnt signaling (Mahmoudi *et al.*, 2010). DOT1L exists in a large complex, known as DOT/COM, which can mediate H3K79me3 of histones that are monoubiquitinated on H2B (Mohan *et al.*, 2010). Grappa, the fly DOT1L homolog, and Bre1, the E3 ligase responsible for H2B ubiquitination, are required for a subset of Wg targets in the developing wing, suggesting that H3K79me3 is the most crucial DOT1L-mediated chromatin mark promoting Wnt target gene transcription (Mohan *et al.*, 2010).

Given the large number of TCF-β-catenin associated factors implicated activation of Wnt targets, it is difficult to imagine them all residing on target gene chromatin simultaneously. It seems likely that factors will associate with WREs and target promoters in a sequential manner (Mosimann *et al.*, 2009). Indeed, there is some evidence for dynamic switching of coregulators during the repressed and activated states at the *c-myc* WRE (Sierra *et al.*, 2006). More comprehensive studies with multiple coregulators will be needed to address this issue.

## 7. Variations of the TCF Transcriptional Switch

Thus far, this review has assumed a model where a single TCF occupies WRE chromatin in the presence and absence of Wnt signaling, nucleating both the repression and activation complexes that mediate that transcriptional switch. In this section, three significant additions to this paradigm will be discussed. Regulation of nuclear POP-1 levels through the Wnt/β-catenin asymmetry pathway is crucial for Wnt signaling in

*C. elegans* and perhaps other systems. While a single *TCF* gene is the norm in most invertebrates, most vertebrates have four *TCF* genes which have become more specialized in function. Finally, while TCF-β-catenin is often thought of as promoting gene activation, there are cases where these proteins also mediate transcriptional repression.

## 7.1. Wnt/β-catenin asymmetry signaling

This pathway was originally discovered in *C. elegans*, where Wnt signaling downregulates nuclear levels of POP-1 (Rocheleau *et al.*, 1997; Thorpe *et al.*, 1997). MOM-4 and LIT-1 are required for this downregulation and encode transforming-growth-factor-β-activated kinase (TAK1) and Nemo-like kinase (NLK), respectively (Meneghini *et al.*, 1999). NLK acts with the β-catenin homolog WRM-1 to phosphosphorylate POP-1 leading to its nuclear export (Lo *et al.*, 2004; Rocheleau *et al.*, 1999; Shin *et al.*, 1999). Wnt signaling also promotes the stabilization and nuclear accumulation of SYS-1 (Huang *et al.*, 2007; Phillips *et al.*, 2007), a protein that is structurally and functionally similar to other β-catenins despite having almost no primary sequence conservation (Kidd *et al.*, 2005; Liu *et al.*, 2008a). The two branches of the Wnt/β-catenin asymmetry pathway result in an increase in nuclear SYS-1 and a decrease in nuclear POP-1, both of which are required for activation of Wnt target genes (Mizumoto and Sawa, 2007; Phillips and Kimble, 2009; Fig. 1.6; Table 1.1). Viewed biochemically, the

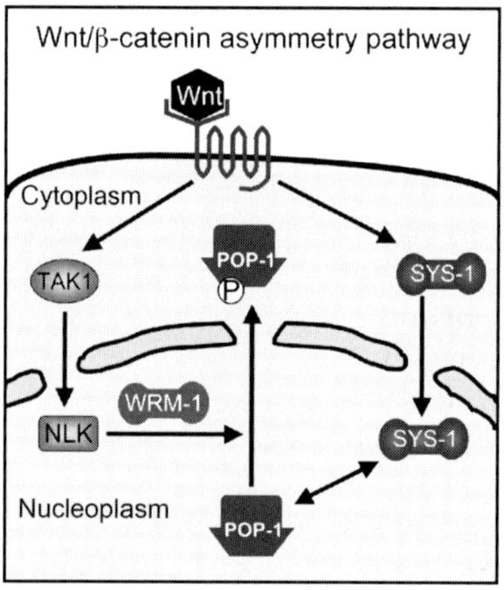


**Figure 1.6** Outline of the Wnt/β-catenin asymmetry pathway in *C. elegans*. Wnt signaling through Fz receptors branches at Dsh (not shown) to generate two signals inside the cell. See text for further explanation. (For color version of this figure, the reader is referred to the Web version of this chapter.)

**Table 1.1** Summary of Wnt/β-catenin asymmetry pathway outputs in different conditions

| Genotype | Wnt | Nuclear SYS-1 levels | Nuclear POP-1 levels | Wnt target gene expression |
|---|---|---|---|---|
| Wild type | Absent | Low | High | Low |
| Wild type | Present | High | Low | High |
| *sys-1* mutant | Present | Absent | Low | Low |
| *NLK/lit-1* mutant | Present | High | High | Low |

In the normal situation, Wnt stimulation causes a concomitant increase and decrease in the nuclear levels of SYS-1 and POP-1, respectively, leading to activation of Wnt targets. SYS-1 is required for this activation. In *lit-1* mutants, nuclear export of POP-1 does not occur, reducing activation of target genes by Wnt (Mizumoto and Sawa, 2007; Phillips and Kimble, 2009).

nuclear efflux of POP-1 is thought to increase the nuclear SYS-1/POP-1 ratio so that the POP-1 bound to WRE chromatin is occupied by SYS-1.

Does this pathway contribute to Wnt signaling outside of nematodes? Wnt stimulation activates TAK1 and NLK activity in mammalian cells (Kanei-Ishii *et al.*, 2004; Smit *et al.*, 2004) where NLK phosphorylates TCFs, which inhibits its ability to bind DNA (Ishitani *et al.*, 1999, 2003) and/or leads to its ubiquitination and degradation (Li *et al.*, 2010b; Yamada *et al.*, 2006). Overexpression of NLK leads to a reduction in Wnt/β-catenin signaling in mammalian cell culture, *Drosophila*, *Xenopus*, and sea urchin embryos (Ishitani *et al.*, 1999, 2003; Rottinger *et al.*, 2006; Yamada *et al.*, 2006; Zeng and Verheyen, 2004).

At first glance, the data from *C. elegans* and other systems seems contradictory, with NLK1 playing a positive role in nematodes and an inhibitory one elsewhere. But the inhibitory role is based almost exclusively on overexpression of NLK, which could result in abnormally high inhibition/degradation of TCF function. Loss of *nemo* (the fly NLK) results in elevated Wg signaling, but NLK may act by regulating Arm stability in this species (Zeng and Verheyen, 2004). In zebrafish, knockdown of Nlk results in a loss of Wnt signaling phenotype, and functional data suggests that NLK acts through inhibition of TCF3a and TCF3b, two fish TCF family members that are associated with transcriptional repression (Thorpe and Moon, 2004). Clearly more loss of function data is needed to better understand the role of NLK in vertebrate Wnt signaling. However, interpretation of *nlk* mutant or knockdown data will depend on the particular Wnt target assayed, that is, the degree to which the target is repressed by TCFs in the absence of signaling. In addition, given the more specialized function of different vertebrate TCFs (discussed in the following section), the TCF repertoire at a particular Wnt target may influence NLK's influence on its expression.

## 7.2. A Wnt-dependent "TCF exchange" on vertebrate WREs

In contrast to most invertebrates, which contain a single *TCF* gene with little alternative splicing, most vertebrates have four *TCF* genes and numerous isoforms (Archbold *et al.*, 2011; Parker *et al.*, 2007). TCF3 isoforms repress Wnt targets (Kim *et al.*, 2000; Liu *et al.*, 2005; Merrill *et al.*, 2004), while LEF1 promotes Wnt target activation (Kratochwil *et al.*, 2002; Liu *et al.*, 2005). While TCF4 often acts as an activator (Korinek *et al.*, 1998; Standley *et al.*, 2006), it has also been linked to repression (Nguyen *et al.*, 2009; Tang *et al.*, 2008). Likewise, TCF1 is associated with both activities (Galceran *et al.*, 1999; Liu *et al.*, 2005; Roose *et al.*, 1999).

Given the division of labor among vertebrate TCFs in regard to transcriptional repression and activation, models where different TCFs occupy WREs in a Wnt-dependent manner have been proposed (Liu *et al.*, 2005; Najdi *et al.*, 2009; Standley *et al.*, 2006; Tang *et al.*, 2008). However, a direct demonstration of this has been lacking, as well as a mechanism that could drive an exchange of TCFs on WRE chromatin. Recently, it has been shown in *Xenopus* and human cells that Wnt/β-catenin signaling promotes the phosphorylation of LEF1, TCF3, and TCF4 by homeodomain interacting protein kinase 2 (HIPK2; Hikasa and Sokol, 2011; Hikasa *et al.*, 2010). These phosphorylated TCFs can no longer occupy WRE chromatin. TCF1 lacks the HIPK2 phosphorylation sites and replaces TCF3 on the *Vent2* WRE in *Xenopus* embryos after Wnt signaling (Hikasa and Sokol, 2011; Fig. 1.7). TCF3 and TCF1 have also recently been shown to have opposing effects in controlling Wnt response in embryonic stem cells (Yi *et al.*, 2011), suggesting that the TCF exchange described at the *Vent2* gene may be a general paradigm.

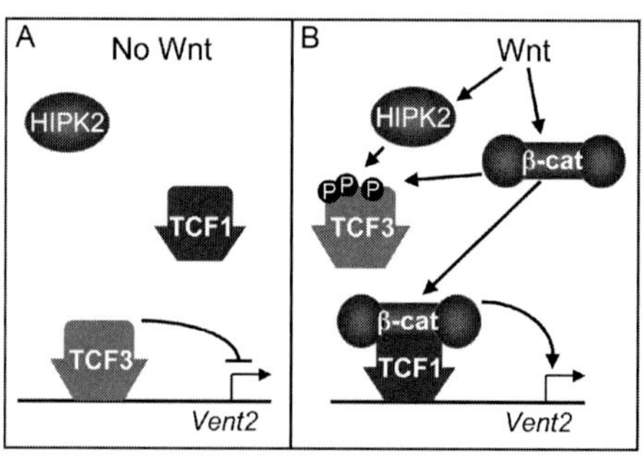


**Figure 1.7** Regulation of a "TCF exchange" on a WRE by HIPK2 and β-catenin. See text for further explanation. (For color version of this figure, the reader is referred to the Web version of this chapter.)

### 7.3. A reverse transcriptional switch for targets repressed by Wnt signaling

While TCF-β-catenin complexes are most often associated with transcriptional activation, there are several examples where these proteins promote direct repression as well (Archbold *et al.*, 2011; Hoverter and Waterman, 2008). In most of these cases, the TCF-binding sites responsible for repression are similar in sequence to those that mediate activation and have no function in the absence of signaling (Delmas *et al.*, 2007; Jamora *et al.*, 2003; Theisen *et al.*, 2007). Why TCF-β-catenin bound to these sites repress rather than activate transcription is not understood.

In contrast to the above examples, several targets in *Drosophila* are regulated by a "reverse switch", where TCF activates their expression in the absence of signaling and then represses in an Arm-dependent manner (Blauwkamp *et al.*, 2008). Detailed examination of the WRE for one of these targets revealed the presence of novel TCF-binding sites (AGAWAW; classic TCF sites have a consensus of CTTTGWWS; W=A/T; S=C/G). Mutation of these novel sites results in reduction of expression in the absence of signaling, and the ability of the WRE to repress the metallothionin promoter is lost (Blauwkamp *et al.*, 2008). Changing the novel TCF sites to classic ones in the WRE converts the transcriptional output from repression to activation (Blauwkamp *et al.*, 2008). This is a dramatic example of allosteric regulation of a transcription factor by the DNA-binding site.

The mechanism by which TCF–Arm represses transcription remains to be determined, but it does not appear to require the N-terminal and C-terminal transactivation domains of Arm. Mutation of the Lgs-binding site and deletion of the C-terminus of Arm results in a protein (called DisArmed) that cannot activate Wg targets, but nonetheless retains the ability to repress targets (Blauwkamp *et al.*, 2008). The DisArmed mutation effectively uncouples Arm transcriptional activation from repression and may be a useful reagent to identify additional targets of direct repression.

## 8. Conclusions and Future Directions

Since a transcriptional switch mechanism was first invoked for TCFs more than a decade ago, it has been a productive working model to gene regulation by Wnt/β-catenin signaling. Although the particular composition of each WRE dictates the importance of TCF repression and activation, it is likely the both mechanisms contribute to some degree on the vast majority of Wnt targets. Given the importance of Wnt/β-catenin signaling in metazoan biology and the need for precision in expression of Wnt targets, it is not surprising that so many factors contribute to repression and activation of these targets. But to understand whether this diversity is important for how Wnt targets are regulated in different contexts, it will be necessary to know which factors directly contribute to regulation for a particular target in a particular cell. Future studies will require a more holistic approach, examining the interplay between multiple coregulators in the same study.

In regard to the variations of the classic transcriptional switch model presented here, a major question is their generality. It seems likely that an exchange of different TCF gene products on WRE in response to Wnt signaling is common in vertebrate systems, where the TCF family members have become more specialized for either repression or activation. Regulation of TCF by the Wnt/β-catenin asymmetry pathway, first revealed with POP-1 in *C. elegans*, clearly occurs in other systems, though likely not as predominant as in nematodes. In the case of direct repression through novel TCF sites, whether this mechanism is more than just an oddity limited to a few *Drosophila* targets will require additional investigation. But this is true to some extent with all studies of Wnt-mediated gene regulation. Our understanding of Wnt/β-catenin signaling has clearly benefited from the model system revolution, where discoveries in one system are readily translated to another. But as the field matures, it is also important to keep in mind that many molecular mechanisms controlling TCF and β-catenin activity have been tinkered with through evolution, so that results obtained in one organism should serve as inspiration for studies in others, not dogma.

## ACKNOWLEDGMENTS

The author thanks C. Bhambhani for critical reading of the manuscript. This work was supported by grants from the NIH (GM08994) and NSF (0950348) to K.M.C.

## REFERENCES

Aasland, R., Gibson, T. J., and Stewart, A. F. (1995). The PHD finger: Implications for chromatin-mediated transcriptional regulation. *Trends Biochem. Sci.* **20,** 56–59.

Adachi, S., Jigami, T., Yasui, T., *et al.* (2004). Role of a BCL9-related beta-catenin-binding protein, B9L, in tumorigenesis induced by aberrant activation of Wnt signaling. *Cancer Res.* **64,** 8496–8501.

Akiyama, H., Lyons, J. P., Mori-Akiyama, Y., *et al.* (2004). Interactions between Sox9 and beta-catenin control chondrocyte differentiation. *Genes Dev.* **18,** 1072–1087.

Andrews, P. G., He, Z., Popadiuk, C., and Kao, K. R. (2009). The transcriptional activity of Pygopus is enhanced by its interaction with cAMP-response-element-binding protein (CREB)-binding protein. *Biochem. J.* **422,** 493–501.

Arce, L., Pate, K. T., and Waterman, M. L. (2009). Groucho binds two conserved regions of LEF-1 for HDAC-dependent repression. *BMC Cancer* **9,** 159.

Archbold, H. C., Yang, Y. X., Chen, L., and Cadigan, K. M. (2011). How do they do Wnt they do?: Regulation of transcription by the Wnt/beta-catenin pathway. *Acta Physiol. (Oxf.)in press.*

Atcha, F. A., Syed, A., Wu, B., *et al.* (2007). A unique DNA binding domain converts TCFs into strong Wnt effectors. *Mol. Cell. Biol.* **27,** 8352–8363.

Barker, N., Hurlstone, A., Musisi, H., *et al.* (2001). The chromatin remodelling factor Brg-1 interacts with beta-catenin to promote target gene activation. *EMBO J.* **20,** 4935–4943.

Barolo, S., and Posakony, J. W. (2002). Three habits of highly effective signaling pathways: Principles of transcriptional control by developmental cell signaling. *Genes Dev.* **16,** 1167–1181.

Baron, R., Rawadi, G., and Roman-Roman, S. (2006). Wnt signaling: A key regulator of bone mass. *Curr. Top. Dev. Biol.* **76,** 103–127.

Bastide, P., Darido, C., Pannequin, J., *et al.* (2007). Sox9 regulates cell proliferation and is required for Paneth cell differentiation in the intestinal epithelium. *J. Cell Biol.* **178,** 635–648.

Bauer, A., Chauvet, S., Huber, O., *et al.* (2000). Pontin52 and reptin52 function as antagonistic regulators of beta-catenin signalling activity. *EMBO J.* **19,** 6121–6130.

Beagle, B., and Johnson, G. V. (2010). AES/GRG5: More than just a dominant-negative TLE/GRG family member. *Dev. Dyn.* **239,** 2795–2805.

Behrens, J., von Kries, J. P., Kuhl, M., *et al.* (1996). Functional interaction of beta-catenin with the transcription factor LEF-1. *Nature* **382,** 638–642.

Belenkaya, T. Y., Han, C., Standley, H. J., *et al.* (2002). pygopus Encodes a nuclear protein essential for wingless/Wnt signaling. *Development* **129,** 4089–4101.

Benchabane, H., Xin, N., Tian, A., *et al.* (2011). Jerky/Earthbound facilitates cell-specific Wnt/Wingless signalling by modulating beta-catenin-TCF activity. *EMBO J.* **30,** 1444–1458.

Bhambhani, C., Chang, J. L., Akey, D. L., and Cadigan, K. M. (2011). The oligomeric state of CtBP determines its role as a transcriptional co-activator and co-repressor of Wingless targets. *EMBO J.* **30,** 2031–2043.

Billin, A. N., Thirlwell, H., and Ayer, D. E. (2000). Beta-catenin-histone deacetylase interactions regulate the transition of LEF1 from a transcriptional repressor to an activator. *Mol. Cell. Biol.* **20,** 6882–6890.

Blauwkamp, T. A., Chang, M. V., and Cadigan, K. M. (2008). Novel TCF-binding sites specify transcriptional repression by Wnt signalling. *EMBO J.* **27,** 1436–1446.

Blythe, S. A., Cha, S. W., Tadjuidje, E., *et al.* (2010). beta-Catenin primes organizer gene expression by recruiting a histone H3 arginine 8 methyltransferase, Prmt2. *Dev. Cell* **19,** 220–231.

Bowerman, B. (1995). Determinants of blastomere identity in the early *C. elegans* embryo. *Bioessays* **17,** 405–414.

Brack, A. S., Murphy-Seiler, F., Hanifi, J., *et al.* (2009). BCL9 is an essential component of canonical Wnt signaling that mediates the differentiation of myogenic progenitors during muscle regeneration. *Dev. Biol.* **335,** 93–105.

Brannon, M., Gomperts, M., Sumoy, L., *et al.* (1997). A beta-catenin/XTcf-3 complex binds to the siamois promoter to regulate dorsal axis specification in Xenopus. *Genes Dev.* **11,** 2359–2370.

Brannon, M., Brown, J. D., Bates, R., *et al.* (1999). XCtBP is a XTcf-3 co-repressor with roles throughout Xenopus development. *Development* **126,** 3159–3170.

Brantjes, H., Roose, J., van De Wetering, M., and Clevers, H. (2001). All Tcf HMG box transcription factors interact with Groucho-related co-repressors. *Nucleic Acids Res.* **29,** 1410–1419.

Brembeck, F. H., Schwarz-Romond, T., Bakkers, J., *et al.* (2004). Essential role of BCL9-2 in the switch between beta-catenin's adhesive and transcriptional functions. *Genes Dev.* **18,** 2225–2230.

Brocardo, M., and Henderson, B. R. (2008). APC shuttling to the membrane, nucleus and beyond. *Trends Cell Biol.* **18,** 587–596.

Brunner, E., Peter, O., Schweizer, L., and Basler, K. (1997). pangolin encodes a Lef-1 homologue that acts downstream of Armadillo to transduce the Wingless signal in Drosophila. *Nature* **385,** 829–833.

Cadigan, K. M., and Peifer, M. (2009). Wnt signaling from development to disease: Insights from model systems. *Cold Spring Harb. Perspect. Biol.* **1,** a002881.

Carrera, I., Janody, F., Leeds, N., *et al.* (2008). Pygopus activates Wingless target gene transcription through the mediator complex subunits Med12 and Med13. *Proc. Natl. Acad. Sci. USA* **105,** 6644–6649.

Cavallo, R. A., Cox, R. T., Moline, M. M., *et al.* (1998). Drosophila Tcf and Groucho interact to repress Wingless signalling activity. *Nature* **395,** 604–608.

Chang, M. V., Chang, J. L., Gangopadhyay, A., *et al.* (2008). Activation of wingless targets requires bipartite recognition of DNA by TCF. *Curr. Biol.* **18,** 1877–1881.

Chen, G., and Courey, A. J. (2000). Groucho/TLE family proteins and transcriptional repression. *Gene* **249,** 1–16.

Chen, G., Nguyen, P. H., and Courey, A. J. (1998). A role for Groucho tetramerization in transcriptional repression. *Mol. Cell. Biol.* **18,** 7259–7268.

Chen, G., Fernandez, J., Mische, S., and Courey, A. J. (1999). A functional interaction between the histone deacetylase Rpd3 and the corepressor groucho in Drosophila development. *Genes Dev.* **13,** 2218–2230.

Chen, J., Luo, Q., Yuan, Y., *et al.* (2010). Pygo2 associates with MLL2 histone methyltransferase and GCN5 histone acetyltransferase complexes to augment Wnt target gene expression and breast cancer stem-like cell expansion. *Mol. Cell. Biol.* **30,** 5621–5635.

Clevers, H. (2006). Wnt/beta-catenin signaling in development and disease. *Cell* **127,** 469–480.

Collins, R. T., and Treisman, J. E. (2000). Osa-containing Brahma chromatin remodeling complexes are required for the repression of wingless target genes. *Genes Dev.* **14,** 3140–3152.

Cuilliere-Dartigues, P., El-Bchiri, J., Krimi, A., *et al.* (2006). TCF-4 isoforms absent in TCF-4 mutated MSI-H colorectal cancer cells colocalize with nuclear CtBP and repress TCF-4-mediated transcription. *Oncogene* **25,** 4441–4448.

Daniel, J. M. (2007). Dancing in and out of the nucleus: p120(ctn) and the transcription factor Kaiso. *Biochim. Biophys. Acta* **1773,** 59–68.

Daniel, J. M., Spring, C. M., Crawford, H. C., *et al.* (2002). The p120(ctn)-binding partner Kaiso is a bi-modal DNA-binding protein that recognizes both a sequence-specific consensus and methylated CpG dinucleotides. *Nucleic Acids Res.* **30,** 2911–2919.

Daniels, D. L., and Weis, W. I. (2002). ICAT inhibits beta-catenin binding to Tcf/Lef-family transcription factors and the general coactivator p300 using independent structural modules. *Mol. Cell* **10,** 573–584.

Daniels, D. L., and Weis, W. I. (2005). Beta-catenin directly displaces Groucho/TLE repressors from Tcf/Lef in Wnt-mediated transcription activation. *Nat. Struct. Mol. Biol.* **12,** 364–371.

de la Roche, M., and Bienz, M. (2007). Wingless-independent association of Pygopus with dTCF target genes. *Curr. Biol.* **17**(6), 556–561.

Deka, J., Wiedemann, N., Anderle, P., *et al.* (2010). Bcl9/Bcl9l are critical for Wnt-mediated regulation of stem cell traits in colon epithelium and adenocarcinomas. *Cancer Res.* **70,** 6619–6628.

Delmas, V., Beermann, F., Martinozzi, S., *et al.* (2007). Beta-catenin induces immortalization of melanocytes by suppressing p16INK4a expression and cooperates with N-Ras in melanoma development. *Genes Dev.* **21,** 2923–2935.

Eroglu, B., Wang, G., Tu, N., *et al.* (2006). Critical role of Brg1 member of the SWI/SNF chromatin remodeling complex during neurogenesis and neural crest induction in zebrafish. *Dev. Dyn.* **235,** 2722–2735.

Fan, M. J., Gruning, W., Walz, G., and Sokol, S. Y. (1998). Wnt signaling and transcriptional control of Siamois in Xenopus embryos. *Proc. Natl. Acad. Sci. USA* **95,** 5626–5631.

Fang, M., Li, J., Blauwkamp, T., *et al.* (2006). C-terminal-binding protein directly activates and represses Wnt transcriptional targets in Drosophila. *EMBO J.* **25,** 2735–2745.

Feng, Y., Lee, N., and Fearon, E. R. (2003). TIP49 regulates beta-catenin-mediated neoplastic transformation and T-cell factor target gene induction via effects on chromatin remodeling. *Cancer Res.* **63,** 8726–8734.

Fiedler, M., Sanchez-Barrena, M. J., Nekrasov, M., *et al.* (2008). Decoding of methylated histone H3 tail by the Pygo-BCL9 Wnt signaling complex. *Mol. Cell* **30,** 507–518.

Galceran, J., Farinas, I., Depew, M. J., *et al.* (1999). Wnt3a−/−-like phenotype and limb deficiency in Lef1(−/−)Tcf1(−/−) mice. *Genes Dev.* **13,** 709–717.

Gay, F., Calvo, D., Lo, M. C., *et al.* (2003). Acetylation regulates subcellular localization of the Wnt signaling nuclear effector POP-1. *Genes Dev.* **17,** 717–722.

Ge, X., Jin, Q., Zhang, F., *et al.* (2009). PCAF acetylates {beta}-catenin and improves its stability. *Mol. Biol. Cell* **20,** 419–427.

Ghogomu, S. M., van Venrooy, S., Ritthaler, M., *et al.* (2006). HIC-5 is a novel repressor of lymphoid enhancer factor/T-cell factor-driven transcription. *J. Biol. Chem.* **281,** 1755–1764.

Graham, T. A., Weaver, C., Mao, F., *et al.* (2000). Crystal structure of a beta-catenin/Tcf complex. *Cell* **103,** 885–896.

Graham, T. A., Clements, W. K., Kimelman, D., and Xu, W. (2002). The crystal structure of the beta-catenin/ICAT complex reveals the inhibitory mechanism of ICAT. *Mol. Cell* **10,** 563–571.

Grewal, S. I., and Moazed, D. (2003). Heterochromatin and epigenetic control of gene expression. *Science* **301,** 798–802.

Grigoryan, T., Wend, P., Klaus, A., and Birchmeier, W. (2008). Deciphering the function of canonical Wnt signals in development and disease: Conditional loss- and gain-of-function mutations of beta-catenin in mice. *Genes Dev.* **22,** 2308–2341.

Gu, B., Sun, P., Yuan, Y., *et al.* (2009). Pygo2 expands mammary progenitor cells by facilitating histone H3 K4 methylation. *J. Cell Biol.* **185,** 811–826.

Haegebarth, A., and Clevers, H. (2009). Wnt signaling, lgr5, and stem cells in the intestine and skin. *Am. J. Pathol.* **174,** 715–721.

Hamada, F., and Bienz, M. (2004). The APC tumor suppressor binds to C-terminal binding protein to divert nuclear beta-catenin from TCF. *Dev. Cell* **7,** 677–685.

Hecht, A., Vleminckx, K., Stemmler, M. P., *et al.* (2000). The p300/CBP acetyltransferases function as transcriptional coactivators of beta-catenin in vertebrates. *EMBO J.* **19,** 1839–1850.

Hikasa, H., and Sokol, S. Y. (2011). Phosphorylation of TCF proteins by homeodomain-interacting protein kinase 2. *J. Biol. Chem.* **286,** 12093–12100.

Hikasa, H., Ezan, J., Itoh, K., *et al.* (2010). Regulation of TCF3 by Wnt-dependent phosphorylation during vertebrate axis specification. *Dev. Cell* **19,** 521–532.

Hoeppner, L. H., Secreto, F. J., Razidlo, D. F., *et al.* (2011). Lef1DeltaN binds beta-catenin and increases osteoblast activity and trabecular bone mass. *J. Biol. Chem.* **286,** 10950–10959.

Hoffmans, R., and Basler, K. (2004). Identification and in vivo role of the Armadillo-Legless interaction. *Development* **131,** 4393–4400.

Hoffmans, R., Stadeli, R., and Basler, K. (2005). Pygopus and legless provide essential transcriptional coactivator functions to armadillo/beta-catenin. *Curr. Biol.* **15,** 1207–1211.

Hong, J. Y., Park, J. I., Cho, K., *et al.* (2010). Shared molecular mechanisms regulate multiple catenin proteins: Canonical Wnt signals and components modulate p120-catenin isoform-1 and additional p120 subfamily members. *J. Cell Sci.* **123,** 4351–4365.

Hoverter, N. P., and Waterman, M. L. (2008). A Wnt-fall for gene regulation: Repression. *Sci. Signal.* **1,** pe43.

Hsu, S. C., Galceran, J., and Grosschedl, R. (1998). Modulation of transcriptional regulation by LEF-1 in response to Wnt-1 signaling and association with beta-catenin. *Mol. Cell. Biol.* **18,** 4807–4818.

Huang, S., Shetty, P., Robertson, S. M., and Lin, R. (2007). Binary cell fate specification during *C. elegans* embryogenesis driven by reiterated reciprocal asymmetry of TCF POP-1 and its coactivator beta-catenin SYS-1. *Development* **134,** 2685–2695.

Iioka, H., Doerner, S. K., and Tamai, K. (2009). Kaiso is a bimodal modulator for Wnt/beta-catenin signaling. *FEBS Lett.* **583,** 627–632.

Ishitani, T., Ninomiya-Tsuji, J., Nagai, S., *et al.* (1999). The TAK1-NLK-MAPK-related pathway antagonizes signalling between beta-catenin and transcription factor TCF. *Nature* **399,** 798–802.

Ishitani, T., Kishida, S., Hyodo-Miura, J., *et al.* (2003). The TAK1-NLK mitogen-activated protein kinase cascade functions in the Wnt-5a/Ca(2+) pathway to antagonize Wnt/beta-catenin signaling. *Mol. Cell. Biol.* **23,** 131–139.

Jamora, C., DasGupta, R., Kocieniewski, P., and Fuchs, E. (2003). Links between signal transduction, transcription and adhesion in epithelial bud development. *Nature* **422,** 317–322.

Jennings, B. H., and Ish-Horowicz, D. (2008). The Groucho/TLE/Grg family of transcriptional co-repressors. *Genome Biol.* **9,** 205.

Jessen, S., Gu, B., and Dai, X. (2008). Pygopus and the Wnt signaling pathway: A diverse set of connections. *Bioessays* **30,** 448–456.

Kanei-Ishii, C., Ninomiya-Tsuji, J., Tanikawa, J., *et al.* (2004). Wnt-1 signal induces phosphorylation and degradation of c-Myb protein via TAK1, HIPK2, and NLK. *Genes Dev.* **18,** 816–829.

Kessler, R., Hausmann, G., and Basler, K. (2009). The PHD domain is required to link Drosophila Pygopus to Legless/beta-catenin and not to histone H3. *Mech. Dev.* **126,** 752–759.

Kidd, A. R., 3rd, Miskowski, J. A., Siegfried, K. R., *et al.* (2005). A beta-catenin identified by functional rather than sequence criteria and its role in Wnt/MAPK signaling. *Cell* **121,** 761–772.

Kim, C. H., Oda, T., Itoh, M., *et al.* (2000). Repressor activity of Headless/Tcf3 is essential for vertebrate head formation. *Nature* **407,** 913–916.

Kim, S., Xu, X., Hecht, A., and Boyer, T. G. (2006). Mediator is a transducer of Wnt/beta-catenin signaling. *J. Biol. Chem.* **281,** 14066–14075.

Kioussi, C., Briata, P., Baek, S. H., *et al.* (2002). Identification of a Wnt/Dvl/beta-Catenin–> Pitx2 pathway mediating cell-type-specific proliferation during development. *Cell* **111,** 673–685.

Knirr, S., and Frasch, M. (2001). Molecular integration of inductive and mesoderm-intrinsic inputs governs even-skipped enhancer activity in a subset of pericardial and dorsal muscle progenitors. *Dev. Biol.* **238,** 13–26.

Koh, S. S., Li, H., Lee, Y. H., *et al.* (2002). Synergistic coactivator function by coactivator-associated arginine methyltransferase (CARM) 1 and beta-catenin with two different classes of DNA-binding transcriptional activators. *J. Biol. Chem.* **277,** 26031–26035.

Korinek, V., Barker, N., Moerer, P., *et al.* (1998). Depletion of epithelial stem-cell compartments in the small intestine of mice lacking Tcf-4. *Nat. Genet.* **19,** 379–383.

Kramps, T., Peter, O., Brunner, E., *et al.* (2002). Wnt/wingless signaling requires BCL9/ legless-mediated recruitment of pygopus to the nuclear beta-catenin-TCF complex. *Cell* **109,** 47–60.

Kratochwil, K., Galceran, J., Tontsch, S., *et al.* (2002). FGF4, a direct target of LEF1 and Wnt signaling, can rescue the arrest of tooth organogenesis in Lef1(−/−) mice. *Genes Dev.* **16,** 3173–3185.

Labalette, C., Renard, C. A., Neuveut, C., *et al.* (2004). Interaction and functional cooperation between the LIM protein FHL2, CBP/p300, and beta-catenin. *Mol. Cell. Biol.* **24,** 10689–10702.

Labalette, C., Nouet, Y., Levillayer, F., *et al.* (2010). Deficiency of the LIM-only protein FHL2 reduces intestinal tumorigenesis in Apc mutant mice. *PLoS One* **5,** e10371.

Lam, N., Chesney, M. A., and Kimble, J. (2006). Wnt signaling and CEH-22/tinman/ Nkx2.5 specify a stem cell niche in *C. elegans*. *Curr. Biol.* **16,** 287–295.

Lee, H. H., and Frasch, M. (2000). Wingless effects mesoderm patterning and ectoderm segmentation events via induction of its downstream target sloppy paired. *Development* **127,** 5497–5508.

Levy, L., Wei, Y., Labalette, C., *et al.* (2004). Acetylation of beta-catenin by p300 regulates beta-catenin-Tcf4 interaction. *Mol. Cell. Biol.* **24,** 3404–3414.

Li, J., and Wang, C. Y. (2008). TBL1-TBLR1 and beta-catenin recruit each other to Wnt target-gene promoter for transcription activation and oncogenesis. *Nat. Cell Biol.* **10,** 160–169.

Li, H., Kim, J. H., Koh, S. S., and Stallcup, M. R. (2004). Synergistic effects of coactivators GRIP1 and beta-catenin on gene activation: Cross-talk between androgen receptor and Wnt signaling pathways. *J. Biol. Chem.* **279,** 4212–4220.

Li, J., Sutter, C., Parker, D. S., *et al.* (2007). CBP/p300 are bimodal regulators of Wnt signaling. *EMBO J.* **26,** 2284–2294.

Li, F. Q., Mofunanya, A., Harris, K., and Takemaru, K. (2008). Chibby cooperates with 14-3-3 to regulate beta-catenin subcellular distribution and signaling activity. *J. Cell Biol.* **181,** 1141–1154.

Li, F. Q., Mofunanya, A., Fischer, V., *et al.* (2010a). Nuclear-cytoplasmic shuttling of Chibby controls beta-catenin signaling. *Mol. Biol. Cell* **21,** 311–322.

Li, M., Wang, H., Huang, T., *et al.* (2010b). TAB2 scaffolds TAK1 and NLK in repressing canonical Wnt signaling. *J. Biol. Chem.* **285,** 13397–13404.

Li, X., Martinez-Ferrer, M., Botta, V., *et al.* (2011a). Epithelial Hic-5/ARA55 expression contributes to prostate tumorigenesis and castrate responsiveness. *Oncogene* **30,** 167–177.

Li, Z., Nie, F., Wang, S., and Li, L. (2011b). Histone H4 Lys 20 monomethylation by histone methylase SET8 mediates Wnt target gene activation. *Proc. Natl. Acad. Sci. USA* **108,** 3116–3123.

Lin, R., Thompson, S., and Priess, J. R. (1995). pop-1 encodes an HMG box protein required for the specification of a mesoderm precursor in early *C. elegans* embryos. *Cell* **83,** 599–609.

Linggi, B. E., Brandt, S. J., Sun, Z. W., and Hiebert, S. W. (2005). Translating the histone code into leukemia. *J. Cell. Biochem.* **96,** 938–950.

Liu, F., van den Broek, O., Destree, O., and Hoppler, S. (2005). Distinct roles for Xenopus Tcf/Lef genes in mediating specific responses to Wnt/beta-catenin signalling in mesoderm development. *Development* **132,** 5375–5385.

Liu, J., Phillips, B. T., Amaya, M. F., *et al.* (2008a). The *C. elegans* SYS-1 protein is a bona fide beta-catenin. *Dev. Cell* **14,** 751–761.

Liu, Y. I., Chang, M. V., Li, H. E., *et al.* (2008b). The chromatin remodelers ISWI and ACF1 directly repress Wingless transcriptional targets. *Dev. Biol.* **323,** 41–52.

Lo, M. C., Gay, F., Odom, R., *et al.* (2004). Phosphorylation by the beta-catenin/MAPK complex promotes 14-3-3-mediated nuclear export of TCF/POP-1 in signal-responsive cells in *C. elegans*. *Cell* **117,** 95–106.

Logan, C. Y., and Nusse, R. (2004). The Wnt signaling pathway in development and disease. *Annu. Rev. Cell Dev. Biol.* **20,** 781–810.

Lyu, J., Jho, E. H., and Lu, W. (2011). Smek promotes histone deacetylation to suppress transcription of Wnt target gene brachyury in pluripotent embryonic stem cells. *Cell Res.* **21,** 911–921.

Ma, H., Nguyen, C., Lee, K. S., and Kahn, M. (2005). Differential roles for the coactivators CBP and p300 on TCF/beta-catenin-mediated survivin gene expression. *Oncogene* **24,** 3619–3631.

MacDonald, B. T., Tamai, K., and He, X. (2009). Wnt/beta-catenin signaling: Components, mechanisms, and diseases. *Dev. Cell* **17,** 9–26.

Maduro, M. F., Lin, R., and Rothman, J. H. (2002). Dynamics of a developmental switch: Recursive intracellular and intranuclear redistribution of *Caenorhabditis elegans* POP-1 parallels Wnt-inhibited transcriptional repression. *Dev. Biol.* **248,** 128–142.

Mahmoudi, T., Boj, S. F., Hatzis, P., *et al.* (2010). The leukemia-associated Mllt10/Af10-Dot1l are Tcf4/beta-catenin coactivators essential for intestinal homeostasis. *PLoS Biol.* **8,** e1000539.

Major, M. B., Roberts, B. S., Berndt, J. D., *et al.* (2008). New regulators of Wnt/beta-catenin signaling revealed by integrative molecular screening. *Sci. Signal.* **1,** ra12.

Malik, S., and Roeder, R. G. (2010). The metazoan Mediator co-activator complex as an integrative hub for transcriptional regulation. *Nat. Rev. Genet.* **11,** 761–772.

Meneghini, M. D., Ishitani, T., Carter, J. C., *et al.* (1999). MAP kinase and Wnt pathways converge to downregulate an HMG-domain repressor in *Caenorhabditis elegans*. *Nature* **399,** 793–797.

Merrill, B. J., Pasolli, H. A., Polak, L., *et al.* (2004). Tcf3: A transcriptional regulator of axis induction in the early embryo. *Development* **131,** 263–274.

Mieszczanek, J., de la Roche, M., and Bienz, M. (2008). A role of Pygopus as an anti-repressor in facilitating Wnt-dependent transcription. *Proc. Natl. Acad. Sci. USA* **105,** 19324–19329.

Miller, T. C., Rutherford, T. J., Johnson, C. M., *et al.* (2010). Allosteric remodelling of the histone H3 binding pocket in the Pygo2 PHD finger triggered by its binding to the B9L/BCL9 co-factor. *J. Mol. Biol.* **401,** 969–984.

Mizumoto, K., and Sawa, H. (2007). Two betas or not two betas: Regulation of asymmetric division by beta-catenin. *Trends Cell Biol.* **17,** 465–473.

Mohan, M., Herz, H. M., Takahashi, Y. H., *et al.* (2010). Linking H3K79 trimethylation to Wnt signaling through a novel Dot1-containing complex (DotCom). *Genes Dev.* **24,** 574–589.

Molenaar, M., van de Wetering, M., Oosterwegel, M., *et al.* (1996). XTcf-3 transcription factor mediates beta-catenin-induced axis formation in Xenopus embryos. *Cell* **86,** 391–399.

Moore, A. C., Amann, J. M., Williams, C. S., *et al.* (2008). Myeloid translocation gene family members associate with T-cell factors (TCFs) and influence TCF-dependent transcription. *Mol. Cell. Biol.* **28,** 977–987.

Mosimann, C., Hausmann, G., and Basler, K. (2006). Parafibromin/Hyrax activates Wnt/Wg target gene transcription by direct association with beta-catenin/Armadillo. *Cell* **125,** 327–341.

Mosimann, C., Hausmann, G., and Basler, K. (2009). Beta-catenin hits chromatin: Regulation of Wnt target gene activation. *Nat. Rev. Mol. Cell Biol.* **10,** 276–286.

Najdi, R., Syed, A., Arce, L., *et al.* (2009). A Wnt kinase network alters nuclear localization of TCF-1 in colon cancer. *Oncogene* **28,** 4133–4146.

Narlikar, G. J., Fan, H. Y., and Kingston, R. E. (2002). Cooperation between complexes that regulate chromatin structure and transcription. *Cell* **108,** 475–487.

Natarajan, L., Witwer, N. E., and Eisenmann, D. M. (2001). The divergent *Caenorhabditis elegans* beta-catenin proteins BAR-1, WRM-1 and HMP-2 make distinct protein interactions but retain functional redundancy in vivo. *Genetics* **159,** 159–172.

Nguyen, H., Merrill, B. J., Polak, L., *et al.* (2009). Tcf3 and Tcf4 are essential for long-term homeostasis of skin epithelia. *Nat. Genet.* **41,** 1068–1075.

Nusse, R. (2008). Wnt signaling and stem cell control. *Cell Res.* **18,** 523–527.

Olson, L. E., Tollkuhn, J., Scafoglio, C., *et al.* (2006). Homeodomain-mediated beta-catenin-dependent switching events dictate cell-lineage determination. *Cell* **125,** 593–605.

Ou, C. Y., LaBonte, M. J., Manegold, P. C., *et al.* (2011). A coactivator role of CARM1 in the dysregulation of beta-catenin activity in colorectal cancer cell growth and gene expression. *Mol. Cancer Res.* **9,** 660–670.

Park, J. I., Kim, S. W., Lyons, J. P., *et al.* (2005). Kaiso/p120-catenin and TCF/beta-catenin complexes coordinately regulate canonical Wnt gene targets. *Dev. Cell* **8,** 843–854.

Park, J. I., Ji, H., Jun, S., *et al.* (2006). Frodo links Dishevelled to the p120-catenin/Kaiso pathway: Distinct catenin subfamilies promote Wnt signals. *Dev. Cell* **11,** 683–695.

Park, J. I., Venteicher, A. S., Hong, J. Y., *et al.* (2009). Telomerase modulates Wnt signalling by association with target gene chromatin. *Nature* **460,** 66–72.

Parker, D. S., Jemison, J., and Cadigan, K. M. (2002). Pygopus, a nuclear PHD-finger protein required for Wingless signaling in Drosophila. *Development* **129,** 2565–2576.

Parker, D. S., Blauwkamp, T., and Cadigan, K. M. (2007). Wnt/beta-catenin-mediated transcriptional regulation. *In* "Advances in Developmental Biology," (P. M. Wassarman, ed.), Vol. 17, pp. 1–61. Elsevier, San Diego.

Parker, D. S., Ni, Y. Y., Chang, J. L., *et al.* (2008). Wingless signaling induces widespread chromatin remodeling of target loci. *Mol. Cell. Biol.* **28,** 1815–1828.

Petersen, C. P., and Reddien, P. W. (2009). Wnt signaling and the polarity of the primary body axis. *Cell* **139,** 1056–1068.

Phillips, B. T., and Kimble, J. (2009). A new look at TCF and beta-catenin through the lens of a divergent *C. elegans* Wnt pathway. *Dev. Cell* **17,** 27–34.

Phillips, B. T., Kidd, A. R., 3rd, King, R., *et al.* (2007). Reciprocal asymmetry of SYS-1/beta-catenin and POP-1/TCF controls asymmetric divisions in *Caenorhabditis elegans*. *Proc. Natl. Acad. Sci. USA* **104,** 3231–3236.

Polakis, P. (2007). The many ways of Wnt in cancer. *Curr. Opin. Genet. Dev.* **17,** 45–51.

Poy, F., Lepourcelet, M., Shivdasani, R. A., and Eck, M. J. (2001). Structure of a human Tcf4-beta-catenin complex. *Nat. Struct. Biol.* **8,** 1053–1057.

Riese, J., Yu, X., Munnerlyn, A., *et al.* (1997). LEF-1, a nuclear factor coordinating signaling inputs from wingless and decapentaplegic. *Cell* **88,** 777–787.

Rocha, P. P., Scholze, M., Bleiss, W., and Schrewe, H. (2010). Med12 is essential for early mouse development and for canonical Wnt and Wnt/PCP signaling. *Development* **137,** 2723–2731.

Rocheleau, C. E., Downs, W. D., Lin, R., *et al.* (1997). Wnt signaling and an APC-related gene specify endoderm in early *C. elegans* embryos. *Cell* **90,** 707–716.

Rocheleau, C. E., Yasuda, J., Shin, T. H., *et al.* (1999). WRM-1 activates the LIT-1 protein kinase to transduce anterior/posterior polarity signals in *C. elegans*. *Cell* **97,** 717–726.

Roose, J., Molenaar, M., Peterson, J., *et al.* (1998). The Xenopus Wnt effector XTcf-3 interacts with Groucho-related transcriptional repressors. *Nature* **395,** 608–612.

Roose, J., Huls, G., van Beest, M., *et al.* (1999). Synergy between tumor suppressor APC and the beta-catenin-TCF4 target TCF1. *Science* **285,** 1923–1926.

Rottbauer, W., Saurin, A. J., Lickert, H., *et al.* (2002). Reptin and pontin antagonistically regulate heart growth in zebrafish embryos. *Cell* **111,** 661–672.

Rottinger, E., Croce, J., Lhomond, G., *et al.* (2006). Nemo-like kinase (NLK) acts downstream of Notch/Delta signalling to downregulate TCF during mesoderm induction in the sea urchin embryo. *Development* **133,** 4341–4353.

Ruzov, A., Hackett, J. A., Prokhortchouk, A., *et al.* (2009a). The interaction of xKaiso with xTcf3: A revised model for integration of epigenetic and Wnt signalling pathways. *Development* **136,** 723–727.

Ruzov, A., Savitskaya, E., Hackett, J. A., *et al.* (2009b). The non-methylated DNA-binding function of Kaiso is not required in early Xenopus laevis development. *Development* **136,** 729–738.

Sampietro, J., Dahlberg, C. L., Cho, U. S., *et al.* (2006). Crystal structure of a beta-catenin/BCL9/Tcf4 complex. *Mol. Cell* **24,** 293–300.

Schwab, K. R., Patterson, L. T., Hartman, H. A., *et al.* (2007). Pygo1 and Pygo2 roles in Wnt signaling in mammalian kidney development. *BMC Biol.* **5,** 15.

Shetty, P., Lo, M. C., Robertson, S. M., and Lin, R. (2005). *C. elegans* TCF protein, POP-1, converts from repressor to activator as a result of Wnt-induced lowering of nuclear levels. *Dev. Biol.* **285,** 584–592.

Shin, T. H., Yasuda, J., Rocheleau, C. E., *et al.* (1999). MOM-4, a MAP kinase kinase kinase-related protein, activates WRM-1/LIT-1 kinase to transduce anterior/posterior polarity signals in *C. elegans*. *Mol. Cell* **4,** 275–280.

Sierra, J., Yoshida, T., Joazeiro, C. A., and Jones, K. A. (2006). The APC tumor suppressor counteracts beta-catenin activation and H3K4 methylation at Wnt target genes. *Genes Dev.* **20,** 586–600.

Smit, L., Baas, A., Kuipers, J., *et al.* (2004). Wnt activates the Tak1/Nemo-like kinase pathway. *J. Biol. Chem.* **279,** 17232–17240.

Song, L. N., and Gelmann, E. P. (2008). Silencing mediator for retinoid and thyroid hormone receptor and nuclear receptor corepressor attenuate transcriptional activation by the beta-catenin-TCF4 complex. *J. Biol. Chem.* **283,** 25988–25999.

Song, H., Hasson, P., Paroush, Z., and Courey, A. J. (2004). Groucho oligomerization is required for repression in vivo. *Mol. Cell. Biol.* **24,** 4341–4350.

Song, H., Spichiger-Haeusermann, C., and Basler, K. (2009). The ISWI-containing NURF complex regulates the output of the canonical Wingless pathway. *EMBO Rep.* **10,** 1140–1146.

Song, H., Goetze, S., Bischof, J., *et al.* (2010). Coop functions as a corepressor of Pangolin and antagonizes Wingless signaling. *Genes Dev.* **24,** 881–886.

Stadeli, R., and Basler, K. (2005). Dissecting nuclear Wingless signalling: Recruitment of the transcriptional co-activator Pygopus by a chain of adaptor proteins. *Mech. Dev.* **122,** 1171–1182.

Standley, H. J., Destree, O., Kofron, M., *et al.* (2006). Maternal XTcf1 and XTcf4 have distinct roles in regulating Wnt target genes. *Dev. Biol.* **289,** 318–328.

Steger, D. J., Lefterova, M. I., Ying, L., *et al.* (2008). DOT1L/KMT4 recruitment and H3K79 methylation are ubiquitously coupled with gene transcription in mammalian cells. *Mol. Cell. Biol.* **28,** 2825–2839.

Sun, Y., Kolligs, F. T., Hottiger, M. O., *et al.* (2000). Regulation of beta-catenin transformation by the p300 transcriptional coactivator. *Proc. Natl. Acad. Sci. USA* **97,** 12613–12618.

Sustmann, C., Flach, H., Ebert, H., *et al.* (2008). Cell-type-specific function of BCL9 involves a transcriptional activation domain that synergizes with beta-catenin. *Mol. Cell. Biol.* **28,** 3526–3537.

Taelman, V. F., Dobrowolski, R., Plouhinec, J. L., *et al.* (2010). Wnt signaling requires sequestration of glycogen synthase kinase 3 inside multivesicular endosomes. *Cell* **143,** 1136–1148.

Tago, K., Nakamura, T., Nishita, M., *et al.* (2000). Inhibition of Wnt signaling by ICAT, a novel beta-catenin-interacting protein. *Genes Dev.* **14,** 1741–1749.
Takemaru, K. I., and Moon, R. T. (2000). The transcriptional coactivator CBP interacts with beta-catenin to activate gene expression. *J. Cell Biol.* **149,** 249–254.
Takemaru, K., Yamaguchi, S., Lee, Y. S., *et al.* (2003). Chibby, a nuclear beta-catenin-associated antagonist of the Wnt/Wingless pathway. *Nature* **422,** 905–909.
Tang, W., Dodge, M., Gundapaneni, D., *et al.* (2008). A genome-wide RNAi screen for Wnt/beta-catenin pathway components identifies unexpected roles for TCF transcription factors in cancer. *Proc. Natl. Acad. Sci. USA* **105,** 9697–9702.
Teo, J. L., and Kahn, M. (2010). The Wnt signaling pathway in cellular proliferation and differentiation: A tale of two coactivators. *Adv. Drug Deliv. Rev.* **62,** 1149–1155.
Theisen, H., Syed, A., Nguyen, B. T., *et al.* (2007). Wingless directly represses DPP morphogen expression via an armadillo/TCF/Brinker complex. *PLoS One* **2,** e142.
Thompson, B., Townsley, F., Rosin-Arbesfeld, R., *et al.* (2002). A new nuclear component of the Wnt signalling pathway. *Nat. Cell Biol.* **4,** 367–373.
Thorpe, C. J., and Moon, R. T. (2004). nemo-like kinase is an essential co-activator of Wnt signaling during early zebrafish development. *Development* **131,** 2899–2909.
Thorpe, C. J., Schlesinger, A., Carter, J. C., and Bowerman, B. (1997). Wnt signaling polarizes an early *C. elegans* blastomere to distinguish endoderm from mesoderm. *Cell* **90,** 695–705.
Tolwinski, N. S., and Wieschaus, E. (2004). Rethinking WNT signaling. *Trends Genet.* **20,** 177–181.
Topol, L., Chen, W., Song, H., *et al.* (2009). Sox9 inhibits Wnt signaling by promoting beta-catenin phosphorylation in the nucleus. *J. Biol. Chem.* **284,** 3323–3333.
Townsley, F. M., Cliffe, A., and Bienz, M. (2004). Pygopus and Legless target Armadillo/beta-catenin to the nucleus to enable its transcriptional co-activator function. *Nat. Cell Biol.* **6,** 626–633.
Valenta, T., Lukas, J., and Korinek, V. (2003). HMG box transcription factor TCF-4's interaction with CtBP1 controls the expression of the Wnt target Axin2/Conductin in human embryonic kidney cells. *Nucleic Acids Res.* **31,** 2369–2380.
Valenta, T., Lukas, J., Doubravska, L., *et al.* (2006). HIC1 attenuates Wnt signaling by recruitment of TCF-4 and beta-catenin to the nuclear bodies. *EMBO J.* **25,** 2326–2337.
van Amerongen, R., and Nusse, R. (2009). Towards an integrated view of Wnt signaling in development. *Development* **136,** 3205–3214.
van de Wetering, M., Cavallo, R., Dooijes, D., *et al.* (1997). Armadillo coactivates transcription driven by the product of the Drosophila segment polarity gene dTCF. *Cell* **88,** 789–799.
Vietor, I., and Huber, L. A. (2007). Role of TIS7 family of transcriptional regulators in differentiation and regeneration. *Differentiation* **75,** 891–897.
Vietor, I., Kurzbauer, R., Brosch, G., and Huber, L. A. (2005). TIS7 regulation of the beta-catenin/Tcf-4 target gene osteopontin (OPN) is histone deacetylase-dependent. *J. Biol. Chem.* **280,** 39795–39801.
Vleminckx, K., Kemler, R., and Hecht, A. (1999). The C-terminal transactivation domain of beta-catenin is necessary and sufficient for signaling by the LEF-1/beta-catenin complex in Xenopus laevis. *Mech. Dev.* **81,** 65–74.
Waltzer, L., and Bienz, M. (1998). Drosophila CBP represses the transcription factor TCF to antagonize Wingless signalling. *Nature* **395,** 521–525.
Willert, K., and Jones, K. A. (2006). Wnt signaling: Is the party in the nucleus? *Genes Dev.* **20,** 1394–1404.
Wohrle, S., Wallmen, B., and Hecht, A. (2007). Differential control of Wnt target genes involves epigenetic mechanisms and selective promoter occupancy by T-cell factors. *Mol. Cell. Biol.* **27,** 8164–8177.

Wolf, D., Rodova, M., Miska, E. A., *et al.* (2002). Acetylation of beta-catenin by CREB-binding protein (CBP). *J. Biol. Chem.* **277,** 25562–25567.

Wright, K. J., and Tjian, R. (2009). Wnt signaling targets ETO coactivation domain of TAF4/TFIID in vivo. *Proc. Natl. Acad. Sci. USA* **106,** 55–60.

Wu, Y., Zhang, Y., Zhang, H., *et al.* (2010). p15RS attenuates Wnt/{beta}-catenin signaling by disrupting {beta}-catenin.TCF4 Interaction. *J. Biol. Chem.* **285,** 34621–34631.

Yamada, M., Ohnishi, J., Ohkawara, B., *et al.* (2006). NARF, an nemo-like kinase (NLK)-associated ring finger protein regulates the ubiquitylation and degradation of T cell factor/lymphoid enhancer factor (TCF/LEF). *J. Biol. Chem.* **281,** 20749–20760.

Yamaguchi, T. P., Takada, S., Yoshikawa, Y., *et al.* (1999). T (Brachyury) is a direct target of Wnt3a during paraxial mesoderm specification. *Genes Dev.* **13,** 3185–3190.

Yang, X., van Beest, M., Clevers, H., *et al.* (2000). decapentaplegic is a direct target of dTcf repression in the Drosophila visceral mesoderm. *Development* **127,** 3695–3702.

Yang, C. K., Kim, J. H., Li, H., and Stallcup, M. R. (2006a). Differential use of functional domains by coiled-coil coactivator in its synergistic coactivator function with beta-catenin or GRIP1. *J. Biol. Chem.* **281,** 3389–3397.

Yang, C. K., Kim, J. H., and Stallcup, M. R. (2006b). Role of the N-terminal activation domain of the coiled-coil coactivator in mediating transcriptional activation by beta-catenin. *Mol. Endocrinol.* **20,** 3251–3262.

Yi, F., Pereira, L., Hoffman, J. A., *et al.* (2011). Opposing effects of Tcf3 and Tcf1 control Wnt stimulation of embryonic stem cell self-renewal. *Nat. Cell Biol.* **13,** 762–770.

Yoda, A., Kouike, H., Okano, H., and Sawa, H. (2005). Components of the transcriptional Mediator complex are required for asymmetric cell division in *C. elegans*. *Development* **132,** 1885–1893.

Yoon, H. G., Chan, D. W., Reynolds, A. B., *et al.* (2003). N-CoR mediates DNA methylation-dependent repression through a methyl CpG binding protein Kaiso. *Mol. Cell* **12,** 723–734.

Zeng, Y. A., and Verheyen, E. M. (2004). Nemo is an inducible antagonist of Wingless signaling during Drosophila wing development. *Development* **131,** 2911–2920.

Zhang, C., Cho, K., Huang, Y., *et al.* (2008). Inhibition of Wnt signaling by the osteoblast-specific transcription factor Osterix. *Proc. Natl. Acad. Sci. USA* **105,** 6936–6941.

CHAPTER TWO

# Grainy Head and Its Target Genes in Epithelial Morphogenesis and Wound Healing

Shenqiu Wang[1] *and* Christos Samakovlis

## Contents



## Abstract

The Grainy head (Grh) family of transcription factors is characterized by a unique DNA-binding domain that binds to a conserved consensus sequence. Nematodes and flies have a single *grh* gene, whereas mice and humans have evolved three genes encoding Grainy head-like (Grhl) factors. We review the biological function of Grh in different animals and the mechanisms modulating its activity. *grh* and *grhl* genes play a remarkably conserved role in epithelial organ development and extracellular barrier repair after tissue damage. Recent studies in flies and vertebrates suggest that Grh factors may be primary

Department of Developmental Biology, Wenner-Gren Institute, Stockholm University, Stockholm, Sweden
[1] Present address: Cancer Genetics Branch, National Human Genome Research Institute, National Institutes of Health, Bethesda, Maryland, USA

*Current Topics in Developmental Biology*, Volume 98
ISSN 0070-2153, DOI: 10.1016/B978-0-12-386499-4.00002-1

determinants of cell adhesion and epithelial tissue formation. Grh proteins can dimerize and act as activators or repressors in different developmental contexts. In flies, tissue-specific, alternative splicing generates different Grh isoforms with different DNA-binding specificities and functions. Grh activity is also modulated by receptor tyrosine kinases: it is phosphorylated by extracellular signal regulated kinase, and this phosphorylation is selectively required for epidermal barrier repair. Two mechanisms have been proposed to explain the repressive function of Grh on target gene transcription. First, Grh can target the Polycomb silencing complex to specific response elements. Second, it can directly compete for DNA binding with transcriptional activators. Understanding the molecular mechanisms of gene regulation by Grh factors is likely to elucidate phylogenetically conserved mechanisms of epithelial cell morphogenesis and regeneration upon tissue damage.

## 1. The Grainy Head (*grh*) Gene Family

*grh* genes encode a family of transcription factors with an isoleucine-rich activation domain, a unique DNA-binding motif and a dimerization domain (Fig. 2.1) (Attardi and Tjian, 1993; Ting *et al.*, 2003b; Uv *et al.*, 1994; Venkatesan *et al.*, 2003; Wilanowski *et al.*, 2002). The first member of the *grh* gene family was identified in *Drosophila* (Bray and Kafatos, 1991; Bray *et al.*, 1989; Dynlacht *et al.*, 1989; Johnson *et al.*, 1989; Nusslein-Volhard and Kluding, 1984), and *grh* genes have subsequently been found in animals as diverse as nematodes and humans (Huang and Miller, 2000; Janicke *et al.*, 2010; Jones *et al.*, 1988; Kim *et al.*, 1987; Lim *et al.*, 1992; Murata *et al.*, 1998; Rodda *et al.*, 2001; Sueyoshi *et al.*, 1995; Venkatesan *et al.*, 2003; Wilanowski *et al.*, 2002; Wu *et al.*, 1988). Phylogenetic analysis subdivides this gene family into two main classes: the Grh-like (Grhl) subfamily and the CP2 subfamily, depending on whether the family members are more closely related to the *Drosophila grh* (also known as *Elf-1/NTF-1*) or to another *Drosophila* gene, *dCP2* (Ting *et al.*, 2003b; Venkatesan *et al.*, 2003; Wilanowski *et al.*, 2002) (Fig. 2.1).

The fly and worm genomes each contain a single *grh* gene. In contrast, vertebrates have evolved multiple *grhl* genes. The zebrafish has four *grhl* genes: *z-grhl1*, *z-grhl2a* and *2b*, and *z-grhl3*. Mammals, both mice and humans, have evolved three Grh homologues: Grhl-1 (also known as Mammalian Grainy head (MGR)/LBP-32/TFCP2L2), Grhl-2 (Brother-of-MGR (BOM)/TFCP2L3), and Grhl-3 (Sister-of-MGR (SOM)/Grainy head-like epithelial activator (GET-1)/TFCP2L4). This group of genes encodes proteins with highly homologous DNA-binding and dimerization domains. All members of the group show a restricted expression pattern during embryogenesis, and many of them play important roles in organogenesis, epidermal development and regeneration after wounding

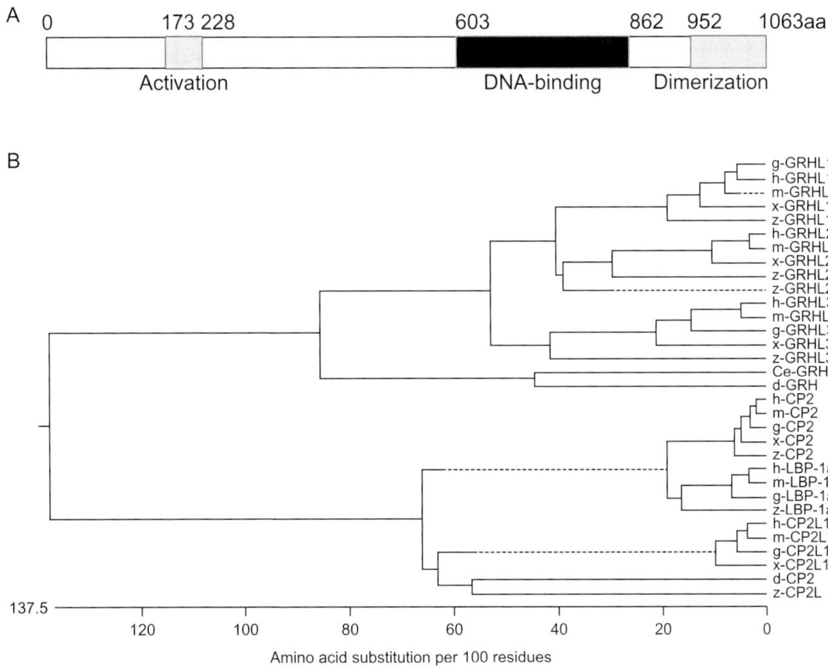


**Figure 2.1** Schematic illustration showing the protein domain organization of *Drosophila* Grh and a phylogenetic tree of published Grh/CP2 proteins. (A) Light gray, dark, and gray boxes indicate the activation domain, DNA-binding motif, and dimerization domain, respectively. The *Drosophila* Grh isoform I is shown here and the numbers indicate amino acids (aa). This domain organization is shared by many Grh proteins. (B) Protein sequence alignments of selected Grh proteins from different species using ClustalW with MegAlign program from DNASTAR Lasergene v8.1.5. The accession numbers for the Grh/CP2 proteins aligned are: AAF60703 (Ce-GRH), NP_610556 (d-CP2), NP_476842 (d-GRH), NP_989715 (g-CP2), XP_422087 (g-CP2L1), XP_426209 (g-GRHL1), XP_417842 (g-GRHL3), XP_426018 (g-LBP-1a), NP_005644 (h-CP2), NP_055368 (h-CP2L1), NP_937825 (h-GRHL1), NP_079191 (h-GRHL2), NP_937816 (h-GRHL3), NP_055332 (h-LBP-1a), NP_258437 (m-CP2), NP_076244 (m-CP2L1), NP_665897 (m-GRHL1), NP_080772 (m-GRHL2), NP_001013778 (m-GRHL3), NP_001076788 (m-LBP-1a), NP_001083962 (x-CP2), NP_001080377 (x-CP2L1), NP_001089071 (x-GRHL1), NP_001011338 (x-GRHL2), NP_001005642 (x-GRHL3), XP_001336482 (z-CP2), XP_001336482 (z-CP2L), XP_001923763 (z-GRHL1), NP_001025263 (z-GRHL2a), NP_001076541 (z-GRHL2b), XP_001332938 (z-GRHL3), NP_001108046 (z-LBP-1a). Abbreviations: Ce, *Caenorhabditis elegans*; d, *Drosophila melanogaster*; g, *Gallus gallus*; h, *Homo sapiens*; m, *Mus musculus*, x, *Xenopus laevis*; z, Zebrafish *Danio rerio*.

(reviewed in Gustavsson *et al.*, 2008; Harden, 2005; Jane *et al.*, 2005; Moussian and Uv, 2005; Stramer and Martin, 2005). Grh factors can function either as transcription activators or repressors, depending on developmental context (Attardi and Tjian, 1993; Bray and Kafatos, 1991; Huang *et al.*, 1995; Liaw *et al.*, 1995).

The second group of genes in the *grh* gene family encodes proteins with higher similarity to dCP2. It includes *CP2*, *LBP-1a*, and *LBP-9* in mammals (Wilanowski *et al.*, 2002). In contrast to the Grh-like subfamily, the CP2-group genes were first identified in vertebrates, and they are widely expressed and involved in diverse cellular processes, such as cell cycle control, growth, and differentiation (Jane *et al.*, 1995; Sueyoshi *et al.*, 1995; Volker *et al.*, 1997; Zhou *et al.*, 2000). An additional distinction between the groups is the ability of the Grh-like factors Grhl-1, Grhl-2, and Grh to interact with each other, but not with members of CP2-like group (Uv *et al.*, 1994; Wilanowski *et al.*, 2002).

It was long believed that *grh* genes were absent in unicellular organisms. However, Traylor-Knowles *et al.* have recently identified Grh homologs in unicellular ancestors of animals, like choanoflagellates and fungi. The fungal *grh* gene cannot be unambiguously assigned to either the *grh* or the *CP2* subfamily. Therefore, although the *grh* gene family originated prior to the metazoan-fungal divergence, the diversification of the Grh/CP2 subfamilies probably occurred in an ancient animal lineage (Traylor-Knowles *et al.*, 2010).

## 2. Grh Function in *Drosophila*

The *Drosophila melanogaster* genome contains a single *grh* gene, which provides a simple and tractable model system for understanding some of the functions of the *grhl* gene family at the molecular level. Grh was originally identified through its ability to bind a *cis*-element of the *Dopa decarboxylase* (*Ddc*) gene and to regulate its larval epidermal expression (Bray *et al.*, 1989; Dynlacht *et al.*, 1989; Johnson *et al.*, 1989). The name *grh* reflects the phenotypes of mutants, which develop a grainy head skeleton and a flimsy epidermal cuticle (Bray and Kafatos, 1991; Nusslein-Volhard and Kluding, 1984). *grh* is predominantly expressed in epithelial tissues derived from the ectoderm, such as the embryonic epidermis, the tracheal airways, the foregut, and the hindgut (Bray and Kafatos, 1991; Hemphala *et al.*, 2003). In addition, Grh is present in the embryonic central nervous system (CNS), larval neuroblasts, and optic lobes, as well as in imaginal discs and the epidermis (Uv *et al.*, 1997).

Grh has been implicated in many aspects of *Drosophila* development, including epidermal barrier formation and reconstruction after wounding (Bray and Kafatos, 1991; Mace *et al.*, 2005), tracheal tube size control (Hemphala *et al.*, 2003), CNS development (Almeida and Bray, 2005; Baumgardt *et al.*, 2009; Cenci and Gould, 2005; Maurange *et al.*, 2008), and the establishment of normal wing hair polarity (Lee and Adler, 2004).

## 2.1. Grh in epidermal barrier maturation

During embryogenesis, a series of morphogenetic events, including epithelial stretching and fusion, forms the continuous embryonic epidermis. Following germ band extension and its subsequent retraction, a large hole remains on the dorsal surface of the embryo. This gap is transiently covered by the amnioserosa, a sheet of large flat cells, which is replaced by lateral epidermal cells that migrate over the amnioserosa to seal the embryo along the dorsal midline (reviewed in Harden, 2002; Jacinto *et al.*, 2002; Martin and Parkhurst, 2004). The cellular processes underlying embryonic dorsal closure resemble the characteristic behavior of cells during epithelial wound closure. Additionally, many of the molecules required for dorsal closure play important roles in epithelial wound healing (reviewed in Grose and Martin, 1999; Martin and Parkhurst, 2004). *grh* mutants show no obvious defects in embryonic dorsal closure. However, Grh overexpression in embryos causes a failure in dorsal closure (Attardi *et al.*, 1993; Narasimha *et al.*, 2008). These dorsal closure defects are either due to ectopic activation of genes encoding septate junction (SJ) proteins in the amnioserosa or due to the overexpression of the SJ proteins in the epidermis (Narasimha *et al.*, 2008). Nevertheless, these experiments suggest that *grh* is redundant for embryonic dorsal closure.

After dorsal closure, the epidermal epithelial cells produce a rigid exoskeleton, the cuticle (Moussian *et al.*, 2006a). The epidermal mature cuticle consists of three morphologically distinct functional layers: the waterproof envelope, the proteinaceous epicuticle, and the chitinaceous procuticle. Chitin, a β1-4 linked polymer of *N*-acetyl-D-glucosamine, is the major constituent and forms a lamellar structure abutting the epidermal cells (Cohen, 2001). The lamellae are built from sheets of chitin microfibrils and are tightly packed to confer stability and elasticity to the exoskeleton (reviewed in Merzendorfer, 2006; Moussian *et al.*, 2005).

The overall structure of the ventrolateral cuticle of *grh* mutant embryos is bulging and distended compared to wild type (Bray and Kafatos, 1991). The specialized structures of the larval cuticle, such as the mouth hooks, are also smaller and discolored. The ventral denticles are frequently disordered, with cases of reversed polarity. Interestingly, when *grh* mutant embryos are removed from the vitelline membrane their cuticle stretches to a much greater extent than the wild-type cuticle. This results in grossly inflated embryonic cuticle preparations. Several other mutants show this characteristic "blimp" cuticle phenotype. They include *krotzkopf verkehrt* (*kkv*), which encodes a chitin synthase; *knickkopf* (*knk*), which encodes a novel protein; and *zeppelin (zep)*, which encodes a protein of unknown function (Ostrowski *et al.*, 2002). The phenotypic similarities of *grh* mutants and mutants in genes affecting cuticle morphogenesis suggest that Grh activates a variety of genes involved in cuticle formation, assembly, and hardening.

Such targets may include the *kkv* gene, which encodes the transmembrane chitin synthase. This synthase links cytosolic UDP-*N*-acetyl-D-glucosamine (UDP-GlcNAc) into chains of repeating GlcNAc residues that extrude from the apical cell surface (reviewed in Merzendorfer, 2006). Chitin can be also modified and degraded by chitinases, which are secreted enzymes that fragment long polysaccharides into shorter chains (reviewed in Arakane *et al.*, 2009; Merzendorfer and Zimoch, 2003). Chitin can also be converted to chitosan by deacetylation. This process is mediated by chitin deacetylases, which remove acetyl groups from chitin and change its biophysical properties (Christodoulidou *et al.*, 1999). Additionally, protein and lipids are cross-linked in the cuticle. *Ddc* and *pale* (or *ple*), which encode Dopa decarboxylase and tyrosine hydroxylase, respectively, are crucial for this cross-linking (reviewed in Wright, 1987b). Both enzymes are components of an enzymatic pathway that converts tyrosine to *N*-acetyldopamine and *N*-β-alanyldopamine, which are subsequently oxidized to quinones by phenol oxidases. The quinones form covalent bonds with histidine residues in the cuticular proteins, leading to cuticular hardening (Schaefer *et al.*, 1987). *Ddc* and *ple* double-mutant larvae develop an almost colorless head skeleton, indicating the role of these proteins in melanin production and cuticle coloration (Mace *et al.*, 2005).

*Ddc* is a direct target of Grh in epidermal cells (Bray and Kafatos, 1991). Loss of *grh* results in a decreased expression of *Ddc* in the epidermis, which may be partly responsible for the weakening of the cuticle and the "blimp" phenotype of *grh* mutants. Reduction in *Ddc* expression may also partially explain the decreased pigmentation of clones containing *grh* mutant cells (Lee and Adler, 2004). It is, however, clear that the reduction of *Ddc* is not the only cause for all of the cuticle-related phenotypes of *grh* mutants. For example, null *Ddc* mutants have discolored mouth hooks and denticle belts, but no other obvious structural defects in their cuticle. *Ddc* mutants, but not *grh* mutants, can survive into larval instars if manually removed from the vitelline membrane (Wright, 1987a). In addition, *Ddc* functions nonautonomously and clones of *Ddc* mutant cells display a less severe pigmentation phenotype than *grh* clones (Lee and Adler, 2004). Therefore, Grh is believed to regulate more genes involved in cuticle formation.

Another direct Grh target in the epidermis is *stitcher* (*stit*), a gene that encodes a putative receptor tyrosine kinase (RTK) of the Ret family (Wang *et al.*, 2009). *stit* expression is initiated in ectodermal tissues prior to dorsal closure and is reduced in the epidermis of *grh* mutants compared to wild type. Conversely, Grh overexpression in the epidermis activates *stit* expression, indicating that Grh controls *stit* during development. The regulatory region of *stit* contains a cluster of four putative conserved Grh-binding elements (Gbes). A genomic DNA fragment encompassing these Gbes directs GFP expression to the epidermis of transgenic embryos, indicating that it contains at least part of the endogenous *stit* enhancer. Grh binds to the Gbe of the *stit* enhancer directly, and the GFP reporter expression is severely reduced in *grh* mutants. Hence, Grh directly activates *stit* during development. *stit* mutants do not show any detectable phenotypes in embryonic development or cuticle maturation, but die as late pupae. This suggests that Stit is not an effector of Grh function during embryonic development (Wang *et al.*, 2009).

Additional candidate *grh* targets during epithelial development are the genes encoding the SJ proteins Fasciclin III (FasIII), Coracle (Cor), and Sinuous (Sinu) (Narasimha *et al.*, 2008). Grh-binding sites are present in the regulatory regions of these genes, and Grh overexpression in the embryonic amnioserosa robustly induces their expression. Further, the protein levels of FasIII and Cor are reduced in *grh* mutant clones in the larva wing discs. However, the expression of these genes is not altered in *grh* mutants during the embryonic stages, suggesting they are probably not responsible for the *grh* epidermal phenotype (Narasimha *et al.*, 2008). Another probable target of Grh is *starry night* (*stan*) (Lee and Adler, 2004). Stan protein levels are directly related to the levels of Grh, both during larval and pupa wing development. Since Grh does not affect Stan protein stability, the results suggest that *grh* regulates *stan* transcription directly. A decrease in *stan* expression may largely account for the PCP defect of *grh* mutants. However, no putative Grh-binding sites have been detected within or around the *stan* gene. This may be due to a large variability in the Grh-binding sites or the fact that *stan* regulation by Grh is indirect. Whether Stan is a *grh* target during epithelial barrier formation remains unknown.

## 2.2. Grh in epidermal wound healing

*Drosophila* has emerged as a key model system to explore the molecular mechanism of epithelial repair. Wound healing in *Drosophila* has been documented at different development stages, including embryonic, larval, and adult (Babcock *et al.*, 2008; Baek *et al.*, 2010; Campos *et al.*, 2010; Galko and Krasnow, 2004; Kim and McGinnis, 2011; Kwon *et al.*, 2010; Mace *et al.*, 2005; Pearson *et al.*, 2009; Ramet *et al.*, 2002; Stramer *et al.*, 2005; Wang *et al.*, 2009; Wood *et al.*, 2002; Wu *et al.*, 2009). Although there are distinct differences, the wound healing processes share some common features in all developmental stages. Grh is crucial for the repair of the protective cuticular layer of the epidermis in wounded *Drosophila* embryos. Upon injury, it activates the expression of cuticle repair genes such as *Ddc* and *ple* in cells surrounding epidermal wounds (Mace *et al.*, 2005; Pearson *et al.*, 2009). Analysis of the regulatory elements required for the expression of both genes upon wounding demonstrates that their activation is dependent on Grh-binding sites. Extracellular signal regulated kinase (ERK) activation also occurs around the wounds and is required for the efficient induction of the

*Ddc* wound response reporter (Mace *et al.*, 2005). The induction of *Ddc* seems to be important for cuticle repair, since *grh* mutants that fail to activate *Ddc* around wounds are defective in cuticle repair, as well as in the displacement of the melanin plug that forms immediately after wounding (Mace *et al.*, 2005). Aseptic wounding of wild-type embryos also induces rapid accumulation of *stit* transcripts at wound sites. The GFP levels driven from the *stit* enhancer are increased around wounds of wild-type embryos and this induction is severely reduced in *grh* mutants. Thus, Grh activates epidermal *stit* expression during development and wound closure. Analysis of *stit* mutants suggested a broader role for Grh in wound healing than simply activating the enzymes involved in the extracellular barrier reconstruction (Wang *et al.*, 2009). Reepithelialization during wound healing relies on an actin cable that forms in the cells along the wound perimeter and operates as a purse-string to draw the wound closed. The epithelial cells at the wound border also migrate, change shape and send out dynamic filopodial protrusions, which are essential for knitting epithelial cells together during embryonic epidermal wound healing (Wood *et al.*, 2002). Thus, the control of the actin cytoskeleton upon wounding is a landmark of the reepithelialization process in the damaged tissues. Live imaging of wounded *stit* mutants showed a delay in reepithelialization and defects in the assembly of the actin ring along the wound perimeter. Stit is also required for the full activation of ERK at the wound sites and the transcriptional activation of the Grh downstream genes *Ddc* and *ple*. Genetic analysis has placed *stit* both downstream and upstream of *grh,* suggesting a positive feedback loop mechanism in wound repair. Activation of Stit (and presumably other receptors) leads to ERK phosphorylation and Grh activation at the wound site. This initiates cuticle-repair gene expression and cytoskeletal rearrangements. Stit also further induces its own transcription and a second tier of responses by a Grh-dependent positive feedback loop. This amplification mechanism ensures efficient epidermal wound repair and relies on transcriptional regulation by Grh without the need of further extracellular stimulation (Wang *et al.*, 2009).

It is presently unknown whether Grh regulates additional genes involved in the reorganization of the actin cytoskeleton and the epidermal cell shape-changes in *Drosophila*. However, keratinocytes from *grhl3* mutant mice show defective actin polymerization and cellular polarity upon wounding. In addition, Grhl3 acts through the small GTPase RhoGEF19 and interacts genetically with planar cell polarity (PCP) genes during epidermal wound healing (Caddy *et al.*, 2010). It will be interesting to see if Grh controls the transcription of cytoskeletal regulators also in *Drosophila*.

## 2.3. Grh in epithelial tube size control

An important late step in the morphogenesis of tubular organs is the final acquisition of distinct and uniform sizes of their branches. The fixed lengths and diameters of tubes dictate the flow rates of the transported gas or liquid and are therefore major determinants of optimal organ function. Genetic analysis in the *Drosophila* airways has revealed several of the cellular mechanisms controlling epithelial tube size (reviewed in Andrew and Ewald, 2010; Casanova, 2007; Ghabrial *et al.*, 2003; Schottenfeld *et al.*, 2010; Swanson and Beitel, 2006; Uv *et al.*, 2003; Wu and Beitel, 2004). The major tracheal airways, the dorsal trunks, consist of a single epithelial cell layer. At the end of embryogenesis, they are surrounded by a basal lamina and an apical specialized cuticular lining forming a characteristic taenidial lining. Like the epidermal cuticle, the spiral taenidial lining is predominantly made of chitin. This apical extracellular matrix (ECM) is thought to provide elasticity to the tubes and a tough, protective barrier against dehydration and pathogens. The lengths and diameters of the different tracheal branches are controlled independently, with the length of branches increasing continuously through development, and the expansion of tube diameter occurring step-wise. Further, neither cell size nor cell number are primary determinants of tube expansion in the embryonic trachea (Beitel and Krasnow, 2000). Analysis of mutants with selective size defects indicates that genes encoding proteins with very diverse molecular functions control tracheal tube size in *Drosophila* embryos.

Grh is the only transcription factor that has been identified among the genes implicated in tracheal tube size control. Grh regulates tube elongation selectively because *grh* mutants show over-elongated and convoluted tubes without any apparent defect in tube diameter or early branch morphogenesis (Hemphala *et al.*, 2003). Labeling *grh* mutant embryos for the adherens junction (AJ) protein E-cadherin (DE-Cad) shows that the apical surface of *grh* tracheal cells is highly irregular and elongated compared to the cobblestone-shaped cells in wild-type embryos. The SJs and the localization of apical basal polarity markers were not affected in *grh* embryos. The cellular defects have been further analyzed by transmission electron microscopy (TEM). Grh mutants show normal morphology along the basal and lateral membrane of the tracheal cells. However, the apical cell domain is excessively overgrown and distorted. Cuticle deposition is also highly abnormal with strong defects in the procuticle layer and the taenidial organization. In addition, the luminal chitin staining in *grh* mutants is decreased compared to wild-type embryos (Luschnig *et al.*, 2006). On the other hand, overexpression of Grh in all tracheal cells throughout embryonic tracheal development is sufficient to prevent luminal extension, and results in a partial failure in branch outgrowth. This phenotype suggests that Grh may also have a redundant function during early tracheal development. Direct *grh* targets in airway size control have not been identified and the molecular mechanism of how *grh* controls tracheal tube size remains unknown. Some clues can be derived from the examination of the differences and similarities between the *grh* mutant phenotypes and the defects of other *Drosophila* mutants with overgrown airway tubes.

The second group of mutants showing overgrown airways involves genes encoding proteins involved in chitin biogenesis or assembly. These genes include *cystic/mummy* (*mmy*), *kkv*, *knk*, *retroactive* (*rtv*), *vermiform* (*verm*), and *serpentine (serp)*. *mmy* encodes the *Drosophila* homologue of UDP-*N*-acetylglucosamine diphosphorylase. This enzyme is required for the production of UDP-*N*-acetylglucosamine, a substrate of chitin synthesis (Araujo *et al.*, 2005; Devine *et al.*, 2005; Tonning *et al.*, 2006). Just before the expansion of the developing tracheal tubes the epithelial cells secrete a luminal chitin matrix. This luminal matrix forms a tightly packed cylindrical structure that grows during tube expansion and disappears shortly before the embryo hatches (Devine *et al.*, 2005; Tonning *et al.*, 2005). Mutations in *mmy*, *kkv*, *knk* or *rtv* result in severe defects in the formation and assembly of the luminal chitin matrix (Araujo *et al.*, 2005; Devine *et al.*, 2005; Moussian *et al.*, 2006b; Tonning *et al.*, 2006, 2005). These defects are accompanied by irregular diametric expansion of the tracheal tubes, which show local constrictions and dilations. In addition, these mutants develop over-elongated and tortuous tracheal branches. The overall apical basal polarity of the tracheal epithelium, the organization of the AJs and the SJ paracellular barrier are not affected by mutations interfering with chitin biosynthesis and assembly. This suggests that the transient apical matrix provides a signal or a template for the uniform expansion of the tracheal tubes. Interestingly, *kkv* mutants show an irregular subapical cytoskeleton (Tonning *et al.*, 2005), suggesting that the chitin cable provides cues for the organization of the apical cell surface and cytoskeleton. *verm* and *serp* encode two related putative luminal chitin deacetylases with a chitin-binding domain. Verm and Serp are required for the structural modification of the apical ECM and the termination of tube elongation (Luschnig *et al.*, 2006; Wang *et al.*, 2006). Collectively, the analysis of the mutants in the "chitin group" of genes suggests that the apical chitin matrix provides either a physical scaffold that defines the shape of the underlying epithelial cells, or it locally signals back to the epithelial cells to adjust their shape in a coordinated manner. The apical ECM defects of mutants in the "chitin group" are not confined to the trachea: they are evident also in the epidermis and other ectodermally derived epithelial tissues (Luschnig *et al.*, 2006; Ostrowski *et al.*, 2002).

The third group of tube size regulators consists of genes encoding proteins associated with the paracellular SJs. SJs are located just basally of the AJs in ectodermally derived epithelial tissues in insects. SJs function as a barrier to control the diffusion of solutes between epithelial cells. This selective paracellular barrier function is analogous to the function of tight junctions (TJs) in vertebrates. Notably, SJs and TJs also share a subset of their constituent proteins, such as claudins (Jane *et al.*, 2005; Wu and Beitel, 2004). Many genes encoding SJ-associated proteins play central roles in tracheal tube size control. These genes include *Na-K ATPase* $\alpha$

(*Atpα*), *melanotransferrin* (*mtf*), *megatrachea* (*mega*), *sinu*, *bulbous/lachesin* (*lac*), *nervana 2* (*nrv2*), *neurexin IV* (*nrx IV*), *neuroglian* (*nrg*), *fasciclin II* (*fasII*), *scribble* (*scrib*), *lethal giant larvae* (*lgl*), *discs large* (*dlg*), *cor*, and *yurt* (*yrt*) (Behr *et al.*, 2003; Beitel and Krasnow, 2000; Genova and Fehon, 2003; Hemphala *et al.*, 2003; Lamb *et al.*, 1998; Laprise *et al.*, 2010; Llimargas *et al.*, 2004; Paul *et al.*, 2007, 2003; Tiklova *et al.*, 2010; Wu *et al.*, 2004, 2007). Mutants lacking these genes show over-elongated and convoluted tracheal tubes, and many of them are also compromised in the paracellular barrier function. Interestingly, *scrib*, *lgl*, *dlg*, *cor*, *and yrt* encode SJ-associated proteins that are also required for the establishment of epithelial apico-basolateral polarity (Laprise *et al.*, 2010). One possible mechanism by which, SJ proteins control tracheal tube size is that of regulating the activity of the apical polarity proteins such as Crumbs (Crb) and atypical protein kinase C (aPKC) (Laprise *et al.*, 2010). Another way to explain the additional functions of the SJ-associated proteins in tube size control derives from their involvement in the assembly of the luminal chitin matrix (Wu and Beitel, 2004). SJs have been directly implicated in the luminal deposition of proteins that modify the tracheal lining. The luminal chitin deacetylases Verm and Serp are retained in tracheal cells of mutants affecting SJ components, which suggests that a function of SJ components is to facilitate the targeting of luminal proteins to the apical cell surface (Wang *et al.*, 2006).

Although both Grh and SJ-associated proteins control tracheal tube elongation, it is likely that Grh regulates tube elongation independently of the functions of SJs, since: (1) *grh* mutants show normal levels and localization of the typical SJ proteins, Nrx and Cor, while mutations in any of the *bona fide* SJ components disrupt the localization of the other components and SJ assembly; (2) double mutants for *grh* and either of the two SJ proteins, *fasII* and *ATPα*, show a more pronounced tracheal tube convolution than single mutants. This suggests that Grh controls tube elongation in parallel to the genes involved in SJ assembly (Hemphala *et al.*, 2003). It is also unlikely that Grh regulates tracheal tube size by antagonizing the expression or the activity of the apical domain determinant Crb, as the levels and localizations of Crb and its negative regulator Cor are intact in *grh* mutants (Hemphala *et al.*, 2003).

Two characteristic features of *grh* mutants are the compromised tracheal taenidial structure and the reduction of the luminal chitin matrix (Hemphala *et al.*, 2003; Luschnig *et al.*, 2006). These features suggest that *grh* controls tracheal tube size at least partially through regulating the expression of genes involved in the chitin biogenesis and maturation.

It will be interesting to characterize the cellular defects of mice deficient in *ghrl* genes, since the expression of these genes is detected in a wide range of tubular tissues, such as the lung and the kidney, undergoing branching morphogenesis (Auden *et al.*, 2006).

## 2.4. Grh in the development of the central nervous system

The *Drosophila* larval CNS is composed of the optic lobes, the central brain (CB), and the ventral nerve cord (VNC). The VNC can be further divided into thoracic and abdominal segments. The neuronal stem-cell-like progenitors called neuroblasts generate the neurons and glia cells in all regions of the *Drosophila* CNS. A fundamental question in the field of neurobiology is how a neural progenitor cell can generate distinct neurons and glia during development. In *Drosophila* embryos, the dividing neuroblasts sequentially express a temporal cascade of transcription factors including Hunchback, Krüppel, Pdm, and Castor (Brody and Odenwald, 2000; Grosskortenhaus *et al.*, 2006; Isshiki *et al.*, 2001; Kambadur *et al.*, 1998). As a result, the neuronal progeny generated from these neuroblasts will have at least four different temporal fates. In addition, a fifth factor, the nuclear receptor Seven-up, is transiently expressed in early embryonic neuroblasts and mediates the switch from Hunchback to Krüppel (Kanai *et al.*, 2005; Mettler *et al.*, 2006). Consequently, mutating or misexpressing these temporal genes results in fate changes of progenies generated by the corresponding neuroblasts (Kanai *et al.*, 2005; Mettler *et al.*, 2006). Castor and Seven-up play multiple roles in many different neuroblast lineages during late neurogenesis (Maurange *et al.*, 2008). Interestingly, Castor also activates opposing temporal feed-forward loops to mediate the neuronal subtype specification within a lineage. Specifically, Castor triggers two different feed-forward loops in its temporal windows. The first loop specifies a unique cell type, whereas the second loop suppresses the first loop to determine an alternate cell fate. *grh* is one of the genes activated in the second loop, and represses Castor to allow the specification of additional cell fates (Baumgardt *et al.*, 2009).

Most embryonic neuroblasts continue dividing during postembryonic stages and generate neurons that will function in the adult fly. Grh is first switched on, possibly activated directly by Castor, in neuroblasts toward the end of embryonic neurogenesis and serves as a terminal embryonic CNS lineage marker (Almeida and Bray, 2005; Baumgardt *et al.*, 2009; Cenci and Gould, 2005; Maurange *et al.*, 2008; Uv *et al.*, 1997). Interestingly, Grh is expressed in postembryonic neuroblasts and their progenitors, the ganglion mother cells (GMC), but not in neurons. Grh regulates the late region-specific patterns of neural proliferation during postembryonic stages. In *grh* mutants, the thoracic neuroblasts divide less and produce fewer progeny than wild type. In contrast, abdominal neural proliferation is increased compared to wild type (Almeida and Bray, 2005; Cenci and Gould, 2005). Mechanistically, Grh provides the competence for the abdominal neuroblasts to undergo Abdominal-A dependent apoptosis. Interestingly, Grh regulates DE-Cad expression in postembryonic neuroblasts, while DE-Cad regulates the proliferation of postembryonic neuroblasts. Therefore, DE-Cad may be at least partially responsible for Grh function in the regional specification of neuroblast proliferation patterns (Almeida and Bray, 2005).

## 2.5. Other Grh functions in epithelial tissues

Apart from its function in embryonic epithelial maturation and tracheal tube size control, Grh is also required for the differentiation of the adult epidermis in *Drosophila* (Lee and Adler, 2004). For example, *grh* mutant clones in the wing show a delay in the formation of wing hairs and often form multiple wing hairs with PCP defects. *grh* mutant clones are also defective in pigmentation and in wing vein specification. Occasionally, *grh* mutant cells are larger than normal in the pupal wing. In addition, it has been suggested that Grh regulates the expression of the DNA replication-related proliferating cell nuclear antigen (*PCNA*) gene in *Drosophila* (Hayashi *et al.*, 1999). The functions of Grh and its related targets in *Drosophila* are summarized in Fig. 2.2.

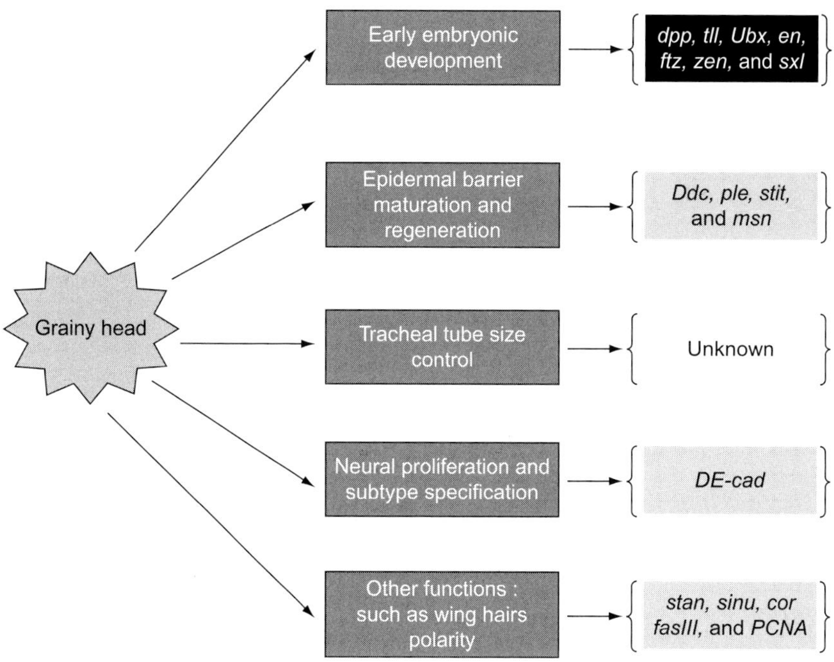


**Figure 2.2** Summary of Grh functions and relevant targets in *Drosophila* development and epidermal wounding healing. Except stan, all genes listed here are directly regulated by Grh. Negatively and positively regulated genes are shown in black and light gray boxes, respectively.

## 3. The Expression and Function of *grhl* Genes in Vertebrates

The mouse *grhl* (*grhl*1–3) genes are expressed in the surface ectoderm and in other epithelial tissues, including the oral cavity, urogenital bladder, and gastrointestinal tract (Auden *et al.*, 2006). However, they each exhibit differential spatiotemporal expression patterns during development. For example, *grhl*-1 expression in the surface ectoderm is dramatically increased from E10.5 to E17.5, while *grhl*-2 and *grhl*-3 expression decrease progressively during the same period. In addition, *grhl*-2 is widely expressed throughout the surface ectoderm at E8.5, whereas *grhl*-3 is only expressed in a small population of nonneural ectoderm cells during this period. The "non-overlapping" expression patterns of *grhl* genes are also detected in non-ectodermal tissues, such as the heart, the lung, and the kidney. The different spatial and temporal expression patterns suggest that the *grhl* genes are not fully redundant with each other during development, despite their extensive sequence identity (Auden *et al.*, 2006).

Indeed, gene targeting in mice has demonstrated that *grhl* genes have related but remarkably different functions during mouse embryogenesis. Mice lacking *grhl*-1 exhibit an abnormal hair coat, defective hair anchoring, and palmoplantar keratoderma (Wilanowski *et al.*, 2008). In contrast, both *grhl*-2 and *grhl*-3 mutant mice show defective epithelial differentiation and impaired neural tube closure (Boglev *et al.*, 2011; Rifat *et al.*, 2010; Yu *et al.*, 2006). Further, the neural tube defects in the *axial defects* mutant mouse were mapped to a mutation causing overexpression of *grhl2* (Brouns *et al.*, 2011).

The *grhl-3* null mice display severe defects in epidermal barrier function (Ting *et al.*, 2005a,b; Yu *et al.*, 2008, 2006, 2009). *Grhl*-3 is also required for efficient wound healing, and eyelid closure (Gustavsson *et al.*, 2008; Hislop *et al.*, 2008; Ting *et al.*, 2005a, 2003a; Yu *et al.*, 2008). Interestingly, *grhl-1* and *grhl-3* function independently of each other, as no dosage-sensitive genetic interaction was observed using mutants in these two genes. In contrast, loss of one allele of *grhl-2* in *grhl-3* null embryos dramatically enhances the eyelid closure defect at birth. In addition, expression of *grhl-2* from the control elements of the *grhl-3* locus can rescue the *grhl-3* null mice wound closure phenotype. However, the skin barrier formation defect remains in these embryos. This is due to an unexpected target specificity of Grhl-2 and Grhl-3. Only Grhl-3 regulates the expression of transglutaminase 1 (TGase1), although Grh-2 and -3 share other targets involved in wound repair. These observations suggest that *grhl* genes have both unique and cooperative roles during epidermal development (Boglev *et al.*, 2011). *grhl* genes are crucial for epidermal development in other vertebrates as well. For example, disruption of *grhl*-1 activity in *Xenopus laevis* results in a severe defect in terminal epidermal differentiation, associated with a dramatically decreased expression of the epidermal keratin gene *XK81A1*(Tao *et al.*, 2005).

*grhl* genes have been implicated in the pathogenesis of several human diseases. A single nucleotide polymorphism found in the human *grhl-2* gene has been associated with age-related hearing loss in a European population, although this association does not exist in Han Chinese (Lin *et al.*, 2011; Peters *et al.*, 2002; Van Laer *et al.*, 2008). Importantly, the wild type allele of human *grhl2* can rescue the inner-ear defects of a Zebrafish *grhl-2b* mutant line, suggesting an evolutionarily conserved function of Grhl-2 in otic development (Han *et al.*, 2011). Human *grhl-2* is also upregulated in human oral squamous cell carcinoma cells and may play a crucial role in telomerase activation during cellular immortalization (Kang *et al.*, 2009). Further, Grhl-2 regulates human telomerase reverse transcriptase gene expression through an epigenetic mechanism and controls the cellular life span in normal human keratinocytes (Chen *et al.*, 2010).

Although the structure of the mammalian skin is very different from the insect epidermis, the function of *grhl* genes in maintaining epidermal integrity has been highly conserved during evolution (reviewed in Harden, 2005; Jane *et al.*, 2005; Moussian and Uv, 2005; Stramer and Martin, 2005). In insects, a single layer of epidermal cells and the extracellular cuticle form the surface barrier. In contrast, the mammalian epidermis is composed of several cell layers. The outermost layer is the stratum corneum, which contains dead granulocytes, encaged in a cornified envelope of cross-linked proteins, keratin cables, and lipids (Jane *et al.*, 2005). In mice, *TGase 1* encodes the enzyme that cross-links components of the stratum corneum and renders it impermeable to water and solutes. Not surprisingly, mice lacking *TGase 1* show defects in the development and maturation of the surface epidermal barriers (Matsuki *et al.*, 1998). The degradation of nuclei and the keratohyalin F-granules in the stratum corneum is incomplete, and consequently the mutants die 4–5h after birth. Homology searches and genetic screens for genes involved in cuticle formation in *Drosophila* had failed to identify homologues of the genes encoding keratin and other structural components of the mammalian barrier. Ddc and Ple have completely different chemical functions than TGase1, but perform similar biological functions. Mouse *TGase 1* is a direct target of *grhl-3* (Ting *et al.*, 2005a) while *grh* directly regulates the epidermal expression of *Ddc* during development in flies (Bray and Kafatos, 1991). In addition, *grh* activates *Ddc* and *ple* expression at wound sites in flies (Mace *et al.*, 2005). Similarly *grhl-3* and *TGase 1* mutant mice show defects in epidermal barrier formation and wound healing (Inada *et al.*, 2000; Matsuki *et al.*, 1998; Ting *et al.*, 2005a). Similarly, knockdown of the nematode (*Caenorhabditis elegans*) *grh* function by RNA interference causes fragile cuticles (Venkatesan *et al.*, 2003). Thus, Grhl transcription factors emerge as phylogenetically conserved regulators of epidermal barrier formation and repair.

It is worth noting that both mouse and *Drosophila* Grh also regulate the expression of genes encoding components of junctional complexes and genes involved in the PCP signaling pathway. The mouse Grhl2 is crucial for the expression of *E-cadherin* and *claudin4*, which encode components of AJ and the TJ complex, respectively (Werth *et al.*, 2010). Grhl2 also regulates E-cadherin expression in the nonneural ectoderm during neural fold fusion (Pyrgaki *et al.*, 2011). In addition, Claudin 1 protein expression is decreased in the epidermis of *grhl-3* mutant mice (Yu *et al.*, 2006). Grhl-3 also directly regulates *RhoGEF19*, which encodes a homolog of the RhoA activator involved in planar polarization during epidermal wound repair in *Xenopus* (Caddy *et al.*, 2010).

Since Grh also controls genes involved in epithelial planar polarization and junctional assembly in flies, these results suggest that some of the important downstream targets of mammalian and fly *grh* genes are also conserved. They also suggest that metazoan Grh factors may have acquired a key role in establishing or maintaining epithelial cell adhesion and planar polarity during tissue morphogenesis.

## 4. Molecular Mechanisms of Grh Function

### 4.1. Upstream regulation by signaling

There is substantial evidence that Grh activity is regulated during development and epidermal wound healing in flies (Fig. 2.3). Grh binds to the torso response elements (tor-RE) of the *tailless* (*tll*) promoter, and the domain of *tll* expression is expanded in embryos lacking maternal Grh. Grh can also be phosphorylated by ERK *in vitro*. These observations suggest that the RTK torso can control Grh activity during early embryogenesis (Liaw *et al.*, 1995). The spatial and temporal expression of the *Gbe-LacZ* reporter in the tracheal airways has provided the first indication that Grh activity is regulated *in vivo* (Hemphala *et al.*, 2003; Uv *et al.*, 1997). The *Gbe-LacZ* construct contains three Grh-binding sites and is selectively upregulated in specialized fusion and terminal cells at the tips of the tracheal branches. Further analysis in the trachea has shown that fibroblast growth factor (FGF) signaling regulates Grh activity. When *branchless* (bnl, a *Drosophila* FGF homolog), or the activated form of Breathless (btl, a *Drosophila* FGF receptor), is ectopically expressed in tracheal cells, *Gbe-LacZ* expression becomes significantly upregulated. The level of Grh protein remains unchanged under these conditions. Bnl signaling is also required for the transcriptional activity of Grh in the trachea, since the expression of *Gbe-LacZ* is both reduced and uniform in *bnl* and *btl* mutant embryos. Importantly, the level of *Gbe-LacZ* expression is not altered by mutations in the *bnl* signaling nuclear effector *pointed*. Because ERK can phosphorylate Grh *in vitro* (Liaw *et al.*,

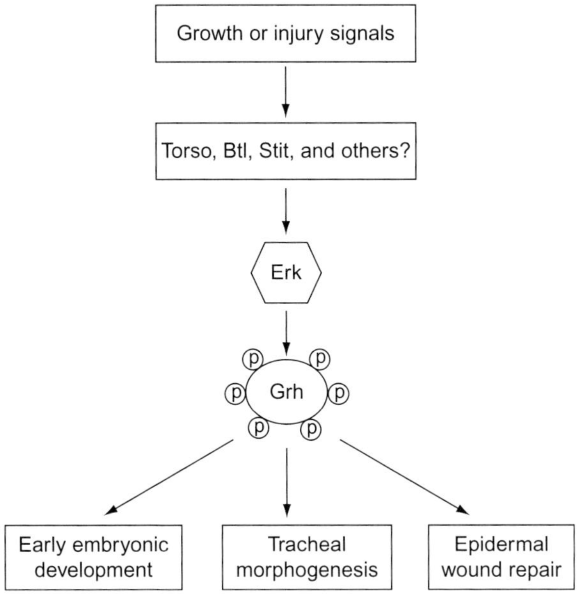


**Figure 2.3** Grh activity is regulated by upstream signaling and ERK during early embryonic development, tracheal morphogenesis, and epidermal wound healing.

1995), it is likely that Grh activity is directly modulated by Bnl-induced phosphorylation during tracheal development.

ERK phosphorylation is also rapidly elevated in epidermal cells upon wounding and contributes to the induction of the *Ddc* wound reporter, which also carries essential Grh-binding sites (Mace *et al.*, 2005). Interestingly, ERK can directly regulate Grh activity through phosphorylation during epidermal wound healing (Kim and McGinnis, 2011). Serine 91 is the principal residue in Grh that is phosphorylated by ERK. However, ERK can also phosphorylate Grh at other serines, such as Serine 88, and some threonines. These phosphorylation sites are required for the Grh-mediated activation of the *Ddc* and *Msn* reporters upon wounding but, intriguingly, these sites are dispensable for Grh function in epidermal barrier maturation. This suggests that Grh is controlled by different mechanisms during wound healing and epidermal development (Kim and McGinnis, 2011). Assuming that both ERK and Grh are rapidly activated around wound sites and that ERK directly phosphorylates Grh, it is reasonable to speculate that wounding triggers a signaling cascade that activates ERK, which in turn phosphorylates the Grh protein and activates Grh-mediated gene expression. However, Bnl signaling is unlikely to account for this, since the *btl* receptor is not present in the epidermis. The Ret family RTK Stit is a candidate receptor for triggering Grh phosphorylation by ERK in the epidermis. Upon wounding, Stit induces ERK phosphorylation along with the Grh-dependent expression of *stit* and cuticle repair genes at the wound sites (Wang *et al.*, 2009). However, other signaling receptors must also exist, since neither ERK activity nor the induction of Grh-dependent wound response genes is abolished in *stit* mutants (Wang *et al.*, 2009). A candidate receptor that may also activate Grh upon wounding is *pvr*. It encodes a RTK related to platelet-derived growth factor and vascular endothelial growth factor receptors (Duchek *et al.*, 2001). Pvr and one of its ligands, Pvf1, are also required for epidermal wound closure in *Drosophila* larvae (Wu *et al.*, 2009). Pvr signaling acts upstream of JNK signaling during *Drosophila* thorax closure, a developmental event that possesses several of the aspects of epithelial wound healing (Ishimaru *et al.*, 2004). Noncanonical Wingless signaling may also regulate Grh activity in flies. Clones of *grh* mutant cells in the wing display PCP defects. In this context, *grh* shows dosage-sensitive genetic interactions with several mutants in genes involved in planar epithelial polarity (Lee and Adler, 2004).

There is increasing evidence indicating that signaling events control the expression levels of Grhl factors in vertebrates. The expression of *Xenopus Grhl-1* is regulated by BMP4 signaling (Tao *et al.*, 2005). The rat *Grhl-3* gene has also been identified as a NGF-dependent transcriptional target of p53 in the ectoderm-derived PC12 pheochromocytoma cells (Brynczka *et al.*, 2007). In addition, tumor necrosis factor alpha can induce the expression of human *Grhl-3* in the mammary carcinoma cell line MCF7 (Guardiola-Serrano *et al.*, 2008). However, there is no direct evidence showing that signaling regulates Grhl factor activity posttranscriptionally. In summary, the results suggest that multiple signaling pathways control the expression and/or the activity of *Grhl* factors during development and pathogenesis.

## 4.2. The partners of Grh

In addition to posttranscriptional modification, Grh activity may also be controlled by its association with other transcription factors. An artificial *lacZ* reporter containing both the Grh binding site and a "Suppressor of hairless" binding site (termed Gbe+Su(H)$_{m8}$) is expressed only in tracheal fusion cells. This localized expression reflects the activation and requirement of Notch signaling in the trachea (Furriols and Bray, 2001), and contrasts with the wide expression of *Gbe-LacZ* in the epidermis and the entire trachea. Removing Su(H) converts the expression of Gbe+Su(H)$_{m8}$ to the same expression pattern as that of *Gbe-LacZ*, which indicates that Su(H) can suppress the Grh-dependent expression of *Gbe-LacZ* when its binding site is combined with Grh-binding sites. Further, ectopic expression of the Notch intracellular domain, which is known to convert Su(H) from a repressor to an activator, dramatically induces the expression of

Gbe+Su(H)$_{m8}$ (Furriols and Bray, 2001). Although the data do not show that Notch signaling regulates Grh activity on endogenous promoters, they suggest that Grh cooperates with other factors, and that its transcriptional activity is modulated by the input of the associated factors.

Consistent with this hypothesis, Grh forms complexes with the Polycomb-group protein Pleiohomeotic (PHO) *in vitro*. This association facilitates the binding of both proteins to their respective targeting site and modulates PHO function (Blastyak *et al.*, 2006). Grh also interacts with another Polycomb protein, Dring, *in vitro* and *in vivo* (Tuckfield *et al.*, 2002). In mice, Grhl-3 cooperates with the LIM-only protein LIMO4, which is also expressed in the developing mammalian epidermis (Hislop *et al.*, 2008; Yu *et al.*, 2006).

In the context of wound responses, both Grh and AP-1 binding sites are present in the wound-responsive enhancers isolated from *Ddc*, *ple*, *msn*, and *kkv*. These results strongly suggest that Grh and AP-1 factors cooperatively regulate common downstream genes, at least during wound healing. This hypothesis is further supported by the fact that both Grh and AP-1 binding sites are essential for the activation of the *Ddc*, *ple*, and *msn* wound reporters (Mace *et al.*, 2005; Pearson *et al.*, 2009).

## 4.3. Competition with other factors for DNA binding

The idea that Grh regulates gene expression by competing with other factors for DNA binding comes from the analysis of the precise timing of gene expression in the *Drosophila* precellular blastoderm (pre-CB) (Harrison *et al.*, 2010). Many promoters of these pre-CB expressed genes contain TAGteam elements. These elements are defined by CAGGTAG and related sequences. Both Grh and the transcriptional activator Zelda (Zld) are TAGteam-binding factors. In addition, Grh competes with Zld for binding to the TAGteam sites in the promoters of *sex lethal* (*sxl*) and *Zerknüllt* (*zen*), although they bind to different TAGteam sequences with varying affinities. Consistently, overexpression of Grh in the early embryo causes defects in cell division, phenocopying Zld depletion (Harrison *et al.*, 2010). Grh binds to repressive elements in the *tll* and *dpp* promoters (Huang *et al.*, 1995; Liaw *et al.*, 1995). Further, mutation of the Grh-binding site can cause an increase in *dpp* expression, and Grh competes with an unidentified activator for binding to sites in the *dpp* promoter (Huang *et al.*, 1995). Therefore, one possible mechanism by which Grh represses gene expression in the pre-CB embryos is through competing with an activator for DNA binding. In addition, as Zld and Grh bind differentially to discrete TAGteam variants, activation at different promoters can be fine-tuned by a combination of the TAGteam sequences present. Thus, Grh, Zld, and the TAGteam elements could combinatorially regulate transcription to achieve the precise timing of gene expression in the pre-CB embryo (Harrison *et al.*, 2010). A summary of the known mechanisms by which Grh controls transcription is shown in Fig. 2.4.

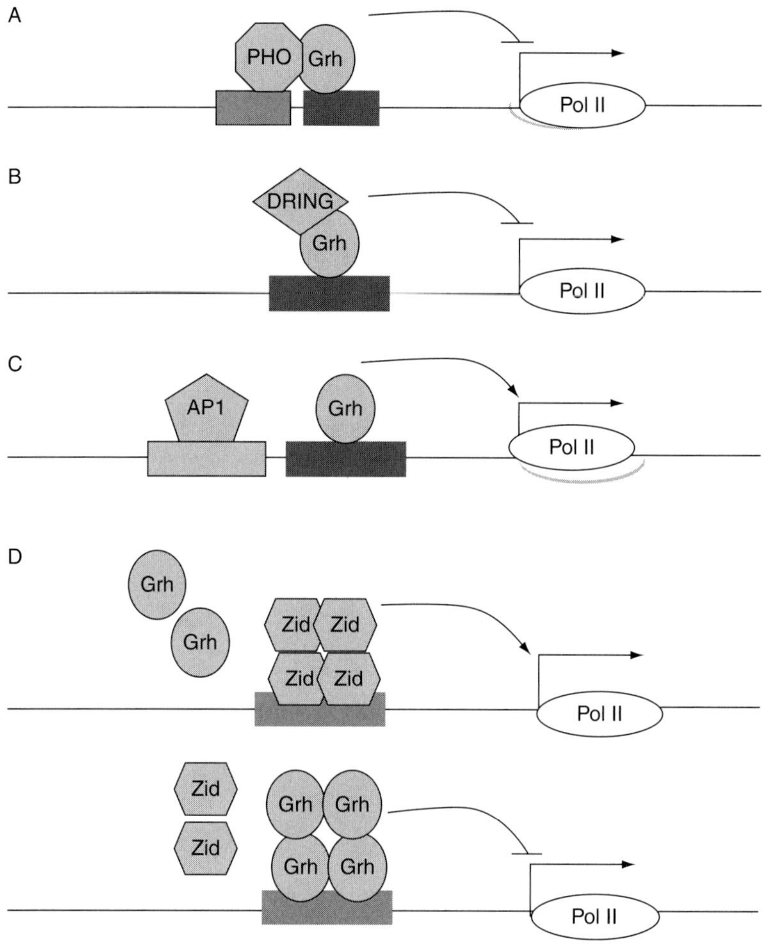


**Figure 2.4** The partners of Grh. (A) Grh can interact physically with the Polycomb-group protein PHO, and this interaction facilitates the binding of both factors to their respective targeting site. Consequently, downstream genes will be repressed in this context. (B) DRING and Grh can form complexes in the promoters of genes repressed by Grh, such as *dpp*. (C) During wound healing, Grh and AP1 associate with their respective binding site and cooperatively induce the expression of wound repair genes. (D) Grh competes with Zld for DNA binding. Both Grh and Zld can bind to the TAGteam elements in the promoters of genes expressed in the precellular blastoderm (upper panel). A high concentration of Zld will compete out Grh for DNA binding, resulting in the activation of gene expression. Conversely, a higher amount of Grh will prevent Zld from DNA binding and repress gene expression (lower panel).

## 4.4. Alternative splicing and tissue-specific Grh functions

A single *grh* gene in *Drosophila* plays several roles, ranging from epidermal maturation to the specification of neuroblasts during CNS development. How does *grh* fulfill its function in such different developmental processes?

One way is through the generation of tissue-specific isoforms by alternative splicing. Alternative splicing of exons is responsible for generating a neuroblast-specific isoform, the O-isoform (Grh-O), of the Grh protein. The O-specific transcript of Grh is present at high levels in neuroblasts but not in the developing optic lobes. In contrast, the N-transcript is expressed in larval epidermis, trachea, and foregut as well as in imaginal discs and optic lobes (Uv *et al.*, 1997). A mutation that selectively abolishes the O-isoform results in pupal and adult lethality, while other *grh* mutations lead to embryonic lethality. Interestingly, reporter genes containing different Grh-binding sites exhibit tissue-specific patterns of expression that correlate with the Grh isoforms, suggesting that the alternative splicing alters the repertoire of target genes controlled in the neuroblasts. The *Ddc* gene contains two Grh-binding sites: be-1 and be-2 (Johnson *et al.*, 1989; Uv *et al.*, 1997). The DNA sequences of be-1 and be-2 are ATAACCGGTTTC and TGAACCGGTCCT and are located 598 and 76 bp upstream of the transcription site of *Ddc* gene, respectively. These two Grh-binding elements can direct Grh-mediated expression of the *lacZ* reporter to different cells. Expression of the be-1 reporter has been detected in the embryonic epidermis and in tracheal cells, whereas the be-2 reporter has been detected only in neuroblasts. The expression of both reporters is greatly reduced in *grh* mutant embryos, suggesting that Grh is responsible for most aspects of their expression (Uv *et al.*, 1997). Therefore, these reporters report faithfully Grh transcriptional activity *in vivo*. These results suggest that Grh can activate the expression of reporter be-2 in a tissue-specific manner, which correlates with the expression pattern of the O-isoform of Grh. Since be-1 is palindromic, while be-2 is not, it has been suggested that be-1 and be-2 represent targets for homodimers and heterodimers of Grh, respectively (Uv *et al.*, 1997).

Importantly, differential splicing of a mammalian *grhl* gene has also been reported. Human *grhl3* generates three different isoforms depending on the usage of two alternative first coding exons and of the second exon. These isoforms are expressed differently, but it remains unknown whether they have different functions (Ting *et al.*, 2003b).

## 5. Conclusions and Future Perspectives

The metazoan Grh factors are phylogenetically conserved regulators of genes involved in epidermal barrier formation. The recent discovery of Grhl factors in fungi and choanoflagelates will prompt to investigate whether Grh also controls extracellular barrier formation and cell wall assembly in unicellular organisms.

Mutations in Grh factors have profound effects in epithelial development and function. The expression of cell junction proteins and cytoskeletal regulators is reduced in several tissues in flies and mice lacking Grh. It will be important to elucidate whether Grh is a direct ancient regulator of cell adhesion, epithelial tissue formation and maintenance.

Grh acts both as an activator and as a repressor of gene expression. Its activity can be modified by signaling, by its association with other transcription factors, and by modulating its expression levels during development. A major challenge will be to define the molecular mechanisms of Grh function in different contexts. How many genes are regulated directly by Grh? How are the tissue-specific and wound-specific functions of Grh controlled?

## ACKNOWLEDGMENTS

We thank the members of the Samakovlis laboratory for suggestions on the text. We also thank Dr. Arun Unni for critical reading of the manuscript. Our work on Grh and wound healing is supported by the Swedish Research Council and the Swedish Cancer Society.

## REFERENCES

Almeida, M. S., and Bray, S. J. (2005). Regulation of post-embryonic neuroblasts by Drosophila Grainyhead. *Mech. Dev.* **122,** 1282–1293.

Andrew, D. J., and Ewald, A. J. (2010). Morphogenesis of epithelial tubes: Insights into tube formation, elongation, and elaboration. *Dev. Biol.* **341,** 34–55.

Arakane, Y., Dixit, R., Begum, K., Park, Y., Specht, C. A., Merzendorfer, H., Kramer, K. J., Muthukrishnan, S., and Beeman, R. W. (2009). Analysis of functions of the chitin deacetylase gene family in Tribolium castaneum. *Insect Biochem. Mol. Biol.* **39,** 355–365.

Araujo, S. J., Aslam, H., Tear, G., and Casanova, J. (2005). mummy/cystic encodes an enzyme required for chitin and glycan synthesis, involved in trachea, embryonic cuticle and CNS development—Analysis of its role in Drosophila tracheal morphogenesis. *Dev. Biol.* **288,** 179–193.

Attardi, L. D., and Tjian, R. (1993). Drosophila tissue-specific transcription factor NTF-1 contains a novel isoleucine-rich activation motif. *Genes Dev.* **7,** 1341–1353.

Attardi, L. D., Von Seggern, D., and Tjian, R. (1993). Ectopic expression of wild-type or a dominant-negative mutant of transcription factor NTF-1 disrupts normal Drosophila development. *Proc. Natl. Acad. Sci. USA* **90,** 10563–10567.

Auden, A., Caddy, J., Wilanowski, T., Ting, S. B., Cunningham, J. M., and Jane, S. M. (2006). Spatial and temporal expression of the Grainyhead-like transcription factor family during murine development. *Gene Expr. Patterns* **6,** 964–970.

Babcock, D. T., Brock, A. R., Fish, G. S., Wang, Y., Perrin, L., Krasnow, M. A., and Galko, M. J. (2008). Circulating blood cells function as a surveillance system for damaged tissue in Drosophila larvae. *Proc. Natl. Acad. Sci. USA* **105,** 10017–10022.

Baek, S. H., Kwon, Y. C., Lee, H., and Choe, K. M. (2010). Rho-family small GTPases are required for cell polarization and directional sensing in Drosophila wound healing. *Biochem. Biophys. Res. Commun.* **394,** 488–492.

Baumgardt, M., Karlsson, D., Terriente, J., Diaz-Benjumea, F. J., and Thor, S. (2009). Neuronal subtype specification within a lineage by opposing temporal feed-forward loops. *Cell* **139,** 969–982.

Behr, M., Riedel, D., and Schuh, R. (2003). The claudin-like megatrachea is essential in septate junctions for the epithelial barrier function in Drosophila. *Dev. Cell* **5,** 611–620.

Beitel, G. J., and Krasnow, M. A. (2000). Genetic control of epithelial tube size in the Drosophila tracheal system. *Development* **127,** 3271–3282.

Blastyak, A., Mishra, R. K., Karch, F., and Gyurkovics, H. (2006). Efficient and specific targeting of Polycomb group proteins requires cooperative interaction between Grainyhead and Pleiohomeotic. *Mol. Cell. Biol.* **26,** 1434–1444.

Boglev, Y., Wilanowski, T., Caddy, J., Parekh, V., Auden, A., Darido, C., Hislop, N. R., Cangkrama, M., Ting, S. B., and Jane, S. M. (2011). The unique and cooperative roles of the Grainy head-like transcription factors in epidermal development reflect unexpected target gene specificity. *Dev. Biol.* **349,** 512–522.

Bray, S. J., and Kafatos, F. C. (1991). Developmental function of Elf-1: An essential transcription factor during embryogenesis in Drosophila. *Genes Dev.* **5,** 1672–1683.

Bray, S. J., Burke, B., Brown, N. H., and Hirsh, J. (1989). Embryonic expression pattern of a family of Drosophila proteins that interact with a central nervous system regulatory element. *Genes Dev.* **3,** 1130–1145.

Brody, T., and Odenwald, W. F. (2000). Programmed transformations in neuroblast gene expression during Drosophila CNS lineage development. *Dev. Biol.* **226,** 34–44.

Brouns, M. R., De Castro, S. C., Terwindt-Rouwenhorst, E. A., Massa, V., Hekking, J. W., Hirst, C. S., Savery, D., Munts, C., Partridge, D., Lamers, W., Kohler, E., van Straaten, H. W., *et al.* (2011). Over-expression of Grhl2 causes spina bifida in the Axial defects mutant mouse. *Hum. Mol. Genet.* **20,** 1536–1546.

Brynczka, C., Labhart, P., and Merrick, B. A. (2007). NGF-mediated transcriptional targets of p53 in PC12 neuronal differentiation. *BMC Genomics* **8,** 139.

Caddy, J., Wilanowski, T., Darido, C., Dworkin, S., Ting, S. B., Zhao, Q., Rank, G., Auden, A., Srivastava, S., Papenfuss, T. A., Murdoch, J. N., Humbert, P. O., *et al.* (2010). Epidermal wound repair is regulated by the planar cell polarity signaling pathway. *Dev. Cell* **19,** 138–147.

Campos, I., Geiger, J. A., Santos, A. C., Carlos, V., and Jacinto, A. (2010). Genetic screen in Drosophila melanogaster uncovers a novel set of genes required for embryonic epithelial repair. *Genetics* **184,** 129–140.

Casanova, J. (2007). The emergence of shape: Notions from the study of the Drosophila tracheal system. *EMBO Rep.* **8,** 335–339.

Cenci, C., and Gould, A. P. (2005). Drosophila Grainyhead specifies late programmes of neural proliferation by regulating the mitotic activity and Hox-dependent apoptosis of neuroblasts. *Development* **132,** 3835–3845.

Chen, W., Dong, Q., Shin, K. H., Kim, R. H., Oh, J. E., Park, N. H., and Kang, M. K. (2010). Grainyhead-like 2 enhances the human telomerase reverse transcriptase gene expression by inhibiting DNA methylation at the 5'-CpG island in normal human keratinocytes. *J. Biol. Chem.* **285,** 40852–40863.

Christodoulidou, A., Briza, P., Ellinger, A., and Bouriotis, V. (1999). Yeast ascospore wall assembly requires two chitin deacetylase isozymes. *FEBS Lett.* **460,** 275–279.

Cohen, E. (2001). Chitin synthesis and inhibition: A revisit. *Pest Manag. Sci.* **57,** 946–950.

Devine, W. P., Lubarsky, B., Shaw, K., Luschnig, S., Messina, L., and Krasnow, M. A. (2005). Requirement for chitin biosynthesis in epithelial tube morphogenesis. *Proc. Natl. Acad. Sci. USA* **102,** 17014–17019.

Duchek, P., Somogyi, K., Jekely, G., Beccari, S., and Rorth, P. (2001). Guidance of cell migration by the Drosophila PDGF/VEGF receptor. *Cell* **107,** 17–26.

Dynlacht, B. D., Attardi, L. D., Admon, A., Freeman, M., and Tjian, R. (1989). Functional analysis of NTF-1, a developmentally regulated Drosophila transcription factor that binds neuronal cis elements. *Genes Dev.* **3,** 1677–1688.

Furriols, M., and Bray, S. (2001). A model Notch response element detects Suppressor of Hairless-dependent molecular switch. *Curr. Biol.* **11,** 60–64.

Galko, M. J., and Krasnow, M. A. (2004). Cellular and genetic analysis of wound healing in Drosophila larvae. *PLoS Biol.* **2,** E239.

Genova, J. L., and Fehon, R. G. (2003). Neuroglian, Gliotactin, and the Na+/K+ ATPase are essential for septate junction function in Drosophila. *J. Cell. Biol.* **161,** 979–989.

Ghabrial, A., Luschnig, S., Metzstein, M. M., and Krasnow, M. A. (2003). Branching morphogenesis of the Drosophila tracheal system. *Annu. Rev. Cell Dev. Biol.* **19,** 623–647.

Grose, R., and Martin, P. (1999). Parallels between wound repair and morphogenesis in the embryo. *Semin. Cell Dev Biol.* **10,** 395–404.

Grosskortenhaus, R., Robinson, K. J., and Doe, C. Q. (2006). Pdm and Castor specify late-born motor neuron identity in the NB7-1 lineage. *Genes Dev.* **20,** 2618–2627.

Guardiola-Serrano, F., Haendeler, J., Lukosz, M., Sturm, K., Melchner, H., and Altschmied, J. (2008). Gene trapping identifies a putative tumor suppressor and a new inducer of cell migration. *Biochem. Biophys. Res. Commun.* **376,** 748–752.

Gustavsson, P., Copp, A. J., and Greene, N. D. (2008). Grainyhead genes and mammalian neural tube closure. *Birth Defects Res. A Clin. Mol. Teratol.* **82,** 728–735.

Han, Y., Mu, Y., Li, X., Xu, P., Tong, J., Liu, Z., Ma, T., Zeng, G., Yang, S., Du, J., and Meng, A. (2011). Grhl2 deficiency impairs otic development and hearing ability in a zebrafish model of the progressive dominant hearing loss DFNA28. *Hum. Mol. Genet.* **20,** 3213–3226.

Harden, N. (2002). Signaling pathways directing the movement and fusion of epithelial sheets: Lessons from dorsal closure in Drosophila. *Differentiation* **70,** 181–203.

Harden, N. (2005). Cell biology. *Of grainy heads and broken skins. Science* **308,** 364–365.

Harrison, M. M., Botchan, M. R., and Cline, T. W. (2010). Grainyhead and Zelda compete for binding to the promoters of the earliest-expressed Drosophila genes. *Dev. Biol.* **345,** 248–255.

Hayashi, Y., Yamagishi, M., Nishimoto, Y., Taguchi, O., Matsukage, A., and Yamaguchi, M. (1999). A binding site for the transcription factor Grainyhead/Nuclear transcription factor-1 contributes to regulation of the Drosophila proliferating cell nuclear antigen gene promoter. *J. Biol. Chem.* **274,** 35080–35088.

Hemphala, J., Uv, A., Cantera, R., Bray, S., and Samakovlis, C. (2003). Grainy head controls apical membrane growth and tube elongation in response to Branchless/FGF signalling. *Development* **130,** 249–258.

Hislop, N. R., Caddy, J., Ting, S. B., Auden, A., Vasudevan, S., King, S. L., Lindeman, G. J., Visvader, J. E., Cunningham, J. M., and Jane, S. M. (2008). Grhl3 and Lmo4 play coordinate roles in epidermal migration. *Dev. Biol.* **321,** 263–272.

Huang, J. D., Dubnicoff, T., Liaw, G. J., Bai, Y., Valentine, S. A., Shirokawa, J. M., Lengyel, J. A., and Courey, A. J. (1995). Binding sites for transcription factor NTF-1/Elf-1 contribute to the ventral repression of decapentaplegic. *Genes Dev.* **9,** 3177–3189.

Huang, N., and Miller, W. L. (2000). Cloning of factors related to HIV-inducible LBP proteins that regulate steroidogenic factor-1-independent human placental transcription of the cholesterol side-chain cleavage enzyme, P450scc. *J. Biol. Chem.* **275,** 2852–2858.

Inada, R., Matsuki, M., Yamada, K., Morishima, Y., Shen, S. C., Kuramoto, N., Yasuno, H., Takahashi, K., Miyachi, Y., and Yamanishi, K. (2000). Facilitated wound healing by activation of the Transglutaminase 1 gene. *Am. J. Pathol.* **157,** 1875–1882.

Ishimaru, S., Ueda, R., Hinohara, Y., Ohtani, M., and Hanafusa, H. (2004). PVR plays a critical role via JNK activation in thorax closure during Drosophila metamorphosis. *EMBO J.* **23,** 3984–3994.

Isshiki, T., Pearson, B., Holbrook, S., and Doe, C. Q. (2001). Drosophila neuroblasts sequentially express transcription factors which specify the temporal identity of their neuronal progeny. *Cell* **106,** 511–521.

Jacinto, A., Woolner, S., and Martin, P. (2002). Dynamic analysis of dorsal closure in Drosophila: From genetics to cell biology. *Dev. Cell* **3,** 9–19.

Jane, S. M., Nienhuis, A. W., and Cunningham, J. M. (1995). Hemoglobin switching in man and chicken is mediated by a heteromeric complex between the ubiquitous transcription factor CP2 and a developmentally specific protein. *EMBO J.* **14,** 97–105.

Jane, S. M., Ting, S. B., and Cunningham, J. M. (2005). Epidermal impermeable barriers in mouse and fly. *Curr. Opin. Genet. Dev.* **15,** 447–453.

Janicke, M., Renisch, B., and Hammerschmidt, M. (2010). Zebrafish grainyhead-like1 is a common marker of different non-keratinocyte epidermal cell lineages, which segregate from each other in a Foxi3-dependent manner. *Int. J. Dev. Biol.* **54,** 837–850.

Johnson, W. A., McCormick, C. A., Bray, S. J., and Hirsh, J. (1989). A neuron-specific enhancer of the Drosophila dopa decarboxylase gene. *Genes Dev.* **3,** 676–686.

Jones, K. A., Luciw, P. A., and Duchange, N. (1988). Structural arrangements of transcription control domains within the 5′-untranslated leader regions of the HIV-1 and HIV-2 promoters. *Genes Dev.* **2,** 1101–1114.

Kambadur, R., Koizumi, K., Stivers, C., Nagle, J., Poole, S. J., and Odenwald, W. F. (1998). Regulation of POU genes by castor and hunchback establishes layered compartments in the Drosophila CNS. *Genes Dev.* **12,** 246–260.

Kanai, M. I., Okabe, M., and Hiromi, Y. (2005). seven-up Controls switching of transcription factors that specify temporal identities of Drosophila neuroblasts. *Dev. Cell* **8,** 203–213.

Kang, X., Chen, W., Kim, R. H., Kang, M. K., and Park, N. H. (2009). Regulation of the hTERT promoter activity by MSH2, the hnRNPs K and D, and GRHL2 in human oral squamous cell carcinoma cells. *Oncogene* **28,** 565–574.

Kim, M., and McGinnis, W. (2011). Phosphorylation of Grainy head by ERK is essential for wound-dependent regeneration but not for development of an epidermal barrier. *Proc. Natl. Acad. Sci. USA* **108,** 650–655.

Kim, C. H., Heath, C., Bertuch, A., and Hansen, U. (1987). Specific stimulation of simian virus 40 late transcription in vitro by a cellular factor binding the simian virus 40 21-base-pair repeat promoter element. *Proc. Natl. Acad. Sci. USA* **84,** 6025–6029.

Kwon, Y. C., Baek, S. H., Lee, H., and Choe, K. M. (2010). Nonmuscle myosin II localization is regulated by JNK during Drosophila larval wound healing. *Biochem. Biophys. Res. Commun.* **393,** 656–661.

Lamb, R. S., Ward, R. E., Schweizer, L., and Fehon, R. G. (1998). Drosophila coracle, a member of the protein 4.1 superfamily, has essential structural functions in the septate junctions and developmental functions in embryonic and adult epithelial cells. Mol. Biol. *Cell* **9,** 3505–3519.

Laprise, P., Paul, S. M., Boulanger, J., Robbins, R. M., Beitel, G. J., and Tepass, U. (2010). Epithelial polarity proteins regulate Drosophila tracheal tube size in parallel to the luminal matrix pathway. *Curr. Biol.* **20,** 55–61.

Lee, H., and Adler, P. N. (2004). The grainy head transcription factor is essential for the function of the frizzled pathway in the Drosophila wing. *Mech. Dev.* **121,** 37–49.

Liaw, G. J., Rudolph, K. M., Huang, J. D., Dubnicoff, T., Courey, A. J., and Lengyel, J. A. (1995). The torso response element binds GAGA and NTF-1/Elf-1, and regulates tailless by relief of repression. *Genes Dev.* **9,** 3163–3176.

Lim, L. C., Swendeman, S. L., and Sheffery, M. (1992). Molecular cloning of the alpha-globin transcription factor CP2. *Mol. Cell. Biol.* **12,** 828–835.

Lin, Y. H., Wu, C. C., Hsu, C. J., Hwang, J. H., and Liu, T. C. (2011). The grainyhead-like 2 gene (GRHL2) single nucleotide polymorphism is not associated with age-related hearing impairment in Han Chinese. *Laryngoscope* **121,** 1303–1307.

Llimargas, M., Strigini, M., Katidou, M., Karagogeos, D., and Casanova, J. (2004). Lachesin is a component of a septate junction-based mechanism that controls tube size and epithelial integrity in the Drosophila tracheal system. *Development* **131,** 181–190.

Luschnig, S., Batz, T., Armbruster, K., and Krasnow, M. A. (2006). serpentine and vermiform encode matrix proteins with chitin binding and deacetylation domains that limit tracheal tube length in Drosophila. *Curr. Biol.* **16,** 186–194.

Mace, K. A., Pearson, J. C., and McGinnis, W. (2005). An epidermal barrier wound repair pathway in Drosophila is mediated by grainy head. *Science* **308,** 381–385.

Martin, P., and Parkhurst, S. M. (2004). Parallels between tissue repair and embryo morphogenesis. *Development* **131,** 3021–3034.

Matsuki, M., Yamashita, F., Ishida-Yamamoto, A., Yamada, K., Kinoshita, C., Fushiki, S., Ueda, E., Morishima, Y., Tabata, K., Yasuno, H., Hashida, M., Iizuka, H., *et al.* (1998). Defective stratum corneum and early neonatal death in mice lacking the gene for transglutaminase 1 (keratinocyte transglutaminase). *Proc. Natl. Acad. Sci. USA* **95,** 1044–1049.

Maurange, C., Cheng, L., and Gould, A. P. (2008). Temporal transcription factors and their targets schedule the end of neural proliferation in Drosophila. *Cell* **133,** 891–902.

Merzendorfer, H. (2006). Insect chitin synthases: A review. *J. Comp. Physiol. B* **176,** 1–15.

Merzendorfer, H., and Zimoch, L. (2003). Chitin metabolism in insects: Structure, function and regulation of chitin synthases and chitinases. *J. Exp. Biol.* **206,** 4393–4412.

Mettler, U., Vogler, G., and Urban, J. (2006). Timing of identity: Spatiotemporal regulation of hunchback in neuroblast lineages of Drosophila by Seven-up and Prospero. *Development* **133,** 429–437.

Moussian, B., and Uv, A. E. (2005). An ancient control of epithelial barrier formation and wound healing. *Bioessays* **27,** 987–990.

Moussian, B., Schwarz, H., Bartoszewski, S., and Nusslein-Volhard, C. (2005). Involvement of chitin in exoskeleton morphogenesis in Drosophila melanogaster. *J. Morphol.* **264,** 117–130.

Moussian, B., Seifarth, C., Muller, U., Berger, J., and Schwarz, H. (2006a). Cuticle differentiation during Drosophila embryogenesis. *Arthropod Struct. Dev.* **35,** 137–152.

Moussian, B., Tang, E., Tonning, A., Helms, S., Schwarz, H., Nusslein-Volhard, C., and Uv, A. E. (2006b). Drosophila Knickkopf and Retroactive are needed for epithelial tube growth and cuticle differentiation through their specific requirement for chitin filament organization. *Development* **133,** 163–171.

Murata, T., Nitta, M., and Yasuda, K. (1998). Transcription factor CP2 is essential for lens-specific expression of the chicken alphaA-crystallin gene. *Genes Cells* **3,** 443–457.

Narasimha, M., Uv, A., Krejci, A., Brown, N. H., and Bray, S. J. (2008). Grainy head promotes expression of septate junction proteins and influences epithelial morphogenesis. *J. Cell. Sci.* **121,** 747–752.

Nusslein-Volhard, C. E. W., and Kluding, H. (1984). Mutations affecting the pattern of the larval cuticle in Drosophila melanogaster I. Zygotic loci on the second chromosome. *Wilhelm Roux's Arch. Dev. Biol.* **193,** 15.

Ostrowski, S., Dierick, H. A., and Bejsovec, A. (2002). Genetic control of cuticle formation during embryonic development of Drosophila melanogaster. *Genetics* **161,** 171–182.

Paul, S. M., Palladino, M. J., and Beitel, G. J. (2007). A pump-independent function of the Na, K-ATPase is required for epithelial junction function and tracheal tube-size control. *Development* **134,** 147–155.

Paul, S. M., Ternet, M., Salvaterra, P. M., and Beitel, G. J. (2003). The Na+/K+ ATPase is required for septate junction function and epithelial tube-size control in the Drosophila tracheal system. *Development* **130,** 4963–4974.

Pearson, J. C., Juarez, M. T., Kim, M., Drivenes, O., and McGinnis, W. (2009). Multiple transcription factor codes activate epidermal wound-response genes in Drosophila. *Proc. Natl. Acad. Sci. USA* **106,** 2224–2229.

Peters, L. M., Anderson, D. W., Griffith, A. J., Grundfast, K. M., San Agustin, T. B., Madeo, A. C., Friedman, T. B., and Morell, R. J. (2002). Mutation of a transcription factor, TFCP2L3, causes progressive autosomal dominant hearing loss, DFNA28. *Hum. Mol. Genet.* **11,** 2877–2885.

Pyrgaki, C., Liu, A., and Niswander, L. (2011). Grainyhead-like 2 regulates neural tube closure and adhesion molecule expression during neural fold fusion. *Dev. Biol.* **353,** 38–49.

Ramet, M., Lanot, R., Zachary, D., and Manfruelli, P. (2002). JNK signaling pathway is required for efficient wound healing in Drosophila. *Dev. Biol.* **241,** 145–156.

Rifat, Y., Parekh, V., Wilanowski, T., Hislop, N. R., Auden, A., Ting, S. B., Cunningham, J. M., and Jane, S. M. (2010). Regional neural tube closure defined by the Grainy head-like transcription factors. *Dev. Biol.* **345,** 237–245.

Rodda, S., Sharma, S., Scherer, M., Chapman, G., and Rathjen, P. (2001). CRTR-1, a developmentally regulated transcriptional repressor related to the CP2 family of transcription factors. *J. Biol. Chem.* **276,** 3324–3332.

Schaefer, J., Kramer, K. J., Garbow, J. R., Jacob, G. S., Stejskal, E. O., Hopkins, T. L., and Speirs, R. D. (1987). Aromatic cross-links in insect cuticle: Detection by solid-state 13C and 15N NMR. *Science* **235,** 1200–1204.

Schottenfeld, J., Song, Y., and Ghabrial, A. S. (2010). Tube continued: Morphogenesis of the Drosophila tracheal system. *Curr. Opin. Cell Biol.* **22,** 633–639.

Stramer, B., and Martin, P. (2005). Cell biology: Master regulators of sealing and healing. *Curr. Biol.* **15,** R425–427.

Stramer, B., Wood, W., Galko, M. J., Redd, M. J., Jacinto, A., Parkhurst, S. M., and Martin, P. (2005). Live imaging of wound inflammation in Drosophila embryos reveals key roles for small GTPases during in vivo cell migration. *J. Cell Biol.* **168,** 567–573.

Sueyoshi, T., Kobayashi, R., Nishio, K., Aida, K., Moore, R., Wada, T., Handa, H., and Negishi, M. (1995). A nuclear factor (NF2d9) that binds to the male-specific P450 (Cyp 2d-9) gene in mouse liver. *Mol. Cell. Biol.* **15,** 4158–4166.

Swanson, L. E., and Beitel, G. J. (2006). Tubulogenesis: An inside job. *Curr. Biol.* **16,** R51–53.

Tao, J., Kuliyev, E., Wang, X., Li, X., Wilanowski, T., Jane, S. M., Mead, P. E., and Cunningham, J. M. (2005). BMP4-dependent expression of Xenopus Grainyhead-like 1 is essential for epidermal differentiation. *Development* **132,** 1021–1034.

Tiklova, K., Senti, K. A., Wang, S., Graslund, A., and Samakovlis, C. (2010). Epithelial septate junction assembly relies on melanotransferrin iron binding and endocytosis in Drosophila. *Nat. Cell. Biol.* **12,** 1071–1077.

Ting, S. B., Wilanowski, T., Auden, A., Hall, M., Voss, A. K., Thomas, T., Parekh, V., Cunningham, J. M., and Jane, S. M. (2003a). Inositol- and folate-resistant neural tube defects in mice lacking the epithelial-specific factor Grhl-3. *Nat. Med.* **9,** 1513–1519.

Ting, S. B., Wilanowski, T., Cerruti, L., Zhao, L. L., Cunningham, J. M., and Jane, S. M. (2003b). The identification and characterization of human Sister-of-Mammalian Grainyhead (SOM) expands the grainyhead-like family of developmental transcription factors. *Biochem. J.* **370,** 953–962.

Ting, S. B., Caddy, J., Hislop, N., Wilanowski, T., Auden, A., Zhao, L. L., Ellis, S., Kaur, P., Uchida, Y., Holleran, W. M., Elias, P. M., Cunningham, J. M., *et al.* (2005a). A homolog of Drosophila grainy head is essential for epidermal integrity in mice. *Science* **308,** 411–413.

Ting, S. B., Caddy, J., Wilanowski, T., Auden, A., Cunningham, J. M., Elias, P. M., Holleran, W. M., and Jane, S. M. (2005b). The epidermis of grhl3-null mice displays altered lipid processing and cellular hyperproliferation. *Organogenesis* **2,** 33–35.

Tonning, A., Hemphala, J., Tang, E., Nannmark, U., Samakovlis, C., and Uv, A. (2005). A transient luminal chitinous matrix is required to model epithelial tube diameter in the Drosophila trachea. *Dev. Cell* **9,** 423–430.

Tonning, A., Helms, S., Schwarz, H., Uv, A. E., and Moussian, B. (2006). Hormonal regulation of mummy is needed for apical extracellular matrix formation and epithelial morphogenesis in Drosophila. *Development* **133,** 331–341.

Traylor-Knowles, N., Hansen, U., Dubuc, T. Q., Martindale, M. Q., Kaufman, L., and Finnerty, J. R. (2010). The evolutionary diversification of LSF and Grainyhead transcription factors preceded the radiation of basal animal lineages. *BMC Evol. Biol.* **10,** 101.

Tuckfield, A., Clouston, D. R., Wilanowski, T. M., Zhao, L. L., Cunningham, J. M., and Jane, S. M. (2002). Binding of the RING polycomb proteins to specific target genes in complex with the grainyhead-like family of developmental transcription factors. *Mol. Cell. Biol.* **22,** 1936–1946.

Uv, A. E., Thompson, C. R., and Bray, S. J. (1994). The Drosophila tissue-specific factor Grainyhead contains novel DNA-binding and dimerization domains which are conserved in the human protein CP2. *Mol. Cell. Biol.* **14,** 4020–4031.

Uv, A. E., Harrison, E. J., and Bray, S. J. (1997). Tissue-specific splicing and functions of the Drosophila transcription factor Grainyhead. *Mol. Cell. Biol.* **17,** 6727–6735.

Uv, A., Cantera, R., and Samakovlis, C. (2003). Drosophila tracheal morphogenesis: Intricate cellular solutions to basic plumbing problems. *Trends Cell. Biol.* **13,** 301–309.

Van Laer, L., Van Eyken, E., Fransen, E., Huyghe, J. R., Topsakal, V., Hendrickx, J. J., Hannula, S., Maki-Torkko, E., Jensen, M., Demeester, K., Baur, M., Bonaconsa, A., *et al.* (2008). The grainyhead like 2 gene (GRHL2), alias TFCP2L3, is associated with age-related hearing impairment. *Hum. Mol. Genet.* **17,** 159–169.

Venkatesan, K., McManus, H. R., Mello, C. C., Smith, T. F., and Hansen, U. (2003). Functional conservation between members of an ancient duplicated transcription factor family. *LSF/Grainyhead. Nucleic Acids Res.* **31,** 4304–4316.

Volker, J. L., Rameh, L. E., Zhu, Q., DeCaprio, J., and Hansen, U. (1997). Mitogenic stimulation of resting T cells causes rapid phosphorylation of the transcription factor LSF and increased DNA-binding activity. *Genes Dev.* **11,** 1435–1446.

Wang, S., Jayaram, S. A., Hemphala, J., Senti, K. A., Tsarouhas, V., Jin, H., and Samakovlis, C. (2006). Septate-junction-dependent luminal deposition of chitin deacetylases restricts tube elongation in the Drosophila trachea. *Curr. Biol.* **16,** 180–185.

Wang, S., Tsarouhas, V., Xylourgidis, N., Sabri, N., Tiklova, K., Nautiyal, N., Gallio, M., and Samakovlis, C. (2009). The tyrosine kinase Stitcher activates Grainy head and epidermal wound healing in Drosophila. *Nat. Cell Biol.* **11,** 890–895.

Werth, M., Walentin, K., Aue, A., Schonheit, J., Wuebken, A., Pode-Shakked, N., Vilianovitch, L., Erdmann, B., Dekel, B., Bader, M., Barasch, J., Rosenbauer, F., *et al.* (2010). The transcription factor grainyhead-like 2 regulates the molecular composition of the epithelial apical junctional complex. *Development* **137,** 3835–3845.

Wilanowski, T., Tuckfield, A., Cerruti, L., O'Connell, S., Saint, R., Parekh, V., Tao, J., Cunningham, J. M., and Jane, S. M. (2002). A highly conserved novel family of mammalian developmental transcription factors related to Drosophila grainyhead. *Mech. Dev.* **114,** 37–50.

Wilanowski, T., Caddy, J., Ting, S. B., Hislop, N. R., Cerruti, L., Auden, A., Zhao, L. L., Asquith, S., Ellis, S., Sinclair, R., Cunningham, J. M., and Jane, S. M. (2008). Perturbed desmosomal cadherin expression in grainy head-like 1-null mice. *EMBO J.* **27,** 886–897.

Wood, W., Jacinto, A., Grose, R., Woolner, S., Gale, J., Wilson, C., and Martin, P. (2002). Wound healing recapitulates morphogenesis in Drosophila embryos. *Nat. Cell. Biol.* **4,** 907–912.

Wright, T. R. (1987a). The genetic and molecular organization of the dense cluster of functionally related, vital genes in the DOPA decarboxylase region of the Drosophila melanogaster genome. *Results Probl. Cell Differ.* **14,** 95–120.

Wright, T. R. (1987b). The genetics of biogenic amine metabolism, sclerotization, and melanization in Drosophila melanogaster. *Adv. Genet.* **24,** 127–222.

Wu, V. M., and Beitel, G. J. (2004). A junctional problem of apical proportions: Epithelial tube-size control by septate junctions in the Drosophila tracheal system. *Curr. Opin. Cell Biol.* **16,** 493–499.

Wu, F. K., Garcia, J. A., Harrich, D., and Gaynor, R. B. (1988). Purification of the human immunodeficiency virus type 1 enhancer and TAR binding proteins EBP-1 and UBP-1. *EMBO J.* **7,** 2117–2130.

Wu, V. M., Schulte, J., Hirschi, A., Tepass, U., and Beitel, G. J. (2004). Sinuous is a Drosophila claudin required for septate junction organization and epithelial tube size control. *J. Cell Biol.* **164,** 313–323.

Wu, V. M., Yu, M. H., Paik, R., Banerjee, S., Liang, Z., Paul, S. M., Bhat, M. A., and Beitel, G. J. (2007). Drosophila Varicose, a member of a new subgroup of basolateral MAGUKs, is required for septate junctions and tracheal morphogenesis. *Development* **134,** 999–1009.

Wu, Y., Brock, A. R., Wang, Y., Fujitani, K., Ueda, R., and Galko, M. J. (2009). A blood-borne PDGF/VEGF-like ligand initiates wound-induced epidermal cell migration in Drosophila larvae. *Curr. Biol.* **19,** 1473–1477.

Yu, Z., Lin, K. K., Bhandari, A., Spencer, J. A., Xu, X., Wang, N., Lu, Z., Gill, G. N., Roop, D. R., Wertz, P., and Andersen, B. (2006). The Grainyhead-like epithelial transactivator Get-1/Grhl3 regulates epidermal terminal differentiation and interacts functionally with LMO4. *Dev. Biol.* **299,** 122–136.

Yu, Z., Bhandari, A., Mannik, J., Pham, T., Xu, X., and Andersen, B. (2008). Grainyhead-like factor Get1/Grhl3 regulates formation of the epidermal leading edge during eyelid closure. *Dev. Biol.* **319,** 56–67.

Yu, Z., Mannik, J., Soto, A., Lin, K. K., and Andersen, B. (2009). The epidermal differentiation-associated Grainyhead gene Get1/Grhl3 also regulates urothelial differentiation. *EMBO J.* **28,** 1890–1903.

Zhou, W., Clouston, D. R., Wang, X., Cerruti, L., Cunningham, J. M., and Jane, S. M. (2000). Induction of human fetal globin gene expression by a novel erythroid factor, NF-E4. *Mol. Cell. Biol.* **20,** 7662–7672.

CHAPTER THREE

# Groucho: A Corepressor with Instructive Roles in Development

Wiam Turki-Judeh *and* Albert J. Courey

## Contents



## Abstract

*Drosophila* Groucho (Gro) is the founding member of a family of metazoan corepressors. Gro mediates repression through interactions with a myriad of DNA-binding repressor proteins to direct the silencing of genes involved

Department of Chemistry & Biochemistry and Molecular Biology Institute, University of California, Los Angeles, California, USA

*Current Topics in Developmental Biology*, Volume 98
ISSN 0070-2153, DOI: 10.1016/B978-0-12-386499-4.00003-3

in many developmental processes, including neurogenesis and patterning of the main body axis, as well as receptor tyrosine kinase/Ras/MAPK, Notch, Wingless (Wg)/Wnt, and Decapentaplegic (Dpp) signaling. Gro mediates repression by multiple molecular mechanisms, depending on the regulatory context. Because Gro is a broadly expressed nuclear factor, whereas its repressor partners display restricted temporal and spatial distribution, it was presumed that this corepressor played permissive rather than instructive roles in development. However, a wide range of studies demonstrates that this is not the case. Gro can sense and integrate many cellular inputs to modulate the expression of variety of genes, making it a versatile corepressor with crucial instructive roles in development and signaling.

## Abbreviations

AES amino-terminal enhancer of split
AH1 and AH2 amphipathic α-helices 1 and 2
CRM *cis*-regulatory module
CtBP C-terminal binding protein
EGFR epidermal growth factor receptor
Grg Groucho-related gene
Gro Groucho
HDAC1 histone deacetylase 1
HES Hairy/Enhancer of split
HIPK2 homeodomain-interacting protein kinase 2
MAPK mitogen-activated protein kinase
RTK receptor tyrosine kinase
TLE transducin-like enhancer of split
TSA Trichostatin A

## 1. The Gro/TLE Family

### 1.1. Initial characterization of Gro

The *Drosophila gro* gene was initially identified through a weak hypomorphic allele ($gro^1$) that results in thick tufts of sensory bristles over the eyes, bearing resemblance to the bushy eyebrows of Groucho Marx (Lindsley and Grell, 1968). While $gro^1$ is viable, stronger mutations in *gro* are lethal and reveal essential maternal and zygotic requirements for Gro in a wide variety of developmental processes.

Proof that Gro is a regulator of transcription came from the finding of a functional interaction with the *Drosophila* Hairy protein, a bHLH protein that represses transcription (Paroush *et al.*, 1994). *hairy,* many of the genes of the E(spl) complex, and *deadpan (dpn)* encode structurally related proteins that together comprise the Hairy/E(spl) (HES) family of transcription factors. HES family factors, which are also found in vertebrates, are characterized by a bHLH DNA-binding domain and a C-terminal WRPW motif (Dawson *et al.*, 1995). Gro interacts directly with HES proteins via the WRPW motif (Fisher *et al.*, 1996; Grbavec and Stifani, 1996; Paroush *et al.*, 1994). The functional relevance of the interaction between Gro and HES family factors was demonstrated by experiments showing that *Drosophila* embryos lacking maternally deposited Gro display misregulation of neurogenesis, sex determination, and embryonic segmentation, that is, many of the processes known to be regulated by HES family factors (Delidakis *et al.*, 1991; Paroush *et al.*, 1994).

Subsequent studies have shown that Gro interacts with a wide variety of repressor partners in addition to HES family proteins, including Dorsal, Runt, Huckebein, Goosecoid, and Brinker (Aronson *et al.*, 1997; Dubnicoff *et al.*, 1997; Fisher *et al.*, 1996; Flores-Saaib *et al.*, 2001; Goldstein *et al.*, 1999; Hasson *et al.*, 2001; Jimenez *et al.*, 1997, 1999; Paroush *et al.*, 1994). While some of these repressors contain a WRPW motif, others contain a different peptide motif mediating the interaction with Gro known as the engrailed homology 1 (eh1) motif, and still others lack a recognizable Gro interaction motif.

## 1.2. Gro homologs

Members of the Gro family are present in all or almost all metazoans (Choudhury *et al.*, 1997; Mallo *et al.*, 1993; Miyasaka *et al.*, 1993; Pflugrad *et al.*, 1997; Schmidt and Sladek, 1993; Sharief *et al.*, 1997; Stifani *et al.*, 1992). A comparison of the predicted protein sequences reveals that most of the family members share a conserved structural organization consisting of five domains, which, from N- to C-terminus, are termed the Q, GP, CcN, SP, and WD-repeat domains (Fig. 3.1). The Q and WD-repeat domains are highly conserved, suggesting a highly conserved function (Chen and Courey, 2000; Jennings and Ish-Horowicz, 2008; Stifani *et al.*, 1992). As will be discussed in subsequent sections, the Q domain mediates self-association and repressor binding, while the WD-repeat domain mediates interactions with repressor proteins. The three central domains (the GP, CcN, and SP domains) are much more weakly conserved. However, as will become clear in later sections, these domains nonetheless have multiple vital functions in repression and its regulation.

While the *Drosophila* genome encodes a single Gro family member, the mouse and human genomes encode four family members each, termed transducin-like Enhancer of Split 1–4 (TLE1–4) in humans (Stifani *et al.*, 1992)

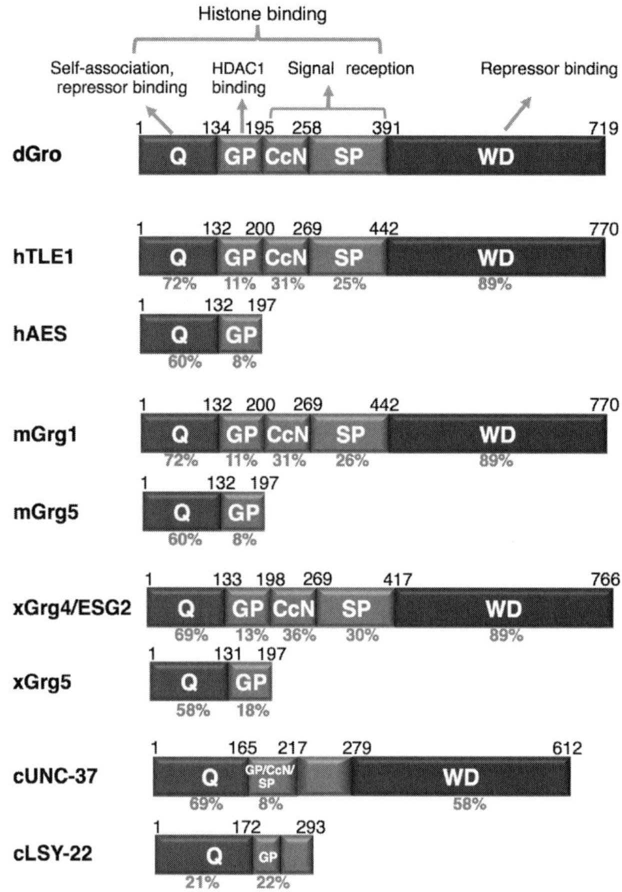


**Figure 3.1** Structural organization and sequence conservation in the Gro/TLE family. A comparison of the amino acid sequence of *Drosophila* Gro (dGro) with those of the human protein hTLE1 and hAES, the mouse proteins mGrg1 and mGrg5, the *Xenopus* proteins xGrg4/ESG2 and xGrg5, and the *C. elegans* proteins cUNC-37 and cLSY-22 reveals that most Gro/TLE family members share a conserved structural organization, which consists of a glutamine-rich (Q) domain, followed by a glycine/proline-rich (GP) domain, a so-called CcN domain, a serine/proline-rich (SP) domain, and a WD-repeat domain. A few of the suspected functions of these domains are indicated above the diagram of dGro. The percentages below the domains indicate sequence identity to the corresponding domains in *Drosophila* Gro. Percent identity was determined from proteins alignments calculated using the ClustalW BLOSUM scoring matrix. The Q and WD-repeat domains are highly conserved, while the central region consisting of GP, CcN, and SP domains is much less well conserved among Gro/TLE family members. The proteins shown have the following NCBI accession numbers: dGro (P16371); hTLE1 (NP_005068); hAES (NP_001121.2), mGrg1 (Q62440), mGrg5 (NP_034477.1), xGrg4/ESG2 (CAA12236), xGrg5 (NP_001083532.1), cUNC-37 (O02482), and cLSY-22 (CAB01968.3). (For color version of this figure, the reader is referred to the Web version of this chapter.)

and Gro-related genes 1–4 (Grg1–4) in mouse (Mallo *et al.*, 1993; Miyasaka *et al.*, 1993). The human and mouse family members are broadly, but specifically, expressed in various developing tissues and are not functionally redundant (Gasperowicz and Otto, 2005; Stifani *et al.*, 1992; Yao *et al.*, 1998). TLE1/Grg1 is involved in myogenesis, hematogenesis, apoptosis, neurogenesis, and eye development (Gao *et al.*, 2001; Imai *et al.*, 1998; Jan *et al.*, 2004; Swingler *et al.*, 2004; Yao *et al.*, 2000; Zhu *et al.*, 2002); TLE2/Grg2 in neural tube formation and osteogenesis (Grbavec *et al.*, 1998; Thirunavukkarasu *et al.*, 1998); TLE3/Grg3 in placental and adipose development (Nakayama *et al.*, 1997; Villanueva *et al.*, 2011); and TLE4/Grg4 in regulating the B cell lineage and heart development (Bajoghli *et al.*, 2007; Eberhard *et al.*, 2000; Linderson *et al.*, 2004). Gro homologs have also been found in frogs (Choudhury *et al.*, 1997; Sharief *et al.*, 1997) and roundworms (Pflugrad *et al.*, 1997).

Although clear Gro/TLE orthologs have not been found outside the metazoans, corepressors exhibiting homology to Gro/TLE proteins have been found in diverse eukaryotic taxa, such as fungi (Treitel and Carlson, 1995) and plants (Liu and Karmarkar, 2008). For example, Tup1, a yeast corepressor, contains a conserved C-terminal WD-repeat domain and this domain is more closely related to the Gro/TLE WD-repeat domain than to WD-repeat domains in other proteins (Flores-Saaib and Courey, 2000). Outside the WD-repeat region, there is no recognizable homology between Tup1 and Gro/TLE proteins, but there is functional conservation (Courey and Jia, 2001). For example, Tup1 contains an N-terminal domain in the same position as the Gro/TLE Q domain that, like the Q domain, mediates essential corepressor self-association (Chen *et al.*, 1998; Flores-Saaib and Courey, 2000; Malave and Dent, 2006; Redd *et al.*, 1997; Tzamarias and Struhl, 1995). Further, as will be discussed later, Gro/TLE binds to hypoacetylated histone tails and the same is true for Tup1 (Edmondson and Roth, 1998; Edmondson *et al.*, 1996; Flores-Saaib and Courey, 2000). As will also be discussed later, Gro/TLE interacts with the histone deacetylase Rpd3/HDAC1, and once again, the same is true for Tup1 (Edmondson *et al.*, 1998; Flores-Saaib and Courey, 2000; Malave and Dent, 2006; Redd *et al.*, 1997). These functional parallels suggest that, despite the poor structural conservation between Gro/TLE and Tup1, the molecular mechanisms used by Gro/TLE to regulate transcription may be similar to those used by Tup1.

## 1.3. An instructional role for Gro/TLE during development

Gro/TLE family proteins have diverse developmental roles mediated by interactions with repressors that regulate embryonic patterning, the transition from proliferation to differentiation, specification of cell identity, and apoptosis (Buscarlet and Stifani, 2007; Chen and Courey, 2000; Fisher and Caudy, 1998; Gasperowicz and Otto, 2005). This is supported by many studies in *Drosophila* that have demonstrated essential roles for Gro in a wide array of developmental processes, including embryonic patterning, sex determination, neuronal development, and imaginal disc patterning (Buscarlet and Stifani, 2007; Chen and Courey, 2000). Gro also plays equally important, although less well understood, roles in vertebrate development (Buscarlet and Stifani, 2007; Fisher and Caudy, 1998; Gasperowicz and Otto, 2005).

As will be repeatedly emphasized in this review, the role of Gro in development is not merely permissive. That is, Gro is not simply a constitutive component of the repression machinery that must be present to allow repression to occur. Rather, it plays many instructive roles in development, by doing such things as converting activators into repressors and dictating the threshold responses to repressor gradients in ways that are highly sensitive to Gro activity levels and that are intensively regulated by signaling pathways.

## 2. Mechanisms of Gro/TLE-Mediated Repression

Despite the widespread importance of Gro/TLE-mediated repression in development and signal transduction, relatively little is known about the mechanisms by which Gro/TLE proteins mediate transcriptional repression. Genetic, molecular, and biochemical data suggest that Gro/TLE proteins direct repression through a variety of mechanisms, depending on the developmental context and the target gene. Biochemical activities of Gro that may underlie many examples of Gro-mediated repression include (1) self-association (oligomerization), (2) direct interaction with histone deacetylase, (3) binding to histones leading to changes in chromatin structure such as increased nucleosome density or chromatin condensation, (4) interactions with the core transcriptional machinery, and (5) masking of activation domains (Fig. 3.2).

### 2.1. Self-association

The highly conserved Gro Q domain mediates homo-oligomerization via two coiled-coil motifs termed amphipathic α-helices 1 and 2 (AH1 and AH2) (Chen *et al.*, 1998; Miyasaka *et al.*, 1993; Pinto and Lobe, 1996). The Q domain forms a predominant homo-tetramer, although other size oligomers are also observed including dimers and larger species that appear to be mulitmers of the tetramer (Chen *et al.*, 1998; Jimenez *et al.*, 1999; Kuo *et al.*, 2011; Palaparti *et al.*, 1997; Pinto and Lobe, 1996; Song *et al.*, 2004). Point mutations that disrupt the hydrophobic faces of AH1 and AH2 eliminate these multimers, converting Gro into a monomer (Chen *et al.*, 1998; Pinto and Lobe, 1996; Song *et al.*, 2004).

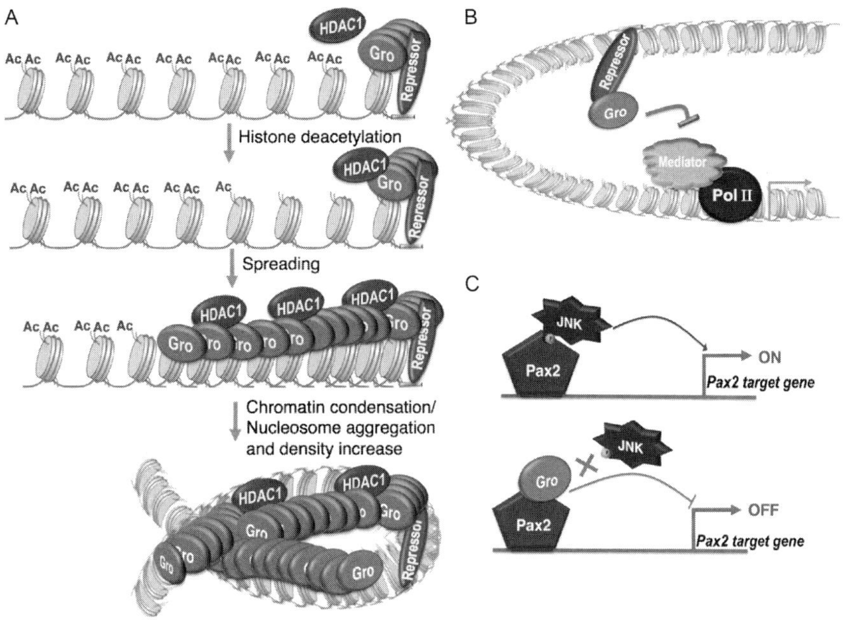


**Figure 3.2** Models for Gro/TLE-mediated repression. Gro/TLE appears to direct repression by multiple mechanisms. (A) A variety of studies indicate that the ability of Gro/TLE proteins to self-associate, recruit histone deacetylase, bind to deacetylated histones, and direct chromatin condensation (or increased nucleosome density) are required for Gro-mediated repression. This multistep model accounts for all of these biochemical functions. (B) Gro/TLE may interact with the Mediator, a component of the basal transcriptional machinery, thereby interfering with the formation or function of the preinitiation complex. (C) Gro/TLE may mask the activation domains of transcriptional activators. In the example shown here, Gro binding prevents phosphorylation of the activation domain in Pax2, thereby preventing Pax2 activation. Additional details regarding these three models and references are given in the text. (See Color Insert.)

A number of lines of evidence indicate that Q domain-mediated Gro self-association makes a significant contribution to Gro-mediated repression. First, repression in both *Drosophila* wing discs and cultured cells is inhibited by point mutations in AH1 and AH2 that prevent self-association. Second, swapping of the Q domain with the heterologous tetramerization domain from the tumor suppressor p53 allows repression in S2 cells, implying that the primary function of the Q domain is to facilitate oligomerization (Chen *et al.*, 1998). Finally, forms of Gro with an intact Q domain, but with mutations in other domains that prevent repression, behave as dominant negatives to prevent repression by endogenous wild-type Gro strongly supporting the idea that Gro oligomerizes *in vivo*. However, oligomerization may not be uniformly required for repression since a mutant *gro* allele encoding an oligomerization deficient form Gro, while lethal, is not a genetic null (Jennings *et al.*, 2008). Thus, Gro/TLE oligomerization may be required for repression of some but not all Gro/TLE target genes, suggesting diverse modes of transcriptional regulation (Jennings and Ish-Horowicz, 2008).

Although it seems clear that self-association is required for many although perhaps not all instances of Gro/TLE-mediated repression, the mechanisms linking homo-oligomerization to repression are still unclear. The self-association properties of Gro/TLE could perhaps allow a silent chromatin structure to spread over large areas of the genome, as proposed for heterochromatic silencing by SIR proteins in budding yeast (Buchberger *et al.*, 2008; Hecht *et al.*, 1995; Hoppe *et al.*, 2002; Huang, 2002; Radman-Livaja *et al.*, 2011; Rusche *et al.*, 2002). In support of this idea, Gro/TLE family proteins bind to chromosomal domains that extend beyond the sites to which Gro/TLE is initially recruited by DNA-bound repressors (Martinez and Arnosti, 2008; Sekiya and Zaret, 2007; Winkler *et al.*, 2010).

## 2.2. Interaction with histone deacetylase

Although the GP domain is poorly conserved, both mammalian and *Drosophila* Gro/TLE make a conserved interaction with the class I histone deacetylase Rpd3/HDAC1 via this domain (Brantjes *et al.*, 2001; Chen *et al.*, 1999). Evidence that this interaction is functional comes from *Drosophila* cell culture studies in which wild-type Rpd3, but not a catalytically inactive mutant, was found to stimulate repression of a Gal4-responsive reporter by a Gal4-Gro fusion protein (Chen *et al.*, 1998). Further, treatment with the histone deacetylase inhibitor trichostatin A (TSA) in both mammalian and *Drosophila* cell culture (Chen *et al.*, 1999; Choi *et al.*, 1999) or RNAi knockdown of *rpd3* in *Drosophila* cell culture (Chen *et al.*, 1999; Winkler *et al.*, 2010) significantly decreases this repression, although neither treatment eliminated Gal4-Gro-mediated repression completely. Thus, in addition to histone deacetylase-dependent mechanisms, Gro may utilize histone deacetylase-independent mechanisms for transcriptional repression. The conclusion that histone deacetylation contributes to Gro-mediated repression is reinforced by chromatin immunoprecipitation (ChIP) assays showing that Gro colocalizes with Rpd3 to the chromatin of target genes, leading to the deacetylation of specific lysine residues in histones H3 and H4, including H3K9, H3K14, H4K5, H4K8, and H4K12 (Winkler *et al.*, 2010).

*In vivo* studies in flies also demonstrate the functional relevance of the interaction between Gro and histone deacetylase. First, a simultaneous twofold reduction in the *gro* and *rpd3* gene dosage leads to synergistic effects on embryonic lethality and pattern formation (Chen *et al.*, 1999). Second, a hypomorphic mutation in *rpd3* disrupts repression by Even-skipped, a Gro-dependent repressor (Mannervik and Levine, 1999). Third, the wing-patterning defects and ectopic repression of Gro target genes caused by overexpression of Gro in the wing disc were significantly reduced when flies were fed either of two different histone deacetylase inhibitors, TSA (which is specific for class I and II histone deacetylases) and HC-toxin (which is specific for the class I enzymes), and this effect was enhanced when the *rpd3* gene dosage was reduced (Winkler *et al.*, 2010).

## 2.3. Chromatin condensation

Gro and related proteins are able to bind to histones, an activity that could contribute to Gro-mediated repression. Such an interaction was first demonstrated for yeast Tup1 (Edmondson and Roth, 1998; Edmondson *et al.*, 1996) and later extended to mammalian and *Drosophila* Gro/TLE (Flores-Saaib and Courey, 2000; Palaparti *et al.*, 1997). Gro binds all four core histones, with highest affinity for histone H3. The interaction is mediated by the N-terminal tail of this histone and is significantly enhanced when the tail is deacetylated.

The notion that histone binding and oligomerization are functionally relevant is supported by a study showing that Gro/TLE binds nucleosomal arrays *in vitro* and condenses the chromatin, thereby reducing the accessibility of the template to the transcriptional machinery (Sekiya and Zaret, 2007). Binding of Gro/TLE to the nucleosomal arrays is dependent on histone tails and nucleosome aggregation requires the Q domain, which, as mentioned above, mediates self-association. Interestingly, binding of Gro to the nucleosomes is not, by itself, sufficient for complete protection of the DNA from DNase I digestion or complete exclusion from the template of activators and components of the general transcriptional. This requires the additional binding of Gro/TLE to a template-bound repressor, which, in turn, leads to the formation of a closed chromatin region spanning three to four nucleosomes and the exclusion of factors such as TBP and RNA polymerase II (Pol II) from the template.

Further evidence that repression might involve significant alterations in chromatin structure is provided by histone H3 ChIP assays on a Gro target in *Drosophila* wing discs. These studies showed that Gro mediated an increase in nucleosomal density, which was reversed in the presence of TSA. This suggests a model in which Gro/TLE directs the formation of a chromatin environment characterized by high-density deacetylated nucleosomes that may block activators and the transcriptional machinery (Winkler *et al.*, 2010).

Putting together the data on Gro self-association, histone deacetylation, and chromatin condensation or increased nucleosome density leads to the following multistep model for Gro-mediated repression (Fig. 3.2A). First, Gro is recruited to the template via an interaction with a DNA-bound repressor. Second, Gro recruits an Rpd3/HDAC1-containing histone deacetylase complex, which catalyzes local histone deacetylation. Third, the ability of Gro to interact both with itself and with deacetylated histones leads to the recruitment of additional Gro. Fourth, Gro association leads to nucleosome aggregation creating a condensed form of chromatin with reduced accessibility to the transcriptional machinery. While this model may explain many instances of Gro-mediated repression, it is important to note that there is at least one documented case in which Gro-mediated repression is not accompanied by a decrease in the accessibility of the template to nucleases (Li and Arnosti, 2011).

## 2.4. Core transcriptional machinery interactions

In addition to chromatin-mediated mechanisms, Gro is likely to repress by other mechanisms as well. For example, Gro may directly interact with the core machinery to inhibit some step in the transcription cycle such as preinitiation complex assembly, promoter release, or elongation (Fig. 3.2B). Along these lines, genetic and biochemical studies in yeast have linked the activity of the Gro-related corepressor Tup1 to several components of the Mediator complex (Gromoller and Lehming, 2000; Kuchin and Carlson, 1998; Papamichos-Chronakis *et al.*, 2000). This complex interacts with the C-terminal domain (CTD) of the largest subunit of RNA Pol II and also makes extensive contact with the core domain (Berk, 1999; Kornberg, 2005; Malik and Roeder, 2010). The Mediator functions as a bridge to relay information from gene-specific regulatory proteins to the basal Pol II transcription machinery and serves as a scaffold for the assembly of a functional preinitiation complex including Pol II and the general transcription factors (Kornberg, 2005).

Evidence that metazoan Gro/TLE homologs may also interact with the Mediator comes from studies of the *Caenorhabditis elegans* Gro/TLE homolog encoded by *unc-37*. In *C. elegans*, the nine pairs of rays in the adult male tail develop from three pairs of embryonic seam cells, V5, V6, and T. Male-specific postembryonic proliferation of V6 begins with the expression of *pal-1* (a homolog of *Drosophila caudal*). Expression of *pal-1* in V6 is partly dependent on a pathway that is repressed normally by the two essential Mediator subunits MED1 and MED12 (also known as SOP3 and SOP1 in *C. elegans*) (Zhang and Emmons, 2002). A mutation in *unc-37* alone has little effect on activation of *pal-1* in V6, thus leading to very few missing V6 rays. The effect is greatly increased, however, when the *unc-37* mutation is combined with mutations in MED1 and MED12 (Zhang and Emmons, 2002). At this point, however, it is not clear if the interaction between the *unc-37* gene product and Mediator subunits is direct or indirect since a physical interaction between these proteins has not been established. For example, it is possible that Gro and the Mediator have separate parallel inputs into *C. elegans* tail development.

## 2.5. Masking of activation domains

Another mechanism of Gro-mediated repression is masking of the activation domains of transcriptional activators (Fig. 3.2C). In general, this involves the binding of Gro to an activator to block coactivator binding or to induce a conformational change in an activation domain that renders it inactive (Arnosti, 2004). The paired domain protein, Pax2, a DNA-binding transcriptional activator that is conserved across eukaryotes, provides a particularly interesting example of masking. Phosphorylation of the Pax2 activation domain by c-jun N-terminal kinase (JNK) potentiates Pax2 activation of target genes (Cai *et al.*, 2003). Gro/TLE binds directly to Pax2 and this interaction inhibits Pax2 phosphorylation by JNK, resulting in a decrease in its transactivation potential (Cai *et al.*, 2003).

## 2.6. Long- versus short-range transcriptional repression

Transcriptional repressors are thought to work via two different mechanisms depending on how far their binding sites are located from the activator sites and/or the core promoter. The two general modes of repression are known as short- and long-range repression (Courey and Jia, 2001; Gray and Levine, 1996b; Mannervik *et al.*, 1999). In short-range repression, the repressors inhibit activators bound within ~100 bp of the repressor binding site. Spatially regulated genes, such as segmentation genes often contain multiple *cis*-regulatory modules (CRMs), each of which directs the formation of a different portion of the transcriptional pattern through combinatorial interactions between multiple activators and repressors, which are, themselves, expressed in spatially restricted patterns. It is critical that repressors bound to one CRM not interfere with the function of another CRM and short-range repression helps to ensure this so-called enhancer autonomy (Gray and Levine, 1996a). Long-range repression, in contrast, involves repressors that are able to render a promoter resistant to the influence of all enhancers including those that are located thousands (or hundreds of thousands) of base pairs away from the binding site of the repressor (Cai *et al.*, 1996; Courey and Jia, 2001; Mannervik *et al.*, 1999).

While we do not understand what determines the distinction between long- and short-range repression, a clue may come from the observation that long-range repressors (e.g., Dorsal, HES family factors) often utilize Gro/TLE as a corepressor, while short-range repressors (e.g., Snail, Knirps) often utilize an alternative corepressor termed C-terminal binding protein (CtBP). However, recent studies on Gro are breaking down this distinction. In particular, these studies suggest that Gro can, in some cases, mediate short-range repression, depending on the DNA-binding repressor and the context of the *cis*-regulatory element. For example, Gro is an essential corepressor for the short-range repressors Sloppy-paired 1 (Slp1) and Knirps (Andrioli *et al.*, 2004; Payankaulam and Arnosti, 2009).

Gro interacts with Slp1 via an eh1 motif and this interaction is functionally relevant because deletion of this region inhibits Slp1 mediated repression of a subset of its target genes (Andrioli *et al.*, 2004). One such target, *runt*, is a pair rule gene expressed in a seven-stripe pattern. These various stripes are under the control of different CRMs. The CRMs that direct formation of stripes 1, 3, and 5 are located near each other, yet the Gro/TLE-Slp1 complex is only able to repress the stripe 1 CRM, allowing for enhancer autonomy (Andrioli *et al.*, 2004).

Additional evidence contravening the idea that Gro is strictly a long-range corepressor is provided by a study showing that Gro physically interacts with the classical short-range repressor Knirps via two N-terminal eh1-like motifs (Payankaulam and Arnosti, 2009). This interaction is functionally relevant because deletions of the two eh1-like motifs not only prevent the Knirps-Gro interaction but also result in very weak repressor activity (Payankaulam and Arnosti, 2009).

At this point, we can only speculate about how it is that Gro can act as a short-range corepressor when it interacts with some targets and a long-range corepressor when it interacts with others. One possibility is that short-range repressors like Knirps and Slp1 induce conformational changes in Gro that inhibit its ability to self-associate and spread, while recruitment by long-range repressors allow such spreading to occur (Payankaulam *et al.*, 2010). Another possibility is that the ability to spread depends on *cis*-regulatory elements that either block or allow spreading. Finally, we note that that long- and short-range repression may be associated with different chromatin modifications (Burks *et al.*, 2009). Perhaps other coregulators interact with Gro in a gene-specific manner to determine the chromatin modification state and thereby determine the range of repression.

In conclusion, Gro/TLE family proteins appear to utilize multiple mechanisms to repress transcription, depending on the regulatory context. The long- and short-range repression categories probably represent an oversimplification. An understanding of the true complexities that determine what mechanism predominates at any given target will require greater knowledge of factors that interact with Gro to modulate its function.

## 2.7. Gro/TLE central domains may be intrinsically disordered regions

As mentioned previously, Gro/TLE family members have a five-domain structure including conserved N- and C-terminal domains (the Q and WD-repeat domains, respectively) and three weakly conserved internal regions (GP, CcN, and SP domains; Fig. 3.1). Sequencing of a large set of lethal Gro/TLE point mutant alleles revealed a large number of mutations that map to the WD-repeat domain and a smaller number that map to the Q domain, demonstrating the functional importance of these two domains

(Jennings *et al.*, 2006, 2008). However, none of the lethal point mutations map to the GP, CcN, or SP domains, an observation that, along with the lack of conservation of these domains, suggests that these internal domains might be dispensable for function (Jennings *et al.*, 2008). Contrary to this suggestion, however, a recent analysis of a series of Gro internal deletion variants suggests that these regions are absolutely critical for repression by Gro and for the regulation of this repression (Turki-Judeh and Courey, unpublished observations).

Rather than reflecting a lack of function, the lack of conservation in the central region could reflect a disordered structure. Like the Gro central domains, regions of intrinsic disorder are typically characterized by a high content of polar and charged amino acids and a low content of bulky hydrophobic amino acids (Chouard, 2011; Dunker *et al.*, 2005). To determine whether the Gro central region is likely to be intrinsically disordered, we applied several protein disorder predictors to the *Drosophila* Gro amino acid sequence (Fig. 3.3) (Li *et al.*, 1999; Radivojac *et al.*, 2003; Romero *et al.*, 1997; Romero *et al.*, 2001; Xue *et al.*, 2010). The predictors produce disorder scores by calculating such parameters as hydropathy and amino acid composition and then comparing these parameters to those calculated from a database of disordered proteins (Dunker *et al.*, 2005; Xue *et al.*, 2010). Essentially, the entire Gro central region displays high

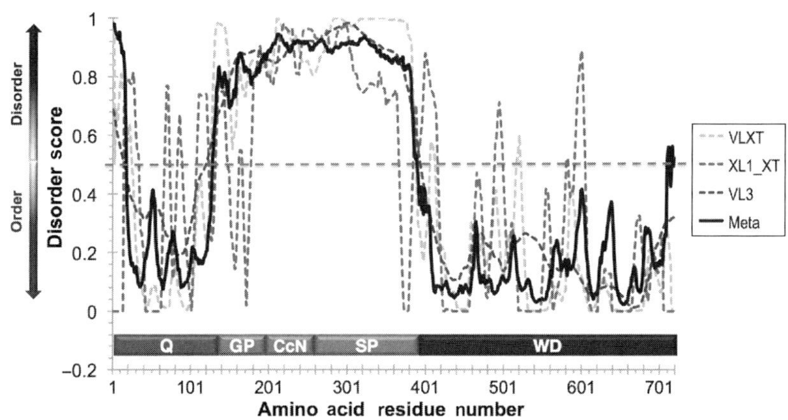


**Figure 3.3** The Gro/TLE central region appears to be intrinsically disordered. Prediction of disorder in *Drosophila* Gro using four disorder prediction algorithms: PONDR-FIT™ (Meta, black solid line) (Xue *et al.*, 2010); PONDR® VLXT (pink dashed line) (Li *et al.*, 1999; Romero *et al.*, 2001); PONDR® XL1_XT (green dashed line) (Romero *et al.*, 1997); and PONDR® VL3 (blue dashed line) (Radivojac *et al.*, 2003). Amino acid residues with scores exceeding 0.5 are likely to be disordered. All four prediction tools strongly suggest that the central region of Gro is disordered, while the Q and WD-repeat domains are ordered. (See Color Insert.)

disorder scores, while the Q and WD-repeat domains appear to be largely ordered. The same is true for all the mammalian Gro family proteins (data not shown). Thus, it seems that the absence of a well-ordered structure in the Gro central domains could lead to a high tolerance of single amino acid changes, thus explaining why lethal point mutations mapping to the GP, CcN, or SP domains have not been identified.

Proteins with regions of intrinsic disorder are not a rare phenomenon (Dunker *et al.*, 2005). In higher eukaryotes, the majority of proteins contain such domains. Many such proteins exist in dynamic ensembles of structural states, thus making structure determination difficult or impossible. Further, such domains often assume extended structures allowing them to contact their binding partners over a large binding surface thereby increasing binding specificity (Dunker *et al.*, 2005). Thus, instead of using large but compactly folded domains that bind only to a few targets, it may sometimes be more efficient to use unstructured domains that can bind to a multitude of proteins, and thereby function as hubs of large regulatory networks (Chouard, 2011; Dunker *et al.*, 2005). In the case of Gro, a disordered central region could account for the position of this protein as a "champion corepressor" (Shilo, 1998) by facilitating a large number of interactions.

While interactions involving disordered domains can be of high specificity, they are generally of low affinity. Such binding may be easily reversed upon posttranslational modification (Dunker *et al.*, 2005). This is in accord with findings showing that the Gro central region is a frequent target for regulation via covalent modification (see below).

## 3. The Regulation of Gro/TLE-Mediated Repression

Gro is not simply a static component of the nuclear machinery that must be present to allow repressors to function. Rather, it is intensively regulated and thus plays an important role in the interpretation of positional and temporal information during development. Mechanisms for regulating Gro function include (1) variation in the strength of the repressor/corepressor interaction; (2) interaction with naturally occurring truncated Gro family proteins; and (3) posttranslational modification of Gro/TLE (Fig. 3.4).

### 3.1. Affinity of the interaction between Gro/TLE and repressors

Gro/TLE function depends on the strength of the Gro/TLE-interaction motif in the repressor protein (Fig. 3.4A). The WD-repeat domain is the primary domain in Gro with responsibility for binding to DNA-bound

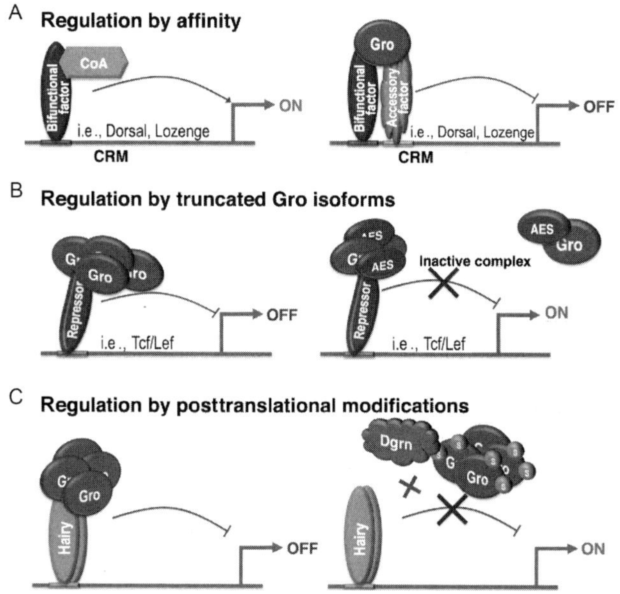


**Figure 3.4** The regulation of Gro/TLE-mediated repression. Gro/TLE function is regulated in a variety of ways. (A) Gro/TLE function depends on the affinity of the Gro/TLE-interaction motif in the repressor protein for Gro. The dual functionality of Lozenge and Dorsal relies on the fact that these two factors have an intrinsically low affinity for Gro. They therefore require assistance from accessory proteins to create a high-affinity platform for Gro/TLE recruitment. (B) The AES proteins, which consist only of the Q and GP domains, inhibit Gro/TLE-mediated transcriptional repression in a dominant negative manner perhaps by forming mixed oligomers with full-length Gro/TLE proteins through the Q domain, resulting in an inactive complex. (C) Gro/TLE is also regulated by posttranslational modification. For example, sumoylated Gro/TLE is targeted by the ubiquitin ligase Degringloade (Dgrn) for inactivation, thereby antagonizing Gro/TLE function. See text for references and details. (See Color Insert.)

repressors. The high-resolution X-ray structure of the WD-repeat domain shows that it assumes a toroidal shape in which each repeat forms a four-stranded antiparallel β-sheet (Pickles *et al.*, 2002). Thus, the structure resembles a seven-bladed propeller, sometimes termed a β-propeller, a structure known to mediate many protein–protein interactions (Li and Roberts, 2001; Pickles *et al.*, 2002).

The β-propeller provides the binding site for two distinct short peptide sequences that, as mentioned previously, are often found in Gro-interacting repressors. These are the WRPW motif and the Engrailed homology-1 (eh1) motif (consensus sequence = FxIxxIL), found in a wide variety of Gro-dependent repressor proteins. Repressors that contain WRPW or its variants include HES, Runt, Ripply1, Bowline, Huckebein, and Brinker, while proteins that contain eh1 motifs include Engrailed, Goosecoid, Pax, Six, Dorsal, Knirps, Slp1, and Hex (Andrioli *et al.*, 2004; Fisher *et al.*, 1996; Flores-Saaib *et al.*, 2001; Goldstein *et al.*, 1999; Jimenez *et al.*, 1997, 1999; Kawamura *et al.*, 2005; Kondow *et al.*, 2006; Paroush *et al.*, 1994; Payankaulam and Arnosti, 2009; Swingler *et al.*, 2004; Tolkunova *et al.*, 1998). Insight into the basis for the interaction between the β-propeller and these short motifs is provided by genetic and structural analysis of the WD-repeat domain bound to the peptides (Jennings *et al.*, 2006). Although the two motifs occupy overlapping sites across the β-propeller central pore, they bind in different conformations. The WRPW motif adopts a β-strand conformation that forms a tight hydrophobic plug, whereas the eh1 motif forms a short α helix (Jennings *et al.*, 2006).

Runt family members, which are involved in embryonic pattern formation, sex determination, eye patterning, and hematopoiesis, contain a motif similar to the WRPW sequence in which the C-terminal Trp is replaced with a Tyr (Aronson *et al.*, 1997; Canon and Banerjee, 2003). This change weakens binding, and as a result, the Runt family protein Lozenge can function as an activator or a repressor and repression depends on additional accessory proteins, such as Cut to assist with Gro recruitment (Fig. 3.4A) (Canon and Banerjee, 2003). In support of these ideas, replacing the C-terminal Tyr with Trp converted Lozenge into a dedicated repressor independent of the presence or absence of Cut (Canon and Banerjee, 2003).

How can a Trp to Tyr conversion lead to such a dramatic change in affinity for Gro/TLE? According to the crystal structure, all four residues of the WRPW motif are fully engaged by the WD-repeat domain, resulting in a strong hydrophobic interaction (Jennings *et al.*, 2006). This suggests that the phenolic hydroxyl group of a C-terminal tyrosine would be buried in the binding pocket, resulting in a loss of a hydrogen bond, consistent with decreased affinity of this motif (Jennings *et al.*, 2006).

The Rel family transcription factor, Dorsal, which directs dorsoventral embryonic patterning, provides another example of a factor in which the regulation of Gro affinity by accessory factors plays a critical role in repression. Dorsal lacks high-affinity Gro/TLE recruitment motifs but contains a highly degenerate and therefore weak eh1-like motif in its CTD that is required for Dorsal-mediated repression (Flores-Saaib *et al.*, 2001). The low-affinity nature of this motif seems to be essential for the ability of Dorsal to function as an activator of some targets and a repressor of others as Dorsal becomes a dedicated repressor when its eh1-like motif is replaced with a WRPW motif (Ratnaparkhi *et al.*, 2006).

The mechanism by which Gro converts Dorsal from an activator to a repressor has been characterized through studies of the *zen* gene. The *zen* locus harbors a CRM termed the ventral repression region, which contains multiple Dorsal binding sites, several of which are closely juxtaposed to sites for the ARID-domain containing transcription factor Dead ringer (Dri) (Valentine *et al.*, 1998). Individually, Dorsal and Dri bind weakly to Gro but together form a high-affinity platform for Gro/TLE recruitment (Fig. 3.4A). Thus, if the Dri sites are mutated, Gro fails to bind and Dorsal functions as an activator rather than as a repressor.

## 3.2. Interactions with truncated Gro family proteins

In addition to full-length Gro/TLE family proteins, truncated members of this family termed amino-terminal enhancer of split (AES) family proteins are often found in vertebrate proteomes and also in the *C. elegans* proteome (Bajoghli, 2007; Beagle and Johnson, 2010; Flowers *et al.*, 2010; Li, 2000). Interactions between full-length Gro/TLE family proteins and their shorter relatives from the AES family may be another way to regulate Gro/TLE-mediated repression (Fig. 3.4B). The AES proteins, which consist only of the Q and GP region, were postulated to inhibit Gro/TLE-mediated transcriptional repression in a dominant negative manner by binding to Gro/TLE through the Q domain. Support for this idea comes from experiments in which coexpression of AES and Gro/TLE resulted in a four- to fivefold reduction of Gro/TLE-mediated repression by the Nkx repressor in COS-7 cells (Muhr *et al.*, 2001). Similarly, AES relieves repression by the Gro/TLE-dependent factor proline-rich homeodomain (PRH). Thus, the balance between full-length Gro/TLE family proteins and AES family proteins seems to be crucial for the proper developmental regulation of transcription (Swingler *et al.*, 2004).

How does this naturally occurring truncated form of Gro modulate Gro/TLE mediate repression? A possible answer to this question is suggested by a study showing that AES cannot bind to Rpd3/HDAC1 even though a truncated form of Gro/TLE containing just the Q and GP domains is able to bind this histone deacetylase. This suggests that the GP domain in AES may lack full functionality, and therefore, the Q domain-mediated interaction between Gro/TLE and AES could disrupt the function of the Gro/TLE oligomer (Brantjes *et al.*, 2001). In addition, since the Q domain of Gro/TLE interacts with some repressor proteins, including Tcf/Lef, AES may antagonize Gro/TLE-mediated repression by interacting directly with repressors (Brantjes *et al.*, 2001; Mallo *et al.*, 1993).

## 3.3. Regulation of Gro/TLE function by posttranslational modification

Posttranslational modification of transcription factors is an effective means by which developmental, metabolic, and mitogenic cues can modify gene expression in a rapid and reversible manner. Given that Gro/TLE functions in conjunction with many repressors in a myriad of developmental processes, it represents an ideal candidate for achieving a coordinated transcriptional response. In accord with this view, many studies have demonstrated that Gro/TLE proteins are substrates for posttranslational modification, including phosphorylation, sumoylation, and poly(ADP-ribosyl)ation.

### 3.3.1. Gro/TLE phosphorylation

The Gro/TLE CcN and SP domains appear to be important targets for phosphorylation by a variety of Ser/Thr kinases. While phosphorylation of the CcN domain can be associated with either enhancement or inhibition of Gro-mediated repression, phosphorylation of the SP domain generally inhibits repression.

The CcN domain contains phosphoacceptor sites for casein kinase II (CK II) and the cyclin-dependent kinase family member cdc2 (Stifani *et al.*, 1992), and phosphorylation by both of these kinases modulates Gro/TLE function. For instance, pharmacological inhibition of cdc2 enhances Gro/TLE-mediated repression (Nuthall *et al.*, 2002b). In addition, the phosphorylation of Ser 239 in the CcN domain enhances Gro/TLE activity by increasing the affinity of Gro for HES family factors and chromatin (Nuthall *et al.*, 2002a, 2004). Further, this phosphorylation event is necessary for cofactor-activated phosphorylation in which Gro/TLE undergoes further phosphorylation as a consequence of its interaction with HES family factors (Buscarlet *et al.*, 2009; Nuthall *et al.*, 2002a, 2004).

The SP domain may function as a negative regulatory domain because phosphorylation of residues in this domain results in decreased Gro/TLE repression activity. For example, phosphorylation of a serine in the SP domain by homeodomain-interacting protein kinase 2 (HIPK2) decreases Gro/TLE-mediated repression by attenuating its binding to a DNA-bound repressor partner and by weakening the interaction between Gro/TLE and Rpd3/HDAC1 (Choi *et al.*, 1999, 2005; Lee *et al.*, 2009). However, these experiments were performed using triple point mutations, only two of which mapped to the SP region, and thus it is not clear from these studies if phosphorylation of any single residue in the SP domain modulates Gro-mediated repression.

The SP domain also contains one of two residues that are phosphorylated by mitogen-activated protein kinase (MAPK) in response to activation of the receptor tyrosine kinase (RTK)/Ras/MAPK signaling pathway, resulting in a decrease of Gro/TLE repressive activity (Cinnamon *et al.*, 2008; Hasson *et al.*, 2005). Although the mechanism by which phosphorylation of the SP domain reduces Gro activity is not clear, a clue may be provided by a study showing that mutation of a phosphoacceptor residue in the SP domain inhibited phosphorylation of Ser 239 in the CcN domain, indirectly resulting in weakened Gro/TLE chromatin association (Buscarlet *et al.*, 2009; Nuthall *et al.*, 2004). This suggests cross talk between the posttranslational modifications much like what is observed for histone tails (Berger, 2007).

It is becoming increasingly clear that downregulation of Gro by RTK/Ras/MAPK signaling plays critical roles in many aspects of development, including patterning of the nonsegmental termini in *Drosophila* embryos. Terminal patterning is directed by the gap genes *huckebein (hkb)* and *tailless* (*tll*), the expression of which is restricted to the embryonic termini by Gro through its interactions with the repressor Capicua (Cic) (Jimenez *et al.*, 2000; Paroush *et al.*, 1997). Terminal expression of *hkb* and *tll* depends on the Torso RTK, which works through the RTK/Ras/MAPK signal transduction pathway to alleviate Gro-mediated repression at the embryonic termini. While this depends in part on the signal-induced degradation of Cic, it also appears that direct phosphorylation of Gro itself attenuates Gro function thereby allowing terminal-specific expression of *hkb* and *tll*. Thus, once again Gro appears to have an active and instructional role in pattern formation (Astigarraga *et al.*, 2007; Cinnamon *et al.*, 2008).

It also appears that phosphorylation of Gro in response to RTK/Ras/MAPK signaling may provide a cellular memory of the RTK signal (Ajuria *et al.*, 2011; Cinnamon *et al.*, 2008; Helman *et al.*, 2011). While the phosphorylated (and therefore activated) state of MAPK can be short-lived, the phosphorylated form of Gro that results from the action of activated MAPK can be relatively stable. For example, the gene *intermediate neuroblasts defective* (*ind*) is initially activated in a portion of the presumptive ventral nerve cord by the epidermal growth factor receptor (EGFR), an RTK that signals via Ras and MAPK. This activation occurs, at least part, by the phosphorylation and resulting inactivation of Gro, which serves to keep *ind* off in the absence of the EGFR signal. Gro phosphorylation and therefore *ind* derepression are maintained for several hours after the end of EGFR signaling and resulting dephosphorylation of MAPK. Similar mechanisms appear to allow Gro phosphorylation to function as a memory for signaling via other RTKs, including Torso and fibroblast growth factor receptor (Helman *et al.*, 2011).

### 3.3.2. Other posttranslational modifications

Gro/TLE is also poly(ADP-ribosy)lated. This covalent modification, in which the enzyme poly(ADP-ribose) polymerase1 (PARP-1) transfers ADP-ribose units from $NAD^+$ onto glutamic acid residues of acceptor proteins, is involved in chromatin decondensation, DNA replication, and repair (Ju *et al.*, 2004). A study in rat neural stem cells not only revealed that Gro/TLE is in a protein complex consisting of PARP-1, nucleophosmin, nucleolin, topoisomerase IIb, and Rad50 but also that activation of PARP-1 by $Ca^{2+}$/calmodulin-dependent kinase II (CaMKIIδ) leads to poly(ADP-ribosy)lation of Gro/TLE, resulting in Gro/TLE dissociation from its repressor partner HES and the relief of repression of its target genes (Ju *et al.*, 2004).

Finally, Gro/TLE is also known to be sumoylated. A study in early fly embryos using mass spectrometry-based proteomics identified Gro/TLE as a conjugation target of the ubiquitin-like protein SUMO, but the biological relevance of this modification was unknown (Nie *et al.*, 2009). Since sumoylation often leads to downregulation of targets, this modification might inhibit Gro/TLE-mediated repression. Support for this idea comes from a recent study showing that sumoylated Gro/TLE is specifically inactivated by the ubiquitin ligase Degringloade (Dgrn) thereby antagonizing Gro/TLE function (Fig. 3.4C). Thus, Gro/TLE sumoylation followed by Dgrn-mediated inactivation may also serve as a molecular switch that regulates Gro recruitment in development (Abed *et al.*, 2011).

## 3.4. Gro as a mediator of interplay between signal transduction pathways

Gro/TLE proteins have roles in many signal transduction pathways in addition to the RTK/Ras/MAPK pathway discussed above (Cinnamon and Paroush, 2008). These include the Notch, decapentaplegic (Dpp), and Wnt/Wingless (Wg) pathways and thus Gro has the potential to mediate cross talk between the RTK/Ras/MAPK signal and these other pathways. Perhaps the best evidence for such interplay comes from studies of the Notch pathway.

Gro plays both negative and positive roles in transducing the Notch signal (Fig. 3.5A). In the absence of Notch signaling, Gro is recruited by the DNA-binding transcription factor Suppressor of Hairless (Su(H)) to Notch target genes such as the genes of the E(spl) complex. This recruitment is mediated by the adapter protein Hairless (H) and the resulting complex additionally includes the corepressor CtBP and leads to silencing of the Notch targets including E(spl) complex genes (Nagel *et al.*, 2005). Upon Notch activation, Notch is cleaved, and the Notch intracellular domain translocates to the nucleus and binds to Su(H). This interaction displaces the Gro/TLE-H-CtBP corepressor complex, resulting in the activation of E(spl) genes. As we have seen, these E(spl) factors are members of the HES family of bHLH domain-containing factors. They contain C-terminal WRPW motifs, by which they recruit Gro to repress proneural genes (Barolo *et al.*, 2002; Hasson and Paroush, 2006; Nagel *et al.*, 2005). Thus, in addition to antagonizing Notch signaling via interactions with Su(H), Gro cooperates with Notch signaling via interactions with E(spl) complex-encoded transcriptional repressors. As mentioned above, activation of RTK/Ras/MAPK signaling results in phosphorylated Gro, inhibiting its function as a repressor. As a result, Gro can mediate cross talk between the Notch and RTK/Ras/MAPK signaling pathways.

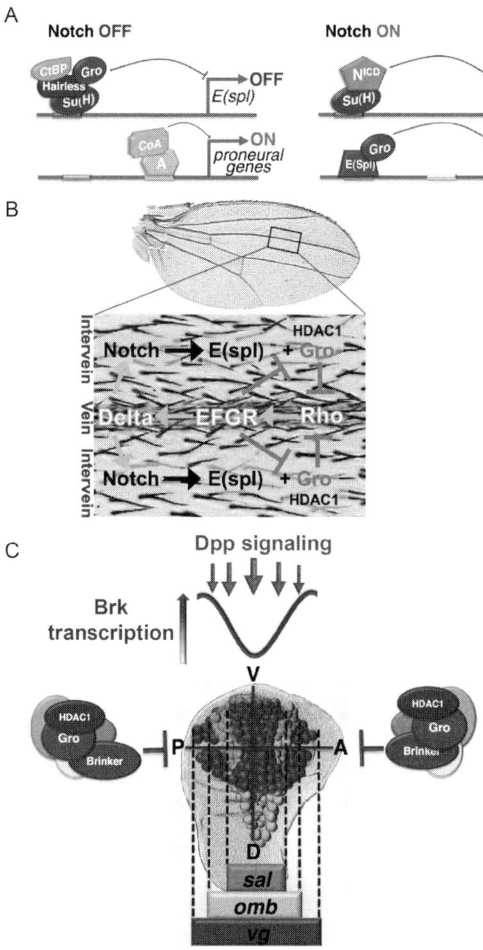


**Figure 3.5** Roles for Gro/TLE family proteins in signaling. (A) Gro plays multiple roles in Notch signaling, working both cooperatively and antagonistically with Notch. In the absence of Notch signaling, Gro is recruited by Suppressor of Hairless (Su(H)) to a corepressor complex at Notch target genes such as the genes of the E(spl) complex. Upon activation of Notch signaling, the Notch intracellular domain translocates to the nucleus displacing the Gro-containing complex from Su(H) and resulting in the activation of E(spl) complex genes. bHLH factors encoded by E(spl) then employ Gro as a corepressor to silence proneural genes. (B) Gro/Notch antagonism is required for wing veination. Wing vein specification is initiated by the activation of *rhomboid* (*rho*) in the wing disc in territories where the veins will develop (the provein region). This leads to activation of EGFR and therefore of the RTK/Ras/MAPK signaling pathway triggering expression of genes that lead to vein formation. (These so-called provein genes have been omitted from the diagram for simplicity). RTK/Ras/MAPK signaling also leads to expression of the Notch ligand Delta in the provein region, which signals to adjacent intervein cells. Notch signaling in the intervein region antagonizes wing veination via the Gro-dependent repression of *rho*, thereby preventing inappropriate expansion of

A good example of the interplay between RTK/Ras/MAPK and Notch signaling may be provided by the events that lead to formation of the *Drosophila* wing veins (Fig. 3.5B). Wing vein specification is initiated by the activation of *rhomboid* (*rho*) in the wing disc in territories where the veins will develop (Guichard *et al.*, 1999). The intramembrane protease encoded by *rho* is then responsible for the processing and activation of the EGFR ligand Spitz leading to the activation of the RTK/Ras/MAPK signaling pathway and the expression of provein genes. EGFR signaling within the provein region leads to expression of the Notch ligand Delta, which in turn results in Notch signaling in cells flanking the provein region (de Celis *et al.*, 1997). Notch signaling in the intervein region serves to antagonize wing veination and prevent inappropriate expansion of veins into the intervein region. As discussed above, Notch signaling requires Gro. Thus, by inactivating Gro, the RTK/Ras/MAPK signal in the provein region serves to prevent the Notch signal from spreading into the provein region, which would block normal wing vein formation. In support of this notion, mutations that convert the MAPK phosphoacceptor residues in Gro to Ala lead to vein loss, while mutation of these residues to phosphomimetic Asp residues leads to excess vein development (Hasson *et al.*, 2005). Finally, the ability of Gro to interfere with wing vein formation is blocked by histone deacetylase inhibitors reinforcing the notion that Gro functions via histone deacetylation (Winkler *et al.*, 2010).

Antagonism between Notch and RTK/Ras/MAPK signaling is not just limited to wing veination but is found in many other developmental contexts, and it is highly likely that Gro will play a role in mediating additional examples of such antagonism (Hasson and Paroush, 2006). Further, as mentioned above, Gro also plays roles in many other signaling pathways including the Wnt/Wg and Dpp pathways where it plays a negative role (Brantjes *et al.*, 2001; Cavallo *et al.*, 1998; Hasson *et al.*, 2001; Roose *et al.*, 1998). Thus, it is very likely that Gro is at the nexus of many signaling pathways thereby playing multiple essential roles in the coordination of signaling during development.

**veins into the intervein region. Gro phosphorylation by RTK/Ras/MAPK signaling in the provein region leads to Gro inactivation. Thus, the RTK/Ras/MAPK signal in the provein region prevents the Notch signal from spreading into the provein region and blocking normal wing vein formation. (C) Anteroposterior patterning in the developing wing results in large part from the repression of *brk* in response to the Dpp signal in the region surrounding the anteroposterior midline. Brk protein acts, in turn, as a Gro-dependent repressor to turn off the expression of genes such as *vg*, *omb*, and *sal* at the anterior and posterior edges of the developing wing. These Brk targets are responsive to different concentrations of Brk, thus leading to different size expression domains. (See Color Insert.)**

## 4. A Final Example of Gro's Instructive Role in Development—Anteroposterior Patterning of the Wing

As the foregoing discussion has made clear, Gro has multiple instructive roles in development: It is actively involved in the interpretation of morphogen gradients and in the reception and interpretation of developmental signals. One final example of a developmental process in which Gro is intimately involved and which illustrates the role of this corepressor in signaling and in determining spatial and temporal boundaries of gene activity is the anteroposterior patterning of the wing imaginal disc as mediated by Dpp signaling (Fig. 3.5C).

*dpp* is expressed in a narrow band along the anteroposterior compartment boundary in the wing disc, from which its secreted protein product spreads toward the anterior and posterior edges of the disc (Muller *et al.*, 2003; Nellen *et al.*, 1996). The resulting Dpp morphogen gradient patterns the wing disc, in large part because the Dpp signal represses *brk* transcription in the region surrounding the midline. Brk acts, in turn, as a Gro-dependent repressor to silence the expression of genes, such as *spalt* (*sal*), *vestigial* (*vg*), and *optomotor blind* (*omb*), at the anterior and posterior edges of the developing wing (Hasson *et al.*, 2001; Sivasankaran *et al.*, 2000). The anteroposterior borders of these target gene expression domains vary since each target is sensitive to a different threshold concentration of Brk.

Once again, as is the case for interpretation of the Cic and Dorsal morphogen gradients, the evidence suggests that the threshold responses to Brk are critically dependent upon levels of Gro activity. At high enough concentrations of Gro, Brk will repress its targets all the way to the midline of the disc where Brk activity is extremely low (Fig. 3.6). Thus, regulatory processes that modulate Gro activity such as Gro phosphorylation and sumoylation could play critical roles in wing patterning by modulating threshold responses to Brk.

At this point, we can only speculate about how the activity of graded transcription factors such as Brk can be so sensitive to Gro concentration and activity levels. One possibility is suggested by the model discussed in the section on the mechanism of Gro-mediated repression in which it was proposed that Gro spreads along chromatin to establish large transcriptionally silent domains. Perhaps these large domains represent cooperative units, the stability of which could be exquisitely sensitive to Gro concentration, thus accounting for the ability of Gro to play an important instructive role in development.

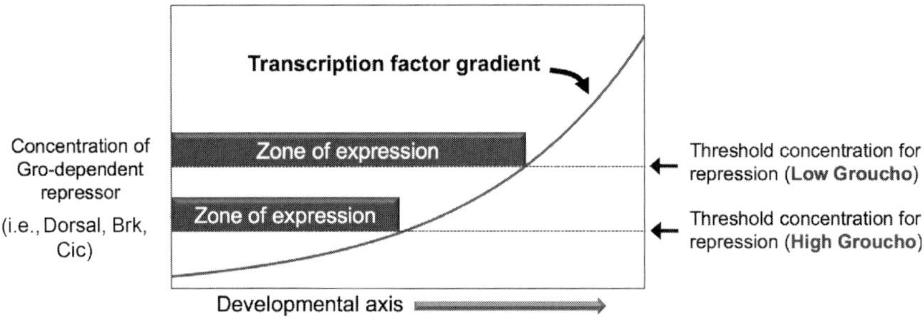


**Figure 3.6** Gro/TLE plays an active role in transcription factor gradient interpretation. Gro/TLE plays a critical role in determining the threshold response of genes to transcription factor gradients. Affinity of a repressor for its binding site along with Gro/TLE activity levels determines the height of a threshold response. When Gro/TLE activity levels are low, higher expression levels of the graded transcription factor is needed to generate repression, whereas when Gro/TLE levels are high, the threshold transcription factor concentration required for repression is lower. (For color version of this figure, the reader is referred to the Web version of this chapter.)

## 5. Concluding Remarks

In conclusion, while early studies on Gro focused on its role as a "universal corepressor" that simply had to be present to allow repression to occur, it now appears that Gro plays an extremely active role in development. Its levels and activity are precisely regulated by multiple mechanisms and such recognition is critical in development. In accord with these conclusions, misregulation of Gro/TLE family members in humans is often linked to cancer, including cervical, colon, lung, brain, pituitary gland, and skin cancers (Allander *et al.*, 2002; Allen *et al.*, 2006; Buscarlet and Stifani, 2007; Rorive *et al.*, 2006; Ruebel *et al.*, 2006; Terry *et al.*, 2007). Moreover, *in vivo* studies show that overexpression of Gro/TLE in transgenic mice induces lung adenocarcinoma (Allen *et al.*, 2006). This is likely because many of the pathways and factors that interact with Gro/TLE have oncogenic properties, including the Wg/Wnt pathway (colorectal cancer), the Notch pathway (acute lymphocytic leukemia), and Runt (acute myeloid leukemia) (Bienz and Clevers, 2000; Lund and van Lohuizen, 2002; Lutterbach *et al.*, 2000; Morin, 1999; Weng *et al.*, 2004; Yamagata *et al.*, 2005). As a result, precisely regulated levels of Gro/TLE activity are essential for normal growth and development, and the illumination of the mechanisms of repression by Gro/TLE may lead to a better understanding of the events in tumorigenesis.

## *ACKNOWLEDGMENTS*

We thank our colleagues in the Courey lab and in the field for many stimulating discussions regarding the mechanisms and regulation of developmental control by Groucho. This work was partially supported by NIH grant GM44522 to A. J. C. and National Research Service Award GM86122 to W. T. J.

## REFERENCES


Abed, M., Barry, K. C., Kenyagin, D., *et al.* (2011). Degringolade, a SUMO-targeted ubiquitin ligase, inhibits Hairy/Groucho-mediated repression. *EMBO J.* **30,** 1289–1301.

Ajuria, L., Nieva, C., Winkler, C., *et al.* (2011). Capicua DNA-binding sites are general response elements for RTK signaling in Drosophila. *Development* **138,** 915–924.

Allander, S. V., Illei, P. B., Chen, Y., *et al.* (2002). Expression profiling of synovial sarcoma by cDNA microarrays: Association of ERBB2, IGFBP2, and ELF3 with epithelial differentiation. *Am. J. Pathol.* **161,** 1587–1595.

Allen, T., van Tuyl, M., Iyengar, P., *et al.* (2006). Grg1 acts as a lung-specific oncogene in a transgenic mouse model. *Cancer Res.* **66,** 1294–1301.

Andrioli, L. P., Oberstein, A. L., Corado, M. S., *et al.* (2004). Groucho-dependent repression by sloppy-paired 1 differentially positions anterior pair-rule stripes in the Drosophila embryo. *Dev. Biol.* **276,** 541–551.

Arnosti, D. N. (2004). Multiple mechanisms of transcriptional repression in eukaryotes. *Handb. Exp. Pharmacol.* **66,** 33–67.

Aronson, B. D., Fisher, A. L., Blechman, K., *et al.* (1997). Groucho-dependent and -independent repression activities of Runt domain proteins. *Mol. Cell. Biol.* **17,** 5581–5587.

Astigarraga, S., Grossman, R., Diaz-Delfin, J., *et al.* (2007). A MAPK docking site is critical for downregulation of Capicua by Torso and EGFR RTK signaling. *EMBO J.* **26,** 668–677.

Bajoghli, B. (2007). Evolution of the Groucho/Tle gene family: Gene organization and duplication events. *Dev. Genes Evol.* **217,** 613–618.

Bajoghli, B., Aghaallaei, N., Soroldoni, D., *et al.* (2007). The roles of Groucho/Tle in left-right asymmetry and Kupffer's vesicle organogenesis. *Dev. Biol.* **303,** 347–361.

Barolo, S., Stone, T., Bang, A. G., *et al.* (2002). Default repression and Notch signaling: Hairless acts as an adaptor to recruit the corepressors Groucho and dCtBP to Suppressor of Hairless. *Genes Dev.* **16,** 1964–1976.

Beagle, B., and Johnson, G. V. (2010). AES/GRG5: More than just a dominant-negative TLE/GRG family member. *Dev. Dyn.* **239,** 2795–2805.

Berger, S. L. (2007). The complex language of chromatin regulation during transcription. *Nature* **447,** 407–412.

Berk, A. J. (1999). Activation of RNA polymerase II transcription. *Curr. Opin. Cell Biol.* **11,** 330–335.

Bienz, M., and Clevers, H. (2000). Linking colorectal cancer to Wnt signaling. *Cell* **103,** 311–320.

Brantjes, H., Roose, J., van De Wetering, M., *et al.* (2001). All Tcf HMG box transcription factors interact with Groucho-related co-repressors. *Nucleic Acids Res.* **29,** 1410–1419.

Buchberger, J. R., Onishi, M., Li, G., *et al.* (2008). Sir3-nucleosome interactions in spreading of silent chromatin in Saccharomyces cerevisiae. *Mol. Cell. Biol.* **28,** 6903–6918.

Burks, P. J., Isaacs, H. V., and Pownall, M. E. (2009). FGF signalling modulates transcriptional repression by Xenopus groucho-related-4. *Biol. Cell* **101,** 301–308.

Buscarlet, M., and Stifani, S. (2007). The 'Marx' of Groucho on development and disease. *Trends Cell Biol.* **17,** 353–361.

Buscarlet, M., Hermann, R., Lo, R., *et al.* (2009). Cofactor-activated phosphorylation is required for inhibition of cortical neuron differentiation by Groucho/TLE1. *PLoS One* **4,** e8107.

Cai, H. N., Arnosti, D. N., and Levine, M. (1996). Long-range repression in the Drosophila embryo. *Proc. Natl. Acad. Sci. USA* **93,** 9309–9314.

Cai, Y., Brophy, P. D., Levitan, I., *et al.* (2003). Groucho suppresses Pax2 transactivation by inhibition of JNK-mediated phosphorylation. *EMBO J.* **22,** 5522–5529.

Canon, J., and Banerjee, U. (2003). In vivo analysis of a developmental circuit for direct transcriptional activation and repression in the same cell by a Runx protein. *Genes Dev.* **17,** 838–843.

Cavallo, R. A., Cox, R. T., Moline, M. M., *et al.* (1998). Drosophila Tcf and Groucho interact to repress Wingless signalling activity. *Nature* **395,** 604–608.
Chen, G., and Courey, A. J. (2000). Groucho/TLE family proteins and transcriptional repression. *Gene* **249,** 1–16.
Chen, G., Nguyen, P. H., and Courey, A. J. (1998). A role for Groucho tetramerization in transcriptional repression. *Mol. Cell. Biol.* **18,** 7259–7268.
Chen, G., Fernandez, J., Mische, S., *et al.* (1999). A functional interaction between the histone deacetylase Rpd3 and the corepressor groucho in Drosophila development. *Genes Dev.* **13,** 2218–2230.
Choi, C. Y., Kim, Y. H., Kwon, H. J., *et al.* (1999). The homeodomain protein NK-3 recruits Groucho and a histone deacetylase complex to repress transcription. *J. Biol. Chem.* **274,** 33194–33197.
Choi, C. Y., Kim, Y. H., Kim, Y. O., *et al.* (2005). Phosphorylation by the DHIPK2 protein kinase modulates the corepressor activity of Groucho. *J. Biol. Chem.* **280,** 21427–21436.
Chouard, T. (2011). Structural biology: Breaking the protein rules. *Nature* **471,** 151–153.
Choudhury, B. K., Kim, J., Kung, H. F., *et al.* (1997). Cloning and developmental expression of Xenopus cDNAs encoding the Enhancer of split groucho and related proteins. *Gene* **195,** 41–48.
Cinnamon, E., and Paroush, Z. (2008). Context-dependent regulation of Groucho/TLE-mediated repression. *Curr. Opin. Genet. Dev.* **18,** 435–440.
Cinnamon, E., Helman, A., Ben-Haroush Schyr, R., *et al.* (2008). Multiple RTK pathways downregulate Groucho-mediated repression in Drosophila embryogenesis. *Development* **135,** 829–837.
Courey, A. J., and Jia, S. (2001). Transcriptional repression: The long and the short of it. *Genes Dev.* **15,** 2786–2796.
Dawson, S. R., Turner, D. L., Weintraub, H., *et al.* (1995). Specificity for the hairy/enhancer of split basic helix-loop-helix (bHLH) proteins maps outside the bHLH domain and suggests two separable modes of transcriptional repression. *Mol. Cell. Biol.* **15,** 6923–6931.
de Celis, J. F., Bray, S., and Garcia-Bellido, A. (1997). Notch signalling regulates veinlet expression and establishes boundaries between veins and interveins in the Drosophila wing. *Development* **124,** 1919–1928.
Delidakis, C., Preiss, A., Hartley, D. A., *et al.* (1991). Two genetically and molecularly distinct functions involved in early neurogenesis reside within the Enhancer of split locus of Drosophila melanogaster. *Genetics* **129,** 803–823.
Dubnicoff, T., Valentine, S. A., Chen, G., *et al.* (1997). Conversion of dorsal from an activator to a repressor by the global corepressor Groucho. *Genes Dev.* **11,** 2952–2957.
Dunker, A. K., Cortese, M. S., Romero, P., *et al.* (2005). Flexible nets. The roles of intrinsic disorder in protein interaction networks. *FEBS J.* **272,** 5129–5148.
Eberhard, D., Jimenez, G., Heavey, B., *et al.* (2000). Transcriptional repression by Pax5 (BSAP) through interaction with corepressors of the Groucho family. *EMBO J.* **19,** 2292–2303.
Edmondson, D. G., and Roth, S. Y. (1998). Interactions of transcriptional regulators with histones. *Methods* **15,** 355–364.
Edmondson, D. G., Smith, M. M., and Roth, S. Y. (1996). Repression domain of the yeast global repressor Tup1 interacts directly with histones H3 and H4. *Genes Dev.* **10,** 1247–1259.
Edmondson, D. G., Zhang, W., Watson, A., *et al.* (1998). In vivo functions of histone acetylation/deacetylation in Tup1p repression and Gcn5p activation. *Cold Spring Harb. Symp. Quant. Biol.* **63,** 459–468.
Fisher, A. L., and Caudy, M. (1998). Groucho proteins: Transcriptional corepressors for specific subsets of DNA-binding transcription factors in vertebrates and invertebrates. *Genes Dev.* **12,** 1931–1940.

Fisher, A. L., Ohsako, S., and Caudy, M. (1996). The WRPW motif of the hairy-related basic helix-loop-helix repressor proteins acts as a 4-amino-acid transcription repression and protein-protein interaction domain. *Mol. Cell. Biol.* **16,** 2670–2677.

Flores-Saaib, R. D., and Courey, A. J. (2000). Analysis of Groucho-histone interactions suggests mechanistic similarities between Groucho- and Tup1-mediated repression. *Nucleic Acids Res.* **28,** 4189–4196.

Flores-Saaib, R. D., Jia, S., and Courey, A. J. (2001). Activation and repression by the C-terminal domain of Dorsal. *Development* **128,** 1869–1879.

Flowers, E. B., Poole, R. J., Tursun, B., *et al.* (2010). The Groucho ortholog UNC-37 interacts with the short Groucho-like protein LSY-22 to control developmental decisions in C. elegans. *Development* **137,** 1799–1805.

Gao, X., Chandra, T., Gratton, M. O., *et al.* (2001). HES6 acts as a transcriptional repressor in myoblasts and can induce the myogenic differentiation program. *J. Cell Biol.* **154,** 1161–1171.

Gasperowicz, M., and Otto, F. (2005). Mammalian Groucho homologs: Redundancy or specificity? *J. Cell. Biochem.* **95,** 670–687.

Goldstein, R. E., Jimenez, G., Cook, O., *et al.* (1999). Huckebein repressor activity in Drosophila terminal patterning is mediated by Groucho. *Development* **126,** 3747–3755.

Gray, S., and Levine, M. (1996a). Short-range transcriptional repressors mediate both quenching and direct repression within complex loci in Drosophila. *Genes Dev.* **10,** 700–710.

Gray, S., and Levine, M. (1996b). Transcriptional repression in development. *Curr. Opin. Cell Biol.* **8,** 358–364.

Grbavec, D., and Stifani, S. (1996). Molecular interaction between TLE1 and the carboxyl-terminal domain of HES-1 containing the WRPW motif. *Biochem. Biophys. Res. Commun.* **223,** 701–705.

Grbavec, D., Lo, R., Liu, Y., *et al.* (1998). Transducin-like Enhancer of split 2, a mammalian homologue of Drosophila Groucho, acts as a transcriptional repressor, interacts with Hairy/Enhancer of split proteins, and is expressed during neuronal development. *Eur. J. Biochem.* **258,** 339–349.

Gromoller, A., and Lehming, N. (2000). Srb7p is a physical and physiological target of Tup1p. *EMBO J.* **19,** 6845–6852.

Guichard, A., Biehs, B., Sturtevant, M. A., *et al.* (1999). rhomboid and Star interact synergistically to promote EGFR/MAPK signaling during Drosophila wing vein development. *Development* **126,** 2663–2676.

Hasson, P., and Paroush, Z. (2006). Crosstalk between the EGFR and other signalling pathways at the level of the global transcriptional corepressor Groucho/TLE. *Br. J. Cancer* **94,** 771–775.

Hasson, P., Muller, B., Basler, K., *et al.* (2001). Brinker requires two corepressors for maximal and versatile repression in Dpp signalling. *EMBO J.* **20,** 5725–5736.

Hasson, P., Egoz, N., Winkler, C., *et al.* (2005). EGFR signaling attenuates Groucho-dependent repression to antagonize Notch transcriptional output. *Nat. Genet.* **37,** 101–105.

Hecht, A., Laroche, T., Strahl-Bolsinger, S., *et al.* (1995). Histone H3 and H4 N-termini interact with SIR3 and SIR4 proteins: A molecular model for the formation of heterochromatin in yeast. *Cell* **80,** 583–592.

Helman, A., Cinnamon, E., Mezuman, S., *et al.* (2011). Phosphorylation of Groucho mediates RTK feedback inhibition and prolonged pathway target gene expression. *Curr. Biol.* **21,** 1102–1110.

Hoppe, G. J., Tanny, J. C., Rudner, A. D., *et al.* (2002). Steps in assembly of silent chromatin in yeast: Sir3-independent binding of a Sir2/Sir4 complex to silencers and role for Sir2-dependent deacetylation. *Mol. Cell. Biol.* **22,** 4167–4180.

Huang, Y. (2002). Transcriptional silencing in Saccharomyces cerevisiae and Schizosaccharomyces pombe. *Nucleic Acids Res.* **30,** 1465–1482.

Imai, Y., Kurokawa, M., Tanaka, K., *et al.* (1998). TLE, the human homolog of groucho, interacts with AML1 and acts as a repressor of AML1-induced transactivation. *Biochem. Biophys. Res. Commun.* **252,** 582–589.

Jan, Y., Matter, M., Pai, J. T., *et al.* (2004). A mitochondrial protein, Bit1, mediates apoptosis regulated by integrins and Groucho/TLE corepressors. *Cell* **116,** 751–762.

Jennings, B. H., and Ish-Horowicz, D. (2008). The Groucho/TLE/Grg family of transcriptional co-repressors. *Genome Biol.* **9,** 205.

Jennings, B. H., Pickles, L. M., Wainwright, S. M., *et al.* (2006). Molecular recognition of transcriptional repressor motifs by the WD domain of the Groucho/TLE corepressor. *Mol. Cell* **22,** 645–655.

Jennings, B. H., Wainwright, S. M., and Ish-Horowicz, D. (2008). Differential in vivo requirements for oligomerization during Groucho-mediated repression. *EMBO Rep.* **9,** 76–83.

Jimenez, G., Paroush, Z., and Ish-Horowicz, D. (1997). Groucho acts as a corepressor for a subset of negative regulators, including Hairy and Engrailed. *Genes Dev.* **11,** 3072–3082.

Jimenez, G., Verrijzer, C. P., and Ish-Horowicz, D. (1999). A conserved motif in goosecoid mediates groucho-dependent repression in Drosophila embryos. *Mol. Cell. Biol.* **19,** 2080–2087.

Jimenez, G., Guichet, A., Ephrussi, A., *et al.* (2000). Relief of gene repression by torso RTK signaling: Role of capicua in Drosophila terminal and dorsoventral patterning. *Genes Dev.* **14,** 224–231.

Ju, B. G., Solum, D., Song, E. J., *et al.* (2004). Activating the PARP-1 sensor component of the groucho/TLE1 corepressor complex mediates a CaMKinase IIdelta-dependent neurogenic gene activation pathway. *Cell* **119,** 815–829.

Kawamura, A., Koshida, S., Hijikata, H., *et al.* (2005). Groucho-associated transcriptional repressor ripply1 is required for proper transition from the presomitic mesoderm to somites. *Dev. Cell* **9,** 735–744.

Kondow, A., Hitachi, K., Ikegame, T., *et al.* (2006). Bowline, a novel protein localized to the presomitic mesoderm, interacts with Groucho/TLE in Xenopus. *Int. J. Dev. Biol.* **50,** 473–479.

Kornberg, R. D. (2005). Mediator and the mechanism of transcriptional activation. *Trends Biochem. Sci.* **30,** 235–239.

Kuchin, S., and Carlson, M. (1998). Functional relationships of Srb10-Srb11 kinase, carboxy-terminal domain kinase CTDK-I, and transcriptional corepressor Ssn6-Tup1. *Mol. Cell. Biol.* **18,** 1163–1171.

Kuo, D., Nie, M., De Hoff, P., *et al.* (2011). A SUMO-Groucho Q domain fusion protein: Characterization and in vivo Ulp1-mediated cleavage. *Protein Expr. Purif.* **76,** 65–71.

Lee, W., Andrews, B. C., Faust, M., *et al.* (2009). Hipk is an essential protein that promotes Notch signal transduction in the Drosophila eye by inhibition of the global co-repressor Groucho. *Dev. Biol.* **325,** 263–272.

Li, S. S. (2000). Structure and function of the Groucho gene family and encoded transcriptional corepressor proteins from human, mouse, rat, Xenopus, Drosophila and nematode. *Proc. Natl. Sci. Counc. Republ. Chin. B* **24,** 47–55.

Li, L. M., and Arnosti, D. N. (2011). Long- and short-range transcriptional repressors induce distinct chromatin states on repressed genes. *Curr. Biol.* **21,** 406–412.

Li, D., and Roberts, R. (2001). WD-repeat proteins: Structure characteristics, biological function, and their involvement in human diseases. *Cell. Mol. Life Sci.* **58,** 2085–2097.

Li, X., Romero, P., Rani, M., *et al.* (1999). Predicting protein disorder for N-, C-, and internal regions. *Genome Inform. Ser. Workshop Genome Inform.* **10,** 30–40.

Linderson, Y., Eberhard, D., Malin, S., *et al.* (2004). Corecruitment of the Grg4 repressor by PU.1 is critical for Pax5-mediated repression of B-cell-specific genes. *EMBO Rep.* **5,** 291–296.

Lindsley, D. L., and Grell, E. H. (1968). Genetic Variations of *Drosophila melanogaster. Carnegie Institution of Washington Publication* **627**.

Liu, Z., and Karmarkar, V. (2008). Groucho/Tup1 family co-repressors in plant development. *Trends Plant Sci.* **13,** 137–144.

Lund, A. H., and van Lohuizen, M. (2002). RUNX: A trilogy of cancer genes. *Cancer Cell* **1,** 213–215.

Lutterbach, B., Westendorf, J. J., Linggi, B., *et al.* (2000). A mechanism of repression by acute myeloid leukemia-1, the target of multiple chromosomal translocations in acute leukemia. *J. Biol. Chem.* **275,** 651–656.

Malave, T. M., and Dent, S. Y. (2006). Transcriptional repression by Tup1-Ssn6. *Biochem. Cell Biol.* **84,** 437–443.

Malik, S., and Roeder, R. G. (2010). The metazoan Mediator co-activator complex as an integrative hub for transcriptional regulation. *Nat. Rev. Genet.* **11,** 761–772.

Mallo, M., Franco del Amo, F., and Gridley, T. (1993). Cloning and developmental expression of Grg, a mouse gene related to the groucho transcript of the Drosophila Enhancer of split complex. *Mech. Dev.* **42,** 67–76.

Mannervik, M., and Levine, M. (1999). The Rpd3 histone deacetylase is required for segmentation of the Drosophila embryo. *Proc. Natl. Acad. Sci. USA* **96,** 6797–6801.

Mannervik, M., Nibu, Y., Zhang, H., *et al.* (1999). Transcriptional coregulators in development. *Science* **284,** 606–609.

Martinez, C. A., and Arnosti, D. N. (2008). Spreading of a corepressor linked to action of long-range repressor hairy. *Mol. Cell. Biol.* **28,** 2792–2802.

Miyasaka, H., Choudhury, B. K., Hou, E. W., *et al.* (1993). Molecular cloning and expression of mouse and human cDNA encoding AES and ESG proteins with strong similarity to Drosophila enhancer of split groucho protein. *Eur. J. Biochem.* **216,** 343–352.

Morin, P. J. (1999). Beta-catenin signaling and cancer. *Bioessays* **21,** 1021–1030.

Muhr, J., Andersson, E., Persson, M., *et al.* (2001). Groucho-mediated transcriptional repression establishes progenitor cell pattern and neuronal fate in the ventral neural tube. *Cell* **104,** 861–873.

Muller, B., Hartmann, B., Pyrowolakis, G., *et al.* (2003). Conversion of an extracellular Dpp/BMP morphogen gradient into an inverse transcriptional gradient. *Cell* **113,** 221–233.

Nagel, A. C., Krejci, A., Tenin, G., *et al.* (2005). Hairless-mediated repression of notch target genes requires the combined activity of Groucho and CtBP corepressors. *Mol. Cell. Biol.* **25,** 10433–10441.

Nakayama, H., Liu, Y., Stifani, S., *et al.* (1997). Developmental restriction of Mash-2 expression in trophoblast correlates with potential activation of the notch-2 pathway. *Dev. Genet.* **21,** 21–30.

Nellen, D., Burke, R., Struhl, G., *et al.* (1996). Direct and long-range action of a DPP morphogen gradient. *Cell* **85,** 357–368.

Nie, M., Xie, Y., Loo, J. A., *et al.* (2009). Genetic and proteomic evidence for roles of Drosophila SUMO in cell cycle control, Ras signaling, and early pattern formation. *PLoS One* **4,** e5905.

Nuthall, H. N., Husain, J., McLarren, K. W., *et al.* (2002a). Role for Hes1-induced phosphorylation in Groucho-mediated transcriptional repression. *Mol. Cell. Biol.* **22,** 389–399.

Nuthall, H. N., Joachim, K., Palaparti, A., *et al.* (2002b). A role for cell cycle-regulated phosphorylation in Groucho-mediated transcriptional repression. *J. Biol. Chem.* **277,** 51049–51057.

Nuthall, H. N., Joachim, K., and Stifani, S. (2004). Phosphorylation of serine 239 of Groucho/TLE1 by protein kinase CK2 is important for inhibition of neuronal differentiation. *Mol. Cell. Biol.* **24,** 8395–8407.

Palaparti, A., Baratz, A., and Stifani, S. (1997). The Groucho/transducin-like enhancer of split transcriptional repressors interact with the genetically defined amino-terminal silencing domain of histone H3. *J. Biol. Chem.* **272,** 26604–26610.

Papamichos-Chronakis, M., Conlan, R. S., Gounalaki, N., *et al.* (2000). Hrs1/Med3 is a Cyc8-Tup1 corepressor target in the RNA polymerase II holoenzyme. *J. Biol. Chem.* **275,** 8397–8403.

Paroush, Z., Finley, R. L., Jr., Kidd, T., *et al.* (1994). Groucho is required for Drosophila neurogenesis, segmentation, and sex determination and interacts directly with hairy-related bHLH proteins. *Cell* **79,** 805–815.

Paroush, Z., Wainwright, S. M., and Ish-Horowicz, D. (1997). Torso signalling regulates terminal patterning in Drosophila by antagonising Groucho-mediated repression. *Development* **124,** 3827–3834.

Payankaulam, S., and Arnosti, D. N. (2009). Groucho corepressor functions as a cofactor for the Knirps short-range transcriptional repressor. *Proc. Natl. Acad. Sci. USA* **106,** 17314–17319.

Payankaulam, S., Li, L. M., and Arnosti, D. N. (2010). Transcriptional repression: Conserved and evolved features. *Curr. Biol.* **20,** R764–R771.

Pflugrad, A., Meir, J. Y., Barnes, T. M., *et al.* (1997). The Groucho-like transcription factor UNC-37 functions with the neural specificity gene unc-4 to govern motor neuron identity in C. elegans. *Development* **124,** 1699–1709.

Pickles, L. M., Roe, S. M., Hemingway, E. J., *et al.* (2002). Crystal structure of the C-terminal WD40 repeat domain of the human Groucho/TLE1 transcriptional corepressor. *Structure* **10,** 751–761.

Pinto, M., and Lobe, C. G. (1996). Products of the grg (Groucho-related gene) family can dimerize through the amino-terminal Q domain. *J. Biol. Chem.* **271,** 33026–33031.

Radivojac, P., Obradovic, Z., Brown, C. J., *et al.* (2003). Prediction of boundaries between intrinsically ordered and disordered protein regions. *Pac. Symp. Biocomput.* **8,** 216–227.

Radman-Livaja, M., Ruben, G., Weiner, A., *et al.* (2011). Dynamics of Sir3 spreading in budding yeast: Secondary recruitment sites and euchromatic localization. *EMBO J.* **30,** 1012–1026.

Ratnaparkhi, G. S., Jia, S., and Courey, A. J. (2006). Uncoupling dorsal-mediated activation from dorsal-mediated repression in the Drosophila embryo. *Development* **133,** 4409–4414.

Redd, M. J., Arnaud, M. B., and Johnson, A. D. (1997). A complex composed of tup1 and ssn6 represses transcription in vitro. *J. Biol. Chem.* **272,** 11193–11197.

Romero, P., Obradovic, Z., and Dunker, K. (1997). Sequence data analysis for long disordered regions prediction in the calcineurin family. *Genome Inform. Ser. Workshop Genome Inform.* **8,** 110–124.

Romero, P., Obradovic, Z., Li, X., *et al.* (2001). Sequence complexity of disordered protein. *Proteins* **42,** 38–48.

Roose, J., Molenaar, M., Peterson, J., *et al.* (1998). The Xenopus Wnt effector XTcf-3 interacts with Groucho-related transcriptional repressors. *Nature* **395,** 608–612.

Rorive, S., Maris, C., Debeir, O., *et al.* (2006). Exploring the distinctive biological characteristics of pilocytic and low-grade diffuse astrocytomas using microarray gene expression profiles. *J. Neuropathol. Exp. Neurol.* **65,** 794–807.

Ruebel, K. H., Leontovich, A. A., Jin, L., *et al.* (2006). Patterns of gene expression in pituitary carcinomas and adenomas analyzed by high-density oligonucleotide arrays, reverse transcriptase-quantitative PCR, and protein expression. *Endocrine* **29,** 435–444.

Rusche, L. N., Kirchmaier, A. L., and Rine, J. (2002). Ordered nucleation and spreading of silenced chromatin in Saccharomyces cerevisiae. *Mol. Biol. Cell* **13,** 2207–2222.

Schmidt, C. J., and Sladek, T. E. (1993). A rat homolog of the Drosophila enhancer of split (groucho) locus lacking WD-40 repeats. *J. Biol. Chem.* **268,** 25681–25686.

Sekiya, T., and Zaret, K. S. (2007). Repression by Groucho/TLE/Grg proteins: Genomic site recruitment generates compacted chromatin in vitro and impairs activator binding in vivo. *Mol. Cell* **28,** 291–303.

Sharief, F. S., Tsoi, S. C., and Li, S. S. (1997). cDNA cloning and genomic organization of enhancer of split groucho gene from nematode Caenorhabditis elegans. *Biochem. Mol. Biol. Int.* **43,** 327–337.

Shilo, B. Z. (1998). Flies over Crete: Drosophila molecular biology. Kolymbari, Crete, July 12–18, 1998. *EMBO J.* **17,** 6769–6771.

Sivasankaran, R., Vigano, M. A., Muller, B., *et al.* (2000). Direct transcriptional control of the Dpp target omb by the DNA binding protein Brinker. *EMBO J.* **19,** 6162–6172.

Song, H., Hasson, P., Paroush, Z., *et al.* (2004). Groucho oligomerization is required for repression in vivo. *Mol. Cell. Biol.* **24,** 4341–4350.

Stifani, S., Blaumueller, C. M., Redhead, N. J., *et al.* (1992). Human homologs of a Drosophila Enhancer of split gene product define a novel family of nuclear proteins. *Nat. Genet.* **2,** 119–127.

Swingler, T. E., Bess, K. L., Yao, J., *et al.* (2004). The proline-rich homeodomain protein recruits members of the Groucho/Transducin-like enhancer of split protein family to co-repress transcription in hematopoietic cells. *J. Biol. Chem.* **279,** 34938–34947.

Terry, J., Saito, T., Subramanian, S., *et al.* (2007). TLE1 as a diagnostic immunohistochemical marker for synovial sarcoma emerging from gene expression profiling studies. *Am. J. Surg. Pathol.* **31,** 240–246.

Thirunavukkarasu, K., Mahajan, M., McLarren, K. W., *et al.* (1998). Two domains unique to osteoblast-specific transcription factor Osf2/Cbfa1 contribute to its transactivation function and its inability to heterodimerize with Cbfbeta. *Mol. Cell. Biol.* **18,** 4197–4208.

Tolkunova, E. N., Fujioka, M., Kobayashi, M., *et al.* (1998). Two distinct types of repression domain in engrailed: One interacts with the groucho corepressor and is preferentially active on integrated target genes. *Mol. Cell. Biol.* **18,** 2804–2814.

Treitel, M. A., and Carlson, M. (1995). Repression by SSN6-TUP1 is directed by MIG1, a repressor/activator protein. *Proc. Natl. Acad. Sci. USA* **92,** 3132–3136.

Tzamarias, D., and Struhl, K. (1995). Distinct TPR motifs of Cyc8 are involved in recruiting the Cyc8-Tup1 corepressor complex to differentially regulated promoters. *Genes Dev.* **9,** 821–831.

Valentine, S. A., Chen, G., Shandala, T., *et al.* (1998). Dorsal-mediated repression requires the formation of a multiprotein repression complex at the ventral silencer. *Mol. Cell. Biol.* **18,** 6584–6594.

Villanueva, C. J., Waki, H., Godio, C., *et al.* (2011). TLE3 is a dual-function transcriptional coregulator of adipogenesis. *Cell Metab.* **13,** 413–427.

Weng, A. P., Ferrando, A. A., Lee, W., *et al.* (2004). Activating mutations of NOTCH1 in human T cell acute lymphoblastic leukemia. *Science* **306,** 269–271.

Winkler, C. J., Ponce, A., and Courey, A. J. (2010). Groucho-mediated repression may result from a histone deacetylase-dependent increase in nucleosome density. *PLoS One* **5,** e10166.

Xue, B., Dunbrack, R. L., Williams, R. W., *et al.* (2010). PONDR-FIT: A meta-predictor of intrinsically disordered amino acids. *Biochim. Biophys. Acta* **1804,** 996–1010.

Yamagata, T., Maki, K., and Mitani, K. (2005). Runx1/AML1 in normal and abnormal hematopoiesis. *Int. J. Hematol.* **82,** 1–8.

Yao, J., Liu, Y., Husain, J., *et al.* (1998). Combinatorial expression patterns of individual TLE proteins during cell determination and differentiation suggest non-redundant functions for mammalian homologs of Drosophila Groucho. *Dev. Growth Differ.* **40,** 133–146.

Yao, J., Liu, Y., Lo, R., *et al.* (2000). Disrupted development of the cerebral hemispheres in transgenic mice expressing the mammalian Groucho homologue transducin-like-enhancer of split 1 in postmitotic neurons. *Mech. Dev.* **93,** 105–115.

Zhang, H., and Emmons, S. W. (2002). Caenorhabditis elegans unc-37/groucho interacts genetically with components of the transcriptional mediator complex. *Genetics* **160,** 799–803.

Zhu, C. C., Dyer, M. A., Uchikawa, M., *et al.* (2002). Six3-mediated auto repression and eye development requires its interaction with members of the Groucho-related family of co-repressors. *Development* **129,** 2835–2849.

CHAPTER FOUR

# *SPARKLING* INSIGHTS INTO ENHANCER STRUCTURE, FUNCTION, AND EVOLUTION

Nicole C. Evans, Christina I. Swanson,[1] *and* Scott Barolo

## Contents



## Abstract

This review focuses on a single *cis*-regulatory element: the *sparkling* eye enhancer of the *Drosophila dPax2* gene. *sparkling* responds to Notch and EGFR signaling, along with other direct regulatory inputs, to drive gene expression that is restricted to cone cells of the developing fly retina. Functional, genetic, biochemical, evolutionary, and bioinformatic analyses have revealed surprising properties of *sparkling*, which may provide new insights into *cis*-regulatory logic and mechanisms of transcriptional activation. These properties include: a very high density of regulatory information and a correspondingly low "junk" content; an unexpectedly complex combinatorial code; tight functional constraints on enhancer organization, paradoxically coupled with high turnover

Department of Cell & Developmental Biology, University of Michigan Medical School, Ann Arbor, Michigan, USA
[1] Present address: Department of Biology, University of North Carolina, Chapel Hill, North Carolina, USA

*Current Topics in Developmental Biology*, Volume 98
ISSN 0070-2153, DOI: 10.1016/B978-0-12-386499-4.00004-5

of DNA sequence and binding site position; a requirement for weak binding of the transcription factor Su(H) to low-affinity sites in order to maintain a cell-type-specific response to Notch signaling; and multiple specialized regulatory sequences conferring functionally distinct activation activities, all of which are required in concert to achieve proper gene expression *in vivo*.

## 1. Introduction

Enhancers, a subclass of *cis*-regulatory elements, are genomic sequences that control gene transcription, both qualitatively and quantitatively, through a wide variety of mechanisms (Blackwood and Kadonaga, 1998; Bulger and Groudine, 2010; Levine, 2010). Enhancers can direct spatially and temporally restricted gene expression, allowing genetically identical cells to take on diverse forms and functions. As the development of multicellular organisms requires precise control of gene expression, it is of no surprise that numerous human diseases and developmental defects are the direct result of enhancer malfunction (e.g., Ghiasvand *et al.*, 2011; Gordon *et al.*, 2009; Kleinjan and van Heyningen, 2005; Kleinjan *et al.*, 2004; Lettice *et al.*, 2003).

As described in detail below, enhancers can generate new patterns of gene expression by integrating multiple positive and negative inputs. These inputs are typically binding sites for sequence-specific transcription factors (TFs). The expression pattern driven by an enhancer is determined in large part by the expression patterns of its regulatory factors: in this way, broad, simple gene expression patterns can give rise to more complex and precise patterns during the development of a tissue. The idea that a "combinatorial code" of regulatory inputs (A+B+C+...) shapes developmental gene expression was perhaps most vividly demonstrated in a 2000 issue of *Cell* that featured three papers describing enhancers of the *Drosophila* genes *even-skipped*, *prospero*, and *dPax2* (Flores *et al.*, 2000; Halfon *et al.*, 2000; Xu *et al.*, 2000). These three signal-regulated enhancers (i.e., enhancers that are directly regulated by TFs under the control of cell–cell signaling pathways) respond to multiple spatial cues, acting through different TF-binding sites clustered within a single regulatory element, to produce novel patterns of gene expression during development.

In the past decade, genomic techniques have revealed a wide array of epigenetic modifications to chromatin, whose presence at a gene locus in some cases correlates with transcriptional activation or repression. Some of these chromatin marks are enriched at enhancers, and some have critical regulatory functions (Barski *et al.*, 2007; He *et al.*, 2010; Heintzman *et al.*, 2007, 2009; Jin *et al.*, 2009; Ong and Corces, 2011; Roh *et al.*, 2007). It is well established that enhancers can often drive proper patterns of gene expression irrespective of their location in the genome, as demonstrated by decades of transgenic reporter experiments in a variety of organisms. The significance of this fact with regard to epigenetics is often overlooked: for most genes, regulatory epigenetic modifications must be largely under the direction of the regulatory DNA sequences they decorate. Of course, information flows in both directions: binding of TFs to specific DNA sites can result in alterations to the state of chromatin, which in turn modulates local TF binding and function. Still, the fact that many enhancers can function properly at new locations in the genome reveals that *cis*-regulatory DNA is usually the primary organizer of gene expression and thus the primary source of developmental pattern.

## 2. The Enhancer: What Is It and What Can It Do?

Eukaryotic enhancers are classically depicted as clusters of transcription factor binding sites (TFBS), found somewhere in the genomic neighborhood of the gene(s) they regulate, where they integrate signals from the cellular environment to direct the timing, levels, and cell-type specificity of gene expression. Enhancers can be located 5′ or 3′ of their target gene and can often be found within introns or UTRs of the transcription unit itself, but they do not always target the promoter nearest to them. *cis*-Regulatory sequences vary greatly in size, typically ranging on average from hundreds of base pairs to a few kilobases. Extremely long or short enhancer sequences have been identified (e.g., Klingler *et al.*, 1996; Schulz *et al.*, 1990), although in many reported cases of large enhancers, little effort has gone toward functionally defining a truly minimal element *in vivo*. The boundaries of *cis*-regulatory elements are difficult to draw with precision: the "minimal" enhancer (i.e., the smallest fragment that is sufficient to generate a given pattern) is often weaker than larger fragments including the minimal element. Highly trimmed sequences sometimes drive a restricted or expanded pattern of gene expression, compared to larger fragments or the gene locus as a whole.

A single gene can be regulated by multiple enhancers, each responsible for a specific domain of the gene's complete expression pattern, a characteristic referred to as enhancer modularity. A classic example of multimodular *cis*-regulation is the *even-skipped* gene of *Drosophila*, whose seven stripes of embryonic expression are controlled by independent enhancers, as are later aspects of gene expression in muscle precursors, the central nervous system, and elsewhere (Goto *et al.*, 1989; Harding *et al.*, 1989; Sackerson *et al.*, 1999; Small *et al.*, 1992). Another informative case study is the sea urchin *endo16* gene, which contains over 30 high-specificity binding sites spread over a 2.3-kb region (Yuh and Davidson, 1996). In other well-documented cases, multiple separate enhancers contribute to a single aspect of a gene's expression pattern (Frankel *et al.*, 2010; Levine, 2010; Perry *et al.*, 2010, 2011; Yao *et al.*, 2008).

## 2.1. How do enhancers activate gene expression?

Enhancers are composed of combinations of protein binding sites, which recruit sequence-specific TFs. These TFs, in turn, recruit non-DNA-binding cofactors, which regulate transcription through a variety of mechanisms, including direct recruitment of RNA polymerase II and the basal transcription machinery, either directly or via Mediator, a large multiunit complex that promotes transcription via assembly of the basal transcription machinery (Malik and Roeder, 2005; Szutorisz *et al.*, 2005; Wang *et al.*, 2005). As mentioned above, enhancers also influence their local chromatin environment via epigenetic changes—for example, by recruiting ATP-dependent nucleosome remodeling complexes or histone acetyltransferases (HATs) or deacetylases (HDACs)—resulting in changes in chromatin structure that stimulate (or inhibit) transcription (Narlikar *et al.*, 2002; Orphanides and Reinberg, 2002). While these biochemical activities are essential for proper transcriptional regulation in many contexts, they have not been shown to be *sufficient* to explain enhancer-mediated gene expression *in vivo*. Growing evidence supports the importance of additional mechanisms of enhancer action, such as long-range enhancer-promoter interactions and the production of regulatory noncoding RNAs (Drissen *et al.*, 2004; Kagey *et al.*, 2010; Kim *et al.*, 2010; Ong and Corces, 2011; Orom *et al.*, 2010; Spilianakis and Flavell, 2004; Tsai *et al.*, 2010; Vakoc *et al.*, 2005; Wang *et al.*, 2011).

## 2.2. Models of enhancer structure

The presence and regulatory significance of enhancer "structure" or "grammar" (i.e., the arrangement and spacing of TFBSs) is currently a topic of active debate in the field (Crocker and Erives, 2008; Crocker *et al.*, 2010; Hare *et al.*, 2008a,b; Kulkarni and Arnosti, 2005; Levine, 2010; Lusk and Eisen, 2010; Papatsenko and Levine, 2007; Papatsenko *et al.*, 2009; Rastegar *et al.*, 2008; Swanson *et al.*, 2010). Two quite different views of enhancer organization, as it relates to function, are the "enhanceosome" model and the "information display" (or "billboard") model (Arnosti and Kulkarni, 2005). In a *cis*-regulatory sequence defined as an enhanceosome, the organization of binding sites within an enhancer is highly constrained, such that only one arrangement results in proper gene expression. Changes in spacing between TFBSs inhibit enhancer activity due to the disruption of local protein–protein interactions, which results in loss of cooperative binding and synergistic activation. Well-studied examples include enhanceosomes of the *IFN-β* and *TCRα* genes (Giese *et al.*, 1995; Thanos and Maniatis, 1995), but this may be a relatively rare class of enhancer.

Conversely, the "information display" model proposes that control of gene transcription is controlled more loosely, by simply acquiring the correct amount of "positive inputs" (in the absence of negative inputs) (Kulkarni and Arnosti, 2005). Note that this view seems to imply only a single class of "activation activity," to which all activating TFs contribute: the total amount of this activity recruited to the enhancer is the critical factor determining gene activation. Under this model, the organization and spacing of TFBSs can be quite flexible, and a high degree of cooperativity among TFs may not be required, since the enhancer presents multiple semiredundant contact surfaces for coactivators or the basal transcription complex. Evidence for this model stems from the well-studied stripe 2 enhancer of the *Drosophila even-skipped* gene (*eve* S2E). While *eve* S2E is fairly well conserved overall, TFBS arrangement shows significant variation among drosophilid and sepsid fly species (Hare *et al.*, 2008b; Ludwig *et al.*, 1998). Additionally, the effects of destroying activator sites within *eve* S2E can be rescued by adding binding sites for a heterologous activator (Arnosti *et al.*, 1996), suggesting that the requirements for transcriptional activation can be extremely flexible.

## 2.3. Persistent questions in the field

Over the past 30 years, enhancers—that is, *cis*-regulatory genomic sequences that stimulate promoter activation—have been identified in all examined forms of life: viruses, bacteria, yeast, and multicellular animals and plants all use this strategy to control gene expression (Banerji *et al.*, 1981; Levine, 2010; Priest *et al.*, 2009). Nevertheless, our knowledge of the basic components and structure of the enhancer remains far from complete. This is a problem for those wishing to understand complex biological systems because, as mentioned above, enhancers are responsible for processing, integrating, and generating complex patterning information. For example, cell–cell signaling pathways (Notch, Hedgehog, BMP, Wnt, etc.) pattern developing tissues and stem cell systems, primarily by directing gene expression via signal-regulated enhancers, yet to date no signal-regulated enhancer has been fully characterized (Barolo and Posakony, 2002; Johnson *et al.*, 2008). Even the extensively studied *eve* S2E is not completely defined with respect to its essential regulatory inputs (Andrioli *et al.*, 2002). Until an enhancer has been characterized to the point that all regulatory sites are defined, the TFs and biochemical activities recruited by those sites are known, and the spatial relationships among those sites (if any) are understood, we cannot fully grasp the relationship between *cis*-regulatory DNA sequence and gene expression patterning. This lack of understanding hampers our ability to mine new enhancers from the genome based on their

DNA sequence or binding site composition, as well as our ability to create custom enhancers as research tools and therapeutic agents.

Further, our basic understanding of the steps involved in the process of transcriptional activation remains incomplete: how many different biochemical activities are required to activate transcription? In which order do they occur? Do individual TFBSs recruit distinct "activation activities," which in combination allow for transcriptional activation, or do all activating binding sites recruit similar activities, which must merely reach a quantitative threshold to trigger activation? If the latter, do different enhancers use different combinations of activation activities, or is there a universal basic recipe? Here, we will highlight recent advances in our understanding of enhancer structure and function, focusing on *sparkling*, a cone-cell-specific eye enhancer of the *Drosophila* gene *dPax2*, as an informative test case.

## 3. The *sparkling* Enhancer of the *dPax2* Gene

The *Drosophila* compound eye consists of approximately 750 simple eyes, or ommatidia, each composed of eight photoreceptors (R1-R8) and four cone cells, surrounded by two primary pigment cells (PPCs), six secondary pigment cells, three tertiary pigment cells, and three mechanosensory bristles. During eye development, which occurs in the eye imaginal disc during larval and pupal stages, sequential EGFR and Notch signaling events recruit undifferentiated retinal cells to the above cell fates (see Voas and Rebay, 2004 for a detailed description of this process). In addition, specification of the R1, R6, R7, cone, and PPC fates requires expression of the Runx-family TF Lozenge (Lz). Because EGFR, Notch, and Lz are all broadly active in the retina, additional activators and repressors must act in concert with these signals to correctly determine cell-type-specific gene expression and differentiation within the eye, as will be discussed below.

*dPax2* expression, which is required for proper cone cell differentiation and maintenance, directly depends on EGFR and Notch signaling, as well as Lz (Flores *et al.*, 2000; Fu and Noll, 1997; Shi and Noll, 2009). The enhancer responsible for cone-cell-specific expression of *dPax2* was identified in the fourth intron of the *dPax2* gene, thanks to spontaneous mutations in that region that impair expression of *dPax2* in cone cells without disrupting its expression or function in other tissues (*dPax2* is also expressed in PPCs and in sheath and shaft cells of the mechanosensory bristle, but these expression domains are under the control of different enhancers; Fu and Noll, 1997; Fu *et al.*, 1998; Johnson *et al.*, 2011). Mutations to *dPax2* that affect cone cells were originally called *sparkling* alleles, so the cone-cell enhancer was named *sparkling* (*spa*) (Fu *et al.*, 1998).

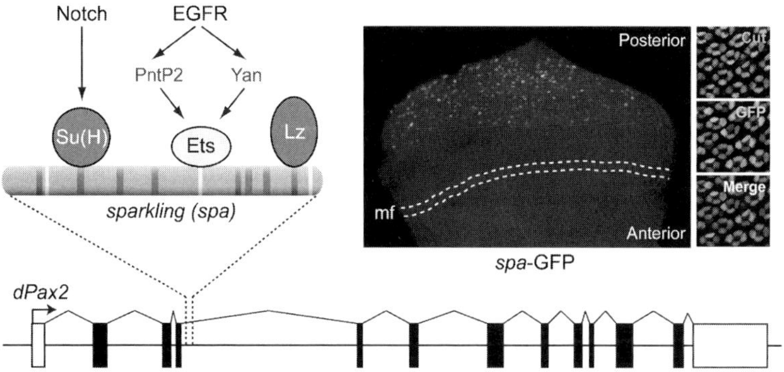


**Figure 4.1** The *sparkling* (*spa*) enhancer of the *Drosophila dPax2 gene* is sufficient to drive cone-cell-specific gene expression *in vivo*. Top left: diagram of the minimal 362-bp *spa* element, with Su(H), Ets, and Lz sites, as determined by Flores *et al.* (2000), indicated with colored stripes. Top right: fluorescence micrograph showing GFP expression, driven by the wild-type *spa* enhancer, in a transgenic third-instar eye imaginal disc. (mf, morphogenic furrow) Smaller images show coexpression of GFP with the cone cell marker Cut in pupae. Bottom: diagram of the structure of the *dPax2* gene, showing the location of *spa* in the fourth intron. (See Color Insert.)

Subsequent transgenic analyses identified a minimal 362bp sequence that is sufficient for cone-cell-specific gene expression (Flores *et al.*, 2000), which we will refer to here as *spa* (Fig. 4.1). This sequence contains five binding sites for the Notch effector Suppressor of Hairless [Su(H)] (Flores *et al.*, 2000). In the absence of Notch signaling, Su(H) confers direct repression on its target genes, while in the presence of active Notch signaling (as occurs in cone cell precursors) it mediates direct activation, thereby acting as a signal-regulated transcriptional switch (Barolo and Posakony, 2002; Barolo *et al.*, 2000; Bray, 2006; Morel and Schweisguth, 2000). *spa* also contains three MGGAW consensus Ets factor binding sites, which are directly bound *in vitro* by one or both of two Ets-family effectors of EGFR/MAPK signaling, Pointed P2 (PntP2), and Yan/Aop (Flores *et al.*, 2000). In the presence of EGFR signaling, PntP2 is phosphorylated and activates gene transcription, while in the absence of signaling, Yan binds to and represses target genes (Brunner *et al.*, 1994; O'Neill *et al.*, 1994; Rebay and Rubin, 1995). In addition, *spa* contains three binding sites for Lz, which is expressed in all undifferentiated progenitor cells in the *Drosophila* eye (Flores *et al.*, 1998). All three of these regulatory inputs were confirmed to be necessary and direct. Genetic ablation of Notch signaling, EGFR signaling, or Lz abolished *dPax2* activation in cone cells, and targeted mutation of the Su(H), Ets, or Lz sites in *spa* abolished its activity in transgenic reporter and rescue assays (Flores *et al.*, 2000).

When the DNA sequence of *spa* is compared across the genomes of other sequenced *Drosophila* species, blocks of conservation are unexpectedly few and short, compared to other enhancers (Fig. 4.2; Swanson *et al.*, 2010, 2011).

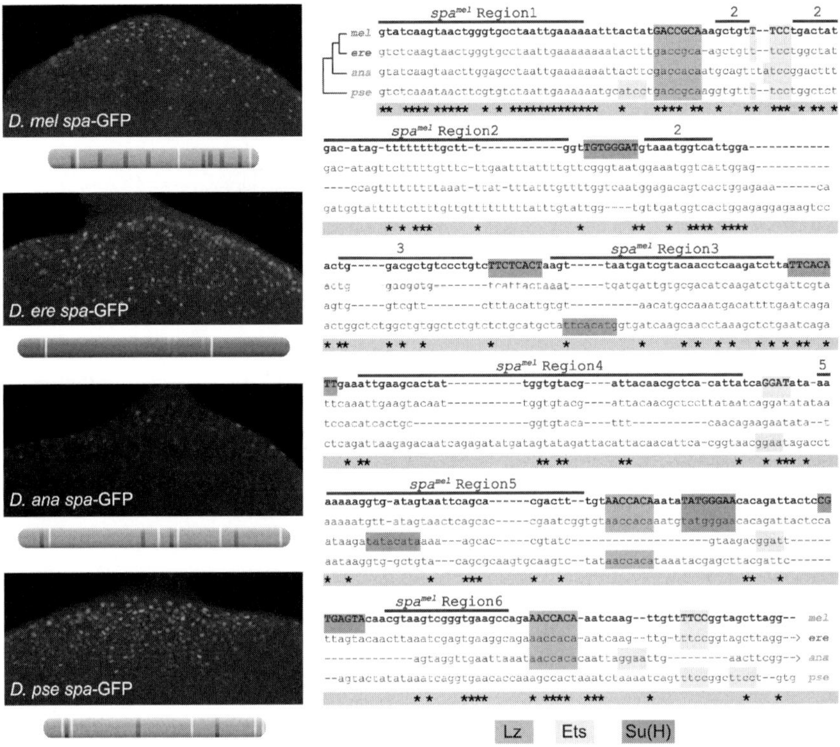


**Figure 4.2** The DNA sequence of *spa* is rapidly evolving. Left: fluorescence images of eye discs from transgenic *D. melanogaster* larvae bearing GFP reporter transgenes driven by *D. melanogaster* (*D. mel spa*) or *spa*-orthologous sequences from *D. erecta* (*D. ere spa*), *D. anannasae* (*D. ana spa*), or *D. pseudoobscura* (*D. pse spa*). Right: sequence alignment of *spa* orthologs from four *Drosophila* species. Predicted and confirmed binding sites for Lz, PntP2/Yan, and Su(H) are indicated with colored boxes. (For interpretation of the references to color in this figure legend, the reader is referred to the Web version of this chapter.)

However, despite very poor sequence conservation, 409bp of orthologous sequence from *D. pseudoobscura* (*D. pse spa*) was capable of driving cone-cell-specific reporter gene expression in transgenic *D. melanogaster (D. mel)* that was indistinguishable from that of *D. mel spa* (Fig. 4.2; Swanson *et al.*, 2010). Similarly, 500bp fragments of *spa*-orthologous sequence from *D. erecta* and *D. ananase* also drove gene expression in *D. mel* cone cells, although with varying different levels of GFP expression (Fig. 4.2). All of these *spa* orthologs contain at least one Lz, one Su(H), and one Ets site, though most individual binding sites were very poorly conserved (Fig. 4.2; Swanson *et al.*, 2011). Therefore, while *spa* sequence is rapidly evolving, its function is conserved. The interesting expression pattern, well-defined regulatory inputs, and evolutionary properties of *spa* inspired the Barolo laboratory to further characterize the *cis*-regulatory logic, structural constraints, and evolutionary history of this enhancer (Johnson *et al.*, 2008; Swanson *et al.*, 2010, 2011).

## 3.1. *sparkling*: Enhanceosome or billboard?

Once Lz, EGFR/Ets, and Notch/Su(H) were identified by Flores *et al.* (2000) as direct regulatory inputs of the *spa* enhancer, it was plausible to imagine that these three regulatory inputs might together be sufficient to generate cone-cell-specific *dPax2* expression. However, this is not the case. Wild-type *spa*-GFP reporter transgenes were active in cone cells of larval eye imaginal discs, but synthetic constructs containing only the Lz, Ets, and Su(H) binding sites from *spa* were incapable of driving gene expression, regardless of the spacing between TFBS (Fig. 4.3A; Johnson *et al.*, 2008; Swanson *et al.*, 2010).

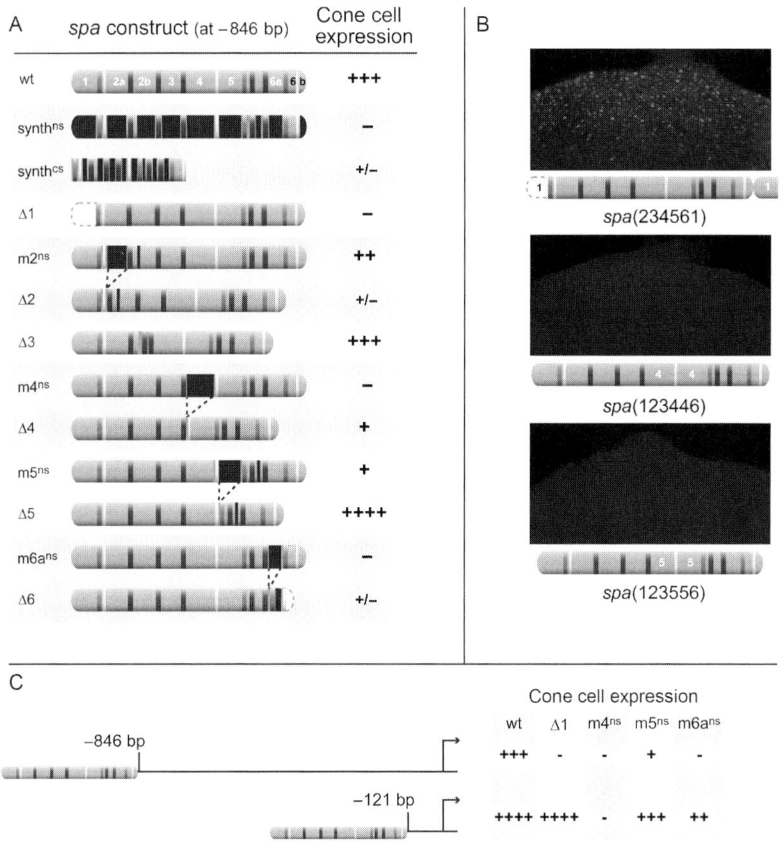


**Figure 4.3** *In vivo* functional dissection of *spa* reveals high complexity, structural constraints, and multiple classes of activation activities. (A) Mutational analysis of *D. mel spa*. "ns" indicates mutations that preserve native spacing within the enhancer; "Δ" indicates sequence deletions. Levels of reporter gene expression in cone cells are indicated *à la* Swanson *et al.* (2010). (B) Relocation and substitution of various regulatory regions within *spa*. (C) Evidence for a region of *spa* that specifically mediates enhancer function at a distance from the promoter. (For color version of this figure, the reader is referred to the Web version of this chapter.)

Therefore, additional regulatory sequences within *spa*, besides the defined TFBSs, must also be necessary for proper *spa* activity *in vivo*. In other words, Lz+Ets+Su(H) is not the complete combinatorial code for *spa*.

To better define the full complement of regulatory sites necessary and, together, sufficient for *spa*'s activity in cone cells, the DNA sequences between the Lz, Ets, and Su(H) sites were systematically mutated. The non-Lz/Ets/Su(H) sequences of *spa* were initially divided in to six regions, each of which was individually deleted. Only one of these, region 3, was found to have no significant effect on *spa* activity *in vivo* (Fig. 4.3A; Swanson *et al.*, 2010). Deletion of region 5 increased GFP expression in cone cells, while the individual deletion of region 1, 2, 4, or 6 resulted in severe loss of gene expression. It is important to note that these deletions resulted in loss of potential novel regulatory site, but they also altered the spacing between sites on either side of the deletion. For this reason, regions 2, 4, 5, and 6 were also subjected to sequence alterations (specifically, every second position was changed to its noncomplementary transversion; A to C, C to A, G to T, T to G) that did not affect the overall spacing or structure of the enhancer. Under these native-spacing conditions, mutating region 2 no longer strongly affected *spa* activity (Fig. 4.3A), suggesting that the length of this region is more functionally significant than its sequence. This is consistent with a structural role for region 2, rather than a role in direct TF recruitment. In contrast to the deletion of region 5, native-spacing mutation of the same region resulted in a severe loss of gene expression. This suggests that the augmented expression associated with the deletion of region 5 might be due to compressed spacing of Ets and Lz sites flanking this region: Ets and Runx factors, including PntP2 and Lz, are well known to interact with one another and to synergistically activate transcription (Dittmer, 2003; Goetz *et al.*, 2000; Jackson Behan *et al.*, 2005; Kim *et al.*, 1999; Liu *et al.*, 2004). The sequence within region 5 must be essential for cone-cell-specific gene expression, which is also the case for regions 4 and 6 as the loss of these regions, either by deletion or native-spacing mutation, abolished *spa* activity (Fig. 4.3A). Similarly, smaller native-spacing mutations within regions 1, 4, 5, and 6 demonstrated that most of the sequence in these regions is required for enhancer function (Swanson *et al.*, 2010). Thus, nearly all of the sequence of *spa* is regulatory, nonredundant, and essential for function *in vivo*.

Perhaps the strongest evidence for structural constraints on the organization of *spa* was seen when the Lz, Ets, and Su(H) sites in *spa* were ablated, and these same sites were restored, but at the 3′ end of the element. No binding sites were ultimately gained or lost, only rearranged relative to other essential sequences. The resultant enhancer construct was active in the developing eye, but in the wrong cell type: expression was lost in cone cells, while ectopic expression was observed in R1 and R6 photoreceptors, which normally do not express *dPax2* (Swanson *et al.*, 2010). Therefore, at least for *spa*, the combinatorial code of TFs binding to the enhancer is not sufficient to specify the proper pattern of gene expression: additional essential patterning information is supplied by the arrangement of regulatory sites. In other words, a given set of inputs (binding sites) can generate multiple possible outputs (expression patterns), and enhancer structure can play an important role in determining the final pattern.

Reasoning from these and other experiments, *spa* seems to behave consistently with an enhanceosome model of action, wherein strict rules of organization and spacing of regulatory sites are enforced, at least in some regions of the enhancer. However, other evidence appears to support an information display model of *spa* function. For example, one essential region of the enhancer (region 1) can be relocated to the opposite end of the enhancer with no apparent effect on enhancer function (Fig. 4.3B; Swanson *et al.*, 2010). This suggests that the regulatory information within *spa* is flexible, at least with respect to the essential contribution of region 1 (more on this region later). The poor sequence conservation of *spa* could also be considered to be consistent with loose organization and therefore with the information display model.

However, evidence strongly suggests that the transcriptional activity of *spa* cannot simply depend on the presence of a sufficient number of activator binding sites, irrespective of their identity. For example, a construct that doubles the number of Lz, Ets, and Su(H) binding sites, but includes *only* those sites, was unable to drive cone-cell-specific gene expression, regardless of spacing between those sites (Swanson *et al.*, 2010; see Fig. 4.5). In other words, the addition of extra Lz/Ets/Su(H) inputs could not functionally compensate for the loss of the novel (i.e., non-Lz/Ets/Su(H)) regulatory inputs, which may mean that these two groups of *spa* regulatory sites provide different transcriptional activation functions. (More evidence supporting this idea will be discussed in the next section.) Similarly, the regulatory contribution of region 4 cannot be substituted with a second copy of region 5, nor can region 5 be functionally replaced by a second copy of region 4 (Fig. 4.3B). It seems that neither a straightforward information display model nor an enhanceosome model accurately represents the structure and function of *spa* (Datta and Small, 2011).

## 3.2. A specialized sequence allowing activation at a distance

Unlike most assays of enhancer function, in which a test sequence is placed immediately upstream of a promoter, most of the reporter experiments described thus far placed the *spa* enhancer at a more distal position, 846 bp 5′ of the transcriptional start site (TSS), forcing the enhancer to act over a moderate distance. (*spa* is located 7 kb downstream of the TSS in the endogenous *dPax2* locus.) This experimental design allowed for the potential identification of subelements within the enhancer that regulate long-range gene transcription, a critical but very poorly understood aspect of transcriptional regulation. After determining which novel sequences within *spa* were essential at −846 bp from the TSS (regions 1, 4, 5, and 6a), enhancers bearing mutations to these sequences were then placed at a more promoter-proximal position of −121 bp. The strength of cone-cell-specific gene expression driven by wild-type *spa* was significantly augmented by moving the enhancer to a more promoter-proximal position (Fig. 4.3C). Individual mutations to most of the essential sites within the enhancer (including the Lz, Ets, or Su(H) sites) affected gene expression regardless of distance from the promoter (Swanson *et al.*, 2010; Fig. 4.3C). However, the loss of expression observed after deletion of region 1, a 40-bp sequence at the 5′ end of the enhancer, was fully rescued when the enhancer was placed adjacent to the promoter. Further, no defects in cell-type patterning were observed in the absence of region 1 when *spa* was placed at −121 bp. Since this region of the enhancer (and only this region) contributed to *spa* function only when the enhancer is placed at a distance from the promoter, it seems plausible that this sequence may facilitate enhancer–promoter interactions. Therefore, region 1 of *spa* was referred to as the "remote control" element, or RCE (Swanson *et al.*, 2010). The *spa* RCE is the only enhancer subelement identified to date, in any organism, that is specifically required for enhancer function at a distance, but completely dispensable for proper expression and pattern formation in a promoter-proximal position. The RCE differs from other long-range elements such as promoter-specific tethering elements or insulator bypass elements, such as those in the *Drosophila AbdB*, *white*, and *cut* loci, as the element is found within the enhancer rather than at the promoter (Calhoun *et al.*, 2002; Celniker and Drewell, 2007; Kostyuchenko *et al.*, 2009; Laney and Biggin, 1997; Misulovin *et al.*, 2008). The function of the RCE does not require special sequences at the *dPax2* promoter, since the all experiments described here were performed using a minimal *Hsp70* promoter. The RCE must be tested more rigorously, in a wider variety of promoter contexts, before its capabilities can be clearly defined, and the identity and function of RCE-binding proteins is an as-yet-unanswered question of potentially high significance. Interestingly, region 1 is the only region of *spa* whose sequence is highly conserved among all sequenced *Drosophila* species (Fig. 4.2). It can also function from either end of the enhancer (Fig. 4.3B). While the limitations of this flexibility have yet to be tested, the unique function of the RCE could be in some way connected to the relatively loose structural constraints pertaining to this region.

### 3.3. Combinatorial control of *sparkling*'s activity

We have already seen that "Lz+Ets+Su(H)" is insufficient to describe the combinatorial code of *spa* activity; additional inputs from regions 1, 4, 5, and 6a are also necessary. While the proteins that bind to each of the essential regions of *spa* have yet to be identified, comparative analysis of *spa* sequences from multiple *Drosophila* species revealed motifs that are required for function in at least two *spa* orthologs (Swanson *et al.*, 2011). Following the example of a highly influential evolutionary analysis of *eve* S2E (Ludwig *et al.*, 2000), chimeras between the 5′ half of *D. mel spa* and the 3′ half of *D. pse spa,* or conversely, the 5′ half of *D. pse spa* and the 3′ half of *D. mel spa*, were assessed in transgenic *D. mel*. The *spa* chimeras provided very different results from those involving *eve* S2E: in the case of *spa*, 5′ *D. mel*+3′ *D. pse* was inactive, while 5′ *D. pse*+3′ *D. mel* drove robust reporter gene expression (Swanson *et al.*, 2011) supporting the idea that there are differences in TFBS composition and arrangement between the two *spa* orthologs. 3′ *D. pse spa* may lack regulatory inputs present in the 3′ half of *D. mel spa*; inputs from these regions may lie (at least partially) in 5′ *D. pse spa*. Detailed mutational analyses of these chimeras were consistent with the hypothesis that regulatory inputs into the 3′ half of *D. mel spa* could be substituted with inputs into the 5′ half of *D. pse spa*, suggesting that *spa* has been significantly reorganized over a relatively short evolutionary period.

MEME motif analysis identified five potential novel motifs present in *D. mel spa* region 4, 5, and 6a, and also within *D. pse spa*, but at noncorresponding positions (Fig. 4.4B). These motifs, named α through ε, are also present in *spa* orthologs from all or most of the 12-sequenced *Drosophila* species. Mutation of these motifs in either *D. mel* or *D. pse spa* impairs enhancer activity (Swanson *et al.*, 2011). Interestingly, the effects of loss of the ε motif in 3′ *D. mel spa* could be rescued by transplanting the 5′ *D. pse* ε input to region 2, providing further evidence for the functional significance of this motif. The ε motif is present in increased copy number in *D. pse spa* relative to *D. mel*, which may account for the ability of *D. pse spa* to function despite its (relatively) fewer Lz and Su(H) sites. A recently derived Ets motif in 5′ *D. pse spa* also seems to make an important compensatory contribution to enhancer function (Swanson *et al.*, 2011). While the identities of the factors that bind these novel motifs and the biochemical activities they recruit are not known, the combinatorial code of of *spa* can now be expanded to Lz+EGFR/Ets+Notch/Su(H)+α+β+γ+δ+ε. However, the sufficiency of these inputs *in vivo* to generate a *spa* expression pattern has not been strictly experimentally tested.

## 3.4. *sparkling* Orthologs reveal evolutionarily labile binding site grammar

As mentioned above, the extent to which "grammar" controls enhancer activity is a vigorously debated issue in the field. As defined by Levine (2010), the term refers to the "phenomenon that spacing and arrangement of binding sites matter for the activity of the enhancer, just like the order of words in a sentence can affect its meaning." Conserved clustering of TFBS

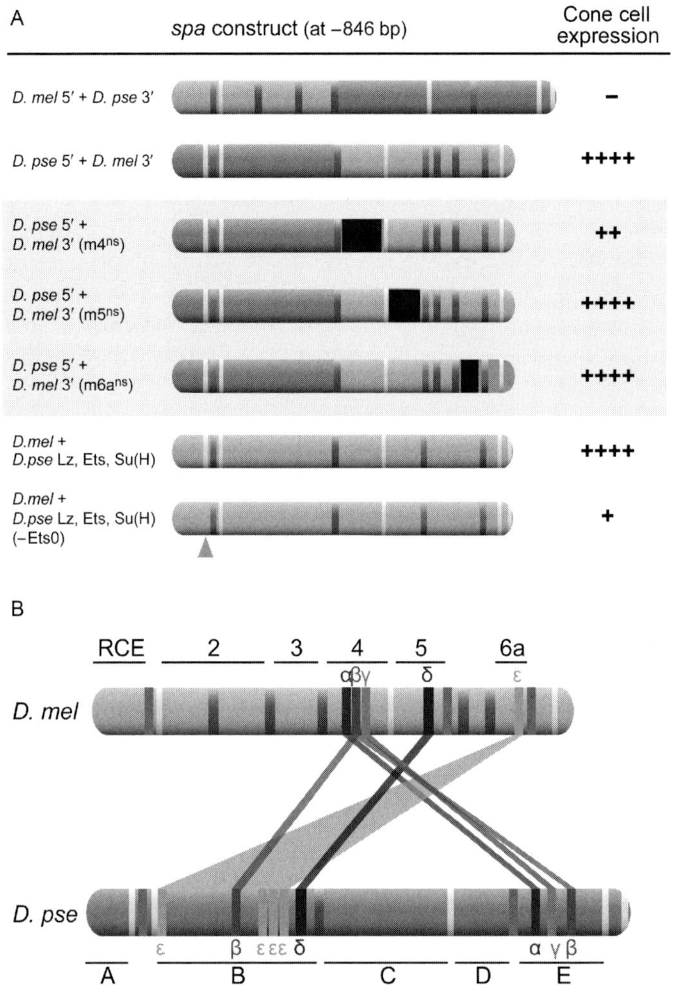


**Figure 4.4** Evolutionary dynamics of the *cis*-regulatory structure of *spa*. (A) Summary of the results of various chimeric enhancer experiments, demonstrating that regulatory information within *spa* has been reorganized since the divergence of *D. mel* and *D. pse*. (B) Diagram summarizing the proposed reorganization of *spa*, with proposed novel regulatory motifs α through ε indicated. (For color version of this figure, the reader is referred to the Web version of this chapter.)

and co-occurrence of TFBS are sometimes used to successfully identify enhancers. This begs the question: are these arrangements functionally significant? A few specific examples of grammar elements have been tested *in vivo*. Activator–activator and activator–repressor binding site spacing has been shown to affect the ability of these proteins to cooperatively bind to

DNA and set thresholds for gene activation (Crocker *et al.*, 2010; Kulkarni and Arnosti, 2005; Papatsenko and Levine, 2007). When nearly 100 early embryonic enhancers were analyzed, linked binding sites of Bicoid-Bicoid, Dorsal-Caudal, Hunchback-Hunchback, and Dorsal-Twist were identified, indicating that grammar control regions are frequently utilized at least in patterning the *Drosophila* embryo (Papatsenko *et al.*, 2009). In contrast to these examples, multiple zebrafish notochord enhancers show no fixed arrangement of essential regulatory sites, and the developmental enhancers for the *even-skipped* stripes appear to lack any conserved arrangements of binding sites (Hare *et al.*, 2008a,b; Rastegar *et al.*, 2008). Additionally, simulation has revealed that there may be an evolutionary tendency to cluster TFBS (Lusk and Eisen, 2010). In this light, linked binding sites may not necessarily be interpreted as evidence for rules of enhancer organization. Given the experimental evidence on both sides of this debate, we are left with the question: does functionally relevant grammar exist in the structure of most enhancers?

Fine-scale structure-function analyses of *spa* have provided evidence for rules of linear organization, yet evolutionary analyses have strongly suggested that enhancer organization is elastic. Despite the high rate of turnover of *spa* sequence, multispecies sequence comparisons revealed a set of binding site configurations with apparently conserved arrangement and spacing. Such linked binding sites are readily apparent for Lz-Ets, as well as Lz-ε and Su(H)-Ets. The existence of conserved Lz-Ets pairs is unsurprising as we know these proteins act cooperatively to activate gene expression. Further, it has been demonstrated that the distance between Lz- and Ets-binding sites is critical for proper *spa* activity. However, in many cases, the positions of these configurations within the enhancer appeared to be rapidly diverging (Swanson *et al.*, 2011). Thus, a lack of significant sequence conservation, and even a lack of binding site conservation, does not necessarily denote an absence of *cis*-regulatory grammar (i.e., structural constraints on enhancer function; Datta and Small, 2011). Additionally, these data suggest that specific arrangement of binding sites are being lost and gained in new locations within the enhancer over evolutionary time, allowing the formation of divergent, yet functional, binding site groupings. In order to rigorously test the proposal that binding site grammar exists within the *spa* lineage enhancers, despite rapid TFBS turnover, further detailed *in vivo* manipulation is required. Do changes in spacing and orientation of identified arrangements affect enhancer function? Do these configurations function in species they do not currently exist in? While *in vivo* verification that these elements are functionally significant is missing from this analysis, all of the species studied have at least one Lz-Ets and Lz-ε grammar element, suggesting that while the position and number of these sites is changing over evolutionary time, some combination of these elements is likely to be necessary for *spa's* ability to drive appropriate *dPax 2* expression. In the context of the extensive debate over enhancer "grammar," *spa* appears at one level to resemble the *eveS2* enhancer, with high levels of sequence divergence and extensive binding site rearrangement. Yet, conserved grammar elements, or words, are readily identifiable throughout the enhancer.

## 3.5. Structural and sequence constraints channel *sparkling* output

Evolutionary analyses of *sparkling* have also led to insights as to how the enhancer achieves cone-cell-specific gene expression. Recall that all of the identified TF inputs into *spa* activity, EGFR/Ets, Notch/Su(H), and Lz, are active in multiple cell types in the developing *Drosophila* retina, yet *spa* drives *dPax2* expression only in cone cells. How does *spa* induce gene expression in this cell type alone? The simplest explanation is that a requirement for additional DNA-binding regulatory factors (such as factors $\alpha$ through $\varepsilon$, perhaps) restricts *spa* activity to cone cells, as discussed above. However, this explanation is not sufficient, as multiple structural rearrangements of *spa* can alter the cell-type specificity of its action. For example, when the Lz, Ets, and Su(H) binding sites were mutated within *spa* and subsequently placed at the 3′ end of *spa* in compressed conformation, gene expression switched from cone cells to R1/R6 photoreceptors (Fig. 4.5A; Swanson *et al.*, 2010). Yet this ectopic gene expression must be due to the new arrangement of TFBS, as restoration of the spacing among the Lz, Ets, and Su(H) sites in this construct resulted in complete loss of activity (Fig. 4.5A). Note that the only difference between these two constructs is the spacing among Lz/Ets/Su(H) sites.

The ectopic R1/R6 activity resulting from this rearrangement requires *spa* regions 1, 4, and 6 as well as Lz and Ets binding sites, but not Su(H) sites, which correlates with genetic data: the R1 and R6 cells respond to EGFR and Lz, but not to Notch signaling (Swanson *et al.*, 2010). Ectopic expression in R1/R6 was also achieved in a different way, by mutating regions 2, 3, 5, and 6b, with all other *spa* sequences and TFBSs in their native configuration (Fig. 4.5A). Restoration of region 5 to this construct abolished R1/R6 expression with no gain in cone cell expression, suggesting that a regulatory site(s) within region 5 represses *spa* activity in photoreceptors. Genetic evidence indicates that ectopic *dPax2* expression, such as that driven by these altered enhancers, would negatively affect the fitness of the fly by disrupting cell-fate specification and differentiation in the eye (Shi and Noll, 2009).

From studies of *spa*'s cell-type specificity, we can also learn more about the structural rules governing this enhancer. For example, two copies of all of *spa*'s Lz/Ets/Su(H) sites in a compressed configuration were shown to drive ectopic expression in photoreceptors, while the same sites in native spacing were incapable of driving gene expression in any cell type (Fig. 4.5A). This is consistent with the idea that unrestrained Lz-Ets

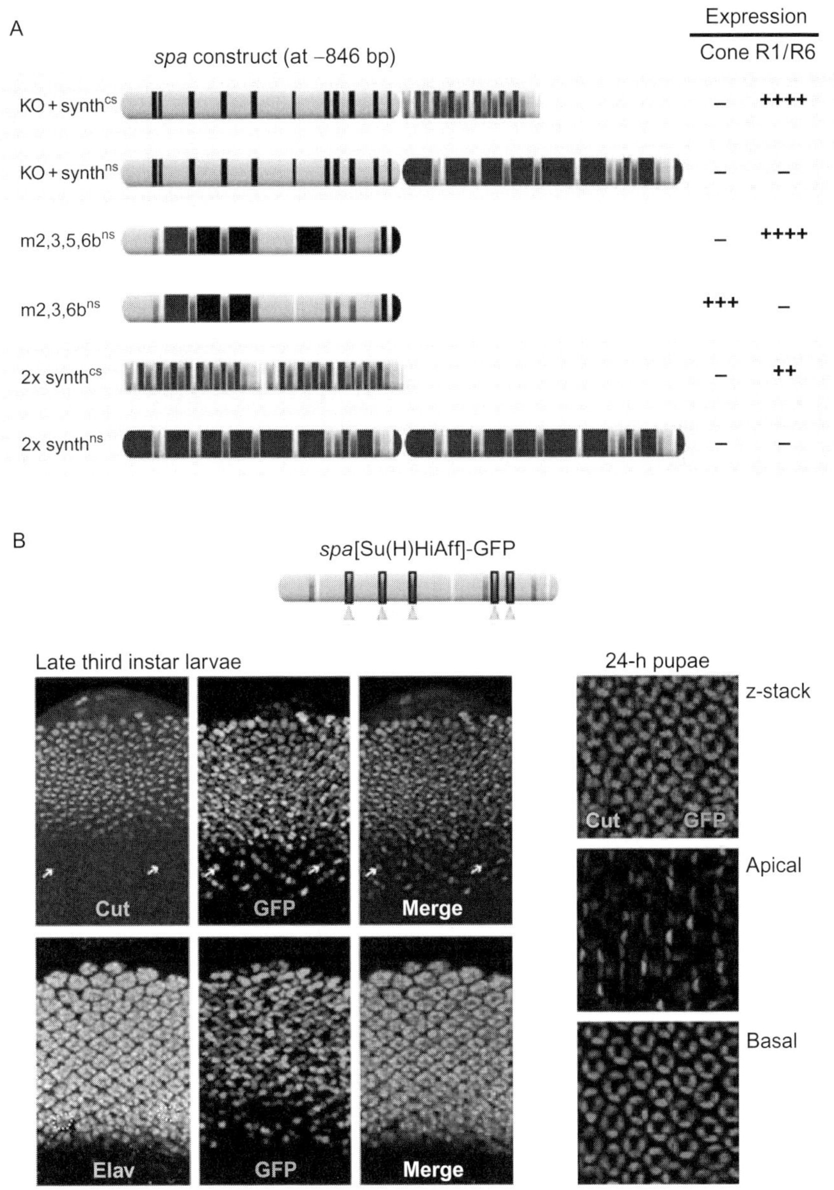


**Figure 4.5** The linear arrangement, spacing, and affinity of binding sites profoundly affects the patterning of gene expression driven by *spa*. (A) Summary of results of experiments testing the effects of binding site organization on cell-type specificity of gene expression. (B) Increasing the affinity of Su(H) binding sites within *spa* creates an enhancer that is oversensitive to Notch signaling, resulting in ectopic gene expression in multiple Notch-responsive cell types. Left: *spa*[Su(H)HiAff]-GFP is expressed ectopically in Cut-negative photoreceptors (arrows) in the larval eye imaginal disc, as shown by coexpression with the neuronal marker Elav. Right: *spa*[Su(H)HiAff]-GFP is also

synergy, which is likely to be promoted by compressed TFBS spacing, results in the ectopic *spa* activity. In a similar vein, the photoreceptor-specific repressor function of region 5 must be only able to work over a short range, as all *spa* sequences (including region 5) are present in the rearranged enhancer construct *spa*(KO+synth$^{cs}$), which was capable of driving robust ectopic R1/R6 gene expression (Fig. 4.5A). Note that the *spa*(KO+synth$^{cs}$) and *spa*(m236b$^{ns}$) constructs both contain region 5, which contains a putative R1/R6 repressor binding site. However, only one confirmation, the wild-type spacing, allows for repression to occur indicating that the presence of a short-range repressor activity. When native-binding site spacing is restored to *spa*(KO+synth$^{cs}$), all reporter expression was lost, suggesting that the activator functions of this enhancer are also short range as both R1/R6 and cone cell expression are absent in this spread-out conformation. Together these data reveal that the activator and repressor activities of *spa* are short range in nature, allowing for precise cell-type-specific output.

### 3.6. Low-affinity binding sites and the Notch response

An entirely different evolutionary strategy appears to make an equally important contribution to the specificity of *spa* activity. In *spa* orthologs in all sequenced *Drosophila* species, the predicted Su(H)-binding sites are nonconsensus and generally of low-predicted affinity. The five confirmed Su(H) sites in *D. mel spa*, and the one predicted site in *D. pse spa*, all deviate by 1–4bp from the high-affinity consensus YGTGRGAAM (Crocker *et al.*, 2010; Flores *et al.*, 2000). Three of the *D. mel* sites and the single *D. pse* site also deviate from the looser, lower-affinity consensus RTGRGAR (Bailey and Posakony, 1995; Nellesen *et al.*, 1999). The conserved property of low affinity for Su(H), even though the individual sites themselves are not conserved, suggested a possible regulatory mechanism. When the five low-affinity Su(H)-binding sites in *D. mel spa* were converted to high-affinity sites, representing a total change of 10bp out of 362bp, levels of cone-cell activation were increased. Possibly more significantly, ectopic gene expression was observed in photoreceptors in larval eye discs, as well as in PPCs in pupal eyes (Fig. 4.3B; Swanson *et al.*, 2011). These data again suggest that strict quantitative control of *spa*'s regulatory inputs is necessary for correct *spa* activity and patterning.

---

**expressed ectopically in primary pigment cells in the 24-h pupal retina, identifiable by lack of Cut staining, apical position within the retina, and characteristic elongated cell shape. (See Color Insert.)**

## 4. Conclusions

One lesson to be learned from the extensive characterization of the *dPax2 sparkling* enhancer by multiple laboratories is the extraordinary amount of information concerning enhancer structure, function, and evolution we can obtain by delving deeply into a single short DNA element. However, the new questions raised by these studies show how much remains to be understood about transcriptional control *in vivo*. *sparkling* cannot be accurately described as either an enhanceosome or a billboard, as it contains both strict structural requirements and flexible binding site arrangements. Its sequence is quite evolutionarily labile, yet stable patterns can be detected among the rapidly shifting regulatory motifs. Mutational analysis, coupled with comparative sequence analysis, revealed that *spa* is crowded with regulatory information, suggesting that the typical enhancer may be more complex than originally anticipated. We have also seen that changes in the structure of the enhancer can result in ectopic gene expression, revealing that combinatorial control alone can be insufficient to determine the cell-type specificity of enhancer function. Furthermore, functional evidence suggests that different sites within an enhancer may mediate distinct, nonsubstitutable regulatory functions, all of which may be required for transcriptional activation *in vivo*. The most important lesson of these studies, we propose, is that the "combinatorial code" view of *cis*-regulatory logic, though accurate as far as it goes, does not adequately address the complexity of the enhancer.

## *ACKNOWLEDGMENTS*

The new research findings presented here were supported by NIH grant GM076509 and ARRA supplement GM07650903S1 to S.B. Experiments were performed as described by Swanson *et al.* (2010, 2011). We thank the members of the Barolo lab for helpful comments on the manuscript.

## *REFERENCES*

Andrioli, L. P., Vasisht, V., Theodosopoulou, E., Oberstein, A., and Small, S. (2002). Anterior repression of a *Drosophila* stripe enhancer requires three position-specific mechanisms. *Development* **129,** 4931–4940.

Arnosti, D. N., and Kulkarni, M. M. (2005). Transcriptional enhancers: Intelligent enhanceosomes or flexible billboards? *J. Cell. Biochem.* **94,** 890–898.

Arnosti, D. N., Barolo, S., Levine, M., and Small, S. (1996). The eve stripe 2 enhancer employs multiple modes of transcriptional synergy. *Development* **122,** 205–214.

Bailey, A. M., and Posakony, J. W. (1995). Suppressor of Hairless directly activates transcription of *Enhancer of split* Complex genes in response to Notch receptor activity. *Genes Dev.* **9,** 2609–2622.

Banerji, J., Rusconi, S., and Schaffer, W. (1981). Expression of a beta-globin gene is enhanced by remote SV 40 DNA sequences. *Cell* **27,** 299–308.

Barolo, S., and Posakony, J. W. (2002). Three habits of highly effective signaling pathways: Principles of transcriptional control by developmental cell signaling. *Genes Dev.* **16,** 1167–1181.

Barolo, S., Walker, R. G., Polyanovsky, A. D., Freschi, G., Keil, T., and Posakony, J. W. (2000). A Notch-independent activity of *Suppressor of Hairless* is required for normal mechanoreceptor physiology. *Cell* **103,** 957–969.

Barski, A., Cuddapah, S., Cui, K., Roh, T. Y., Schones, D. E., Wang, Z., Wei, G., Chepelev, I., and Zhao, K. (2007). High-resolution profiling of histone methylations in the human genome. *Cell* **129,** 823–837.

Blackwood, E. M., and Kadonaga, J. T. (1998). Going the distance: A current view of enhancer action. *Science* **281,** 60–63.

Bray, S. J. (2006). Notch signalling: A simple pathway becomes complex. *Nat. Rev. Mol. Cell Biol.* **7,** 678–689.

Brunner, D., Ducker, K., Oellers, N., Hafen, E., Scholz, H., and Klambt, C. (1994). The ETS domain protein pointed-P2 is a target of MAP kinase in the sevenless signal transduction pathway. *Nature* **370,** 386–389.

Bulger, M., and Groudine, M. (2010). Enhancers: The abundance and function of regulatory sequences beyond promoters. *Dev. Biol.* **339,** 250–257.

Calhoun, V. C., Stathopoulos, A., and Levine, M. (2002). Promoter-proximal tethering elements regulate enhancer-promoter specificity in the Drosophila Antennapedia complex. *Proc. Natl. Acad. Sci. USA* **99,** 9243–9247.

Celniker, S. E., and Drewell, R. A. (2007). Chromatin looping mediates boundary element promoter interactions. *Bioessays* **29,** 7–10.

Crocker, J., and Erives, A. (2008). A closer look at the *eve* stripe 2 enhancers of *Drosophila* and *Themira*. *PLoS Genet.* **4,** e1000276.

Crocker, J., Potter, N., and Erives, A. (2010). Dynamic evolution of precise regulatory encodings creates the clustered site signature of enhancers. *Nat. Commun.* **1,** 99.

Datta, R. R., and Small, S. (2011). Gene regulation: Piecing together the puzzle of enhancer evolution. *Curr. Biol.* **21,** R542–R543.

Dittmer, J. (2003). The biology of the *Ets1* proto-oncogene. *Mol. Cancer* **2,** 29.

Drissen, R., Palstra, R. J., Gillemans, N., Splinter, E., Grosveld, F., Philipsen, S., and de Laat, W. (2004). The active spatial organization of the beta-globin locus requires the transcription factor EKLF. *Genes Dev.* **18,** 2485–2490.

Flores, G. V., Daga, A., Kalhor, H. R., and Banerjee, U. (1998). Lozenge is expressed in pluripotent precursor cells and patterns multiple cell types in the Drosophila eye through the control of cell-specific transcription factors. *Development* **125,** 3681–3687.

Flores, G. V., Duan, H., Yan, H., Nagaraj, R., Fu, W., Zou, Y., Noll, M., and Banerjee, U. (2000). Combinatorial signaling in the specification of unique cell fates. *Cell* **103,** 75–85.

Frankel, N., Davis, G. K., Vargas, D., Wang, S., Payre, F., and Stern, D. L. (2010). Phenotypic robustness conferred by apparently redundant transcriptional enhancers. *Nature* **466,** 490–493.

Fu, W., and Noll, M. (1997). The *Pax2* homolog *sparkling* is required for development of cone and pigment cells in the *Drosophila* eye. *Genes Dev.* **11,** 2066–2078.

Fu, W., Duan, H., Frei, E., and Noll, M. (1998). *shaven* and *sparkling* are mutations in separate enhancers of the *Drosophila Pax2* homolog. *Development* **125,** 2943–2950.

Ghiasvand, N. M., Rudolph, D. D., Mashayekhi, M., Brzezinski, J. A.t., Goldman, D., and Glaser, T. (2011). Deletion of a remote enhancer near ATOH7 disrupts retinal neurogenesis, causing NCRNA disease. *Nat. Neurosci.* **14,** 578–586.

Giese, K., Kingsley, C., Kirshner, J. R., and Grosschedl, R. (1995). Assembly and function of a *TCRα* enhancer complex is dependent on LEF-1-induced DNA bending and multiple protein-protein interactions. *Genes Dev.* **9,** 995–1008.

Goetz, T. L., Gu, T. L., Speck, N. A., and Graves, B. J. (2000). Auto-inhibition of Ets-1 is counteracted by DNA binding cooperativity with core-binding factor alpha2. *Mol. Cell. Biol.* **20,** 81–90.

Gordon, C. T., Tan, T. Y., Benko, S., Fitzpatrick, D., Lyonnet, S., and Farlie, P. G. (2009). Long-range regulation at the *SOX9* locus in development and disease. *J. Med. Genet.* **46,** 649–656.

Goto, T., Macdonald, P., and Maniatis, T. (1989). Early and late periodic patterns of *even-skipped* expression are controlled by distinct regulatory elements that respond to different spatial cues. *Cell* **57,** 413–422.

Halfon, M. S., Carmena, A., Gisselbrecht, S., Sackerson, C. M., Jimenez, F., Baylies, M. K., and Michelson, A. M. (2000). Ras pathway specificity is determined by the integration of multiple signal-activated and tissue-restricted transcription factors. *Cell* **103,** 63–74.

Harding, K., Hoey, T., Warrior, R., and Levine, M. (1989). Autoregulatory and gap gene response elements of the *even-skipped* promoter of *Drosophila. EMBO J.* **8,** 1205–1212.

Hare, E. E., Peterson, B. K., and Eisen, M. B. (2008a). A careful look at binding site reorganization in the *even-skipped* enhancers of *Drosophila* and sepsids. *PLoS Genet.* **4,** e1000268.

Hare, E. E., Peterson, B. K., Iyer, V. N., Meier, R., and Eisen, M. B. (2008b). Sepsid *even-skipped* enhancers are functionally conserved in *Drosophila* despite lack of sequence conservation. *PLoS Genet.* **4,** e1000106.

He, H. H., Meyer, C. A., Shin, H., Bailey, S. T., Wei, G., Wang, Q., Zhang, Y., Xu, K., Ni, M., Lupien, M., *et al.* (2010). Nucleosome dynamics define transcriptional enhancers. *Nat. Genet.* **42,** 343–347.

Heintzman, N. D., Stuart, R. K., Hon, G., Fu, Y., Ching, C. W., Hawkins, R. D., Barrera, L. O., Van Calcar, S., Qu, C., Ching, K. A., *et al.* (2007). Distinct and predictive chromatin signatures of transcriptional promoters and enhancers in the human genome. *Nat. Genet.* **39,** 311–318.

Heintzman, N. D., Hon, G. C., Hawkins, R. D., Kheradpour, P., Stark, A., Harp, L. F., Ye, Z., Lee, L. K., Stuart, R. K., Ching, C. W., *et al.* (2009). Histone modifications at human enhancers reflect global cell-type-specific gene expression. *Nature* **459,** 108–112.

Jackson Behan, K., Fair, J., Singh, S., Bogwitz, M., Perry, T., Grubor, V., Cunningham, F., Nichols, C. D., Cheung, T. L., Batterham, P., *et al.* (2005). Alternative splicing removes an Ets interaction domain from Lozenge during *Drosophila* eye development. *Dev. Genes Evol.* **215,** 423–435.

Jin, C., Zang, C., Wei, G., Cui, K., Peng, W., Zhao, K., and Felsenfeld, G. (2009). H3.3/H2A.Z double variant-containing nucleosomes mark 'nucleosome-free regions' of active promoters and other regulatory regions. *Nat. Genet.* **41,** 941–945.

Johnson, L. A., Zhao, Y., Golden, K., and Barolo, S. (2008). Reverse-engineering a transcriptional enhancer: A case study in Drosophila. *Tissue Eng. Part A* **14,** 1549–1559.

Johnson, S. A., Harmon, K. J., Smiley, S. G., Still, F. M., and Kavaler, J. (2011). Discrete regulatory regions control early and late expression of *D-Pax2* during external sensory organ development. *Dev. Dyn.* **240,** 1769–1778.

Kagey, M. H., Newman, J. J., Bilodeau, S., Zhan, Y., Orlando, D. A., van Berkum, N. L., Ebmeier, C. C., Goossens, J., Rahl, P. B., Levine, S. S., *et al.* (2010). Mediator and cohesin connect gene expression and chromatin architecture. *Nature* **467,** 430–435.

Kim, W. Y., Sieweke, M., Ogawa, E., Wee, H. J., Englmeier, U., Graf, T., and Ito, Y. (1999). Mutual activation of Ets-1 and AML1 DNA binding by direct interaction of their autoinhibitory domains. *EMBO J.* **18,** 1609–1620.

Kim, T. K., Hemberg, M., Gray, J. M., Costa, A. M., Bear, D. M., Wu, J., Harmin, D. A., Laptewicz, M., Barbara-Haley, K., Kuersten, S., *et al.* (2010). Widespread transcription at neuronal activity-regulated enhancers. *Nature* **465,** 182–187.

Kleinjan, D. A., and van Heyningen, V. (2005). Long-range control of gene expression: Emerging mechanisms and disruption in disease. *Am. J. Hum. Genet.* **76,** 8–32.

Kleinjan, D. A., Seawright, A., Childs, A. J., and van Heyningen, V. (2004). Conserved elements in *Pax6* intron 7 involved in (auto)regulation and alternative transcription. *Dev. Biol.* **265,** 462–477.

Klingler, M., Soong, J., Butler, B., and Gergen, J. P. (1996). Disperse versus compact elements for the regulation of *runt* stripes in *Drosophila. Dev. Biol.* **177,** 73–84.

Kostyuchenko, M., Savitskaya, E., Koryagina, E., Melnikova, L., Karakozova, M., and Georgiev, P. (2009). Zeste can facilitate long-range enhancer-promoter communication and insulator bypass in Drosophila melanogaster. *Chromosoma* **118,** 665–674.

Kulkarni, M. M., and Arnosti, D. N. (2005). *cis*-regulatory logic of short-range transcriptional repression in *Drosophila melanogaster. Mol. Cell. Biol.* **25,** 3411–3420.

Laney, J. D., and Biggin, M. D. (1997). Zeste-mediated activation by an enhancer is independent of cooperative DNA binding in vivo. *Proc. Natl. Acad. Sci. USA* **94,** 3602–3604.

Lettice, L. A., Heaney, S. J., Purdie, L. A., Li, L., de Beer, P., Oostra, B. A., Goode, D., Elgar, G., Hill, R. E., and de Graaff, E. (2003). A long-range *Shh* enhancer regulates expression in the developing limb and fin and is associated with preaxial polydactyly. *Hum. Mol. Genet.* **12,** 1725–1735.

Levine, M. (2010). Transcriptional enhancers in animal development and evolution. *Curr. Biol.* **20,** R754–R763.

Liu, H., Holm, M., Xie, X. Q., Wolf-Watz, M., and Grundstrom, T. (2004). AML1/Runx1 recruits calcineurin to regulate granulocyte macrophage colony-stimulating factor by Ets1 activation. *J. Biol. Chem.* **279,** 29398–29408.

Ludwig, M. Z., Patel, N. H., and Kreitman, M. (1998). Functional analysis of eve stripe 2 enhancer evolution in Drosophila: Rules governing conservation and change. *Development* **125,** 949–958.

Ludwig, M. Z., Bergman, C., Patel, N. H., and Kreitman, M. (2000). Evidence for stabilizing selection in a eukaryotic enhancer element. *Nature* **403,** 564–567.

Lusk, R. W., and Eisen, M. B. (2010). Evolutionary mirages: Selection on binding site composition creates the illusion of conserved grammars in Drosophila enhancers. *PLoS Genet.* **6,** e1000829.

Malik, S., and Roeder, R. G. (2005). Dynamic regulation of pol II transcription by the mammalian Mediator complex. *Trends Biochem. Sci.* **30,** 256–263.

Misulovin, Z., Schwartz, Y. B., Li, X. Y., Kahn, T. G., Gause, M., MacArthur, S., Fay, J. C., Eisen, M. B., Pirrotta, V., Biggin, M. D., *et al.* (2008). Association of cohesin and Nipped-B with transcriptionally active regions of the Drosophila melanogaster genome. *Chromosoma* **117,** 89–102.

Morel, V., and Schweisguth, F. (2000). Repression by Suppressor of Hairless and activation by Notch are required to define a single row of *single-minded* expressing cells in the *Drosophila* embryo. *Genes Dev.* **14,** 377–388.

Narlikar, G. J., Fan, H. Y., and Kingston, R. E. (2002). Cooperation between complexes that regulate chromatin structure and transcription. *Cell* **108,** 475–487.

Nellesen, D. T., Lai, E. C., and Posakony, J. W. (1999). Discrete enhancer elements mediate selective responsiveness of *Enhancer of split* Complex genes to common transcriptional activators. *Dev. Biol.* **213,** 33–53.

O'Neill, E. M., Rebay, I., Tjian, R., and Rubin, G. M. (1994). The activities of two Ets-related transcription factors required for *Drosophila* eye development are modulated by the Ras/MAPK pathway. *Cell* **78,** 137–147.

Ong, C. T., and Corces, V. G. (2011). Enhancer function: New insights into the regulation of tissue-specific gene expression. *Nat. Rev. Genet.* **12,** 283–293.
Orom, U. A., Derrien, T., Beringer, M., Gumireddy, K., Gardini, A., Bussotti, G., Lai, F., Zytnicki, M., Notredame, C., Huang, Q., *et al.* (2010). Long noncoding RNAs with enhancer-like function in human cells. *Cell* **143,** 46–58.
Orphanides, G., and Reinberg, D. (2002). A unified theory of gene expression. *Cell* **108,** 439–451.
Papatsenko, D., and Levine, M. (2007). A rationale for the enhanceosome and other evolutionarily constrained enhancers. *Curr. Biol.* **17,** R955–R957.
Papatsenko, D., Goltsev, Y., and Levine, M. (2009). Organization of developmental enhancers in the Drosophila embryo. *Nucleic Acids Res.* **37,** 5665–5677.
Perry, M. W., Boettiger, A. N., Bothma, J. P., and Levine, M. (2010). Shadow enhancers foster robustness of Drosophila gastrulation. *Curr. Biol.* **20,** 1562–1567.
Perry, M. W., Boettiger, A. N., and Levine, M. (2011). Multiple enhancers ensure precision of gap gene-expression patterns in the Drosophila embryo. *Proc. Natl. Acad. Sci. USA* **108,** 13570–13575.
Priest, H. D., Filichkin, S. A., and Mockler, T. C. (2009). Cis-regulatory elements in plant cell signaling. *Curr. Opin. Plant Biol.* **12,** 643–649.
Rastegar, S., Hess, I., Dickmeis, T., Nicod, J. C., Ertzer, R., Hadzhiev, Y., Thies, W. G., Scherer, G., and Strahle, U. (2008). The words of the regulatory code are arranged in a variable manner in highly conserved enhancers. *Dev. Biol.* **318,** 366–377.
Rebay, I., and Rubin, G. M. (1995). Yan functions as a general inhibitor of differentiation and is negatively regulated by activation of the Ras1/MAPK pathway. *Cell* **81,** 857–866.
Roh, T. Y., Wei, G., Farrell, C. M., and Zhao, K. (2007). Genome-wide prediction of conserved and nonconserved enhancers by histone acetylation patterns. *Genome Res.* **17,** 74–81.
Sackerson, C., Fujioka, M., and Goto, T. (1999). The *even-skipped* locus is contained in a 16-kb chromatin domain. *Dev. Biol.* **211,** 39–52.
Schulz, R. A., Xie, X. L., and Miksch, J. L. (1990). *cis*-acting sequences required for the germ line expression of the *Drosophila gonadal* gene. *Dev. Biol.* **140,** 455–458.
Shi, Y., and Noll, M. (2009). Determination of cell fates in the R7 equivalence group of the *Drosophila* eye by the concerted regulation of D-Pax2 and TTK88. *Dev. Biol.* **331,** 68–77.
Small, S., Blair, A., and Levine, M. (1992). Regulation of *even-skipped* stripe 2 in the *Drosophila* embryo. *EMBO J.* **11,** 4047–4057.
Spilianakis, C. G., and Flavell, R. A. (2004). Long-range intrachromosomal interactions in the T helper type 2 cytokine locus. *Nat. Immunol.* **5,** 1017–1027.
Swanson, C. I., Evans, N. C., and Barolo, S. (2010). Structural rules and complex regulatory circuitry constrain expression of a Notch- and EGFR-regulated eye enhancer. *Dev. Cell* **18,** 359–370.
Swanson, C. I., Schwimmer, D. B., and Barolo, S. (2011). Rapid evolutionary rewiring of a structurally constrained eye enhancer. *Curr. Biol.* **21,** 1186–1196.
Szutorisz, H., Dillon, N., and Tora, L. (2005). The role of enhancers as centres for general transcription factor recruitment. *Trends Biochem. Sci.* **30,** 593–599.
Thanos, D., and Maniatis, T. (1995). Virus induction of human *IFN-β* gene expression requires the assembly of an enhanceosome. *Cell* **83,** 1091–1100.
Tsai, M. C., Manor, O., Wan, Y., Mosammaparast, N., Wang, J. K., Lan, F., Shi, Y., Segal, E., and Chang, H. Y. (2010). Long noncoding RNA as modular scaffold of histone modification complexes. *Science* **329,** 689–693.
Vakoc, C. R., Letting, D. L., Gheldof, N., Sawado, T., Bender, M. A., Groudine, M., Weiss, M. J., Dekker, J., and Blobel, G. A. (2005). Proximity among distant regulatory elements at the beta-globin locus requires GATA-1 and FOG-1. *Mol. Cell* **17,** 453–462.

Voas, M. G., and Rebay, I. (2004). Signal integration during development: Insights from the *Drosophila* eye. *Dev. Dyn.* **229,** 162–175.

Wang, G., Balamotis, M. A., Stevens, J. L., Yamaguchi, Y., Handa, H., and Berk, A. J. (2005). Mediator requirement for both recruitment and postrecruitment steps in transcription initiation. *Mol. Cell* **17,** 683–694.

Wang, K. C., Yang, Y. W., Liu, B., Sanyal, A., Corces-Zimmerman, R., Chen, Y., Lajoie, B. R., Protacio, A., Flynn, R. A., Gupta, R. A., *et al.* (2011). A long noncoding RNA maintains active chromatin to coordinate homeotic gene expression. *Nature* **472,** 120–124.

Xu, C., Kauffmann, R. C., Zhang, J., Kladny, S., and Carthew, R. W. (2000). Overlapping activators and repressors delimit transcriptional response to receptor tyrosine kinase signals in the *Drosophila* eye. *Cell* **103,** 87 97.

Yao, L. C., Phin, S., Cho, J., Rushlow, C., Arora, K., and Warrior, R. (2008). Multiple modular promoter elements drive graded *brinker* expression in response to the Dpp morphogen gradient. *Development* **135,** 2183–2192.

Yuh, C. H., and Davidson, E. H. (1996). Modular cis-regulatory organization of Endo16, a gut-specific gene of the sea urchin embryo. *Development* **122,** 1069–1082.

CHAPTER FIVE

# Computational Strategies for the Genome-Wide Identification of *cis*-Regulatory Elements and Transcriptional Targets

Stein Aerts

## Contents



## Abstract

Transcription factors (TFs) are key proteins that decode the information in our genome to express a precise and unique set of proteins and RNA molecules in each cell type in our body. These factors play a pivotal role in all biological processes, including the determination of a cell's fate during development and the maintenance of a cell's physiological function. To achieve this, a TF binds to specific DNA sequences in the noncoding part of the genome, recruits chromatin modifiers and cofactors, and directs the transcription initiation rate of its "target genes." Therefore, a key challenge in deciphering a transcriptional switch is to identify the direct target genes of the master regulators that control the switch, the *cis*-regulatory elements implementing (auto-)regulatory loops,

Laboratory of Computational Biology, Center for Human Genetics, Katholieke Universiteit (K.U.) Leuven, Leuven, Belgium

*Current Topics in Developmental Biology*, Volume 98
ISSN 0070-2153, DOI: 10.1016/B978-0-12-386499-4.00005-7

and the target genes of all the TFs in the downstream regulatory network. A better knowledge of a TF's targetome during specification and differentiation of a particular cell type will generate mechanistic insight into its developmental program. Here, I review computational strategies and methods to predict transcriptional targets by genome-wide searches for TF binding sites using position weight matrices, motif clusters, phylogenetic footprinting, chromatin binding and accessibility data, enhancer classification, motif enrichment, and gene expression signatures.

## 1. Introduction

During development, a plethora of *cis*-regulatory modules (CRMs) interpret the activity of all transcription factors (TFs) in a cell and yield the appropriate genomic response by translating bound TFs into a particular gene expression profile of that cell. Transcriptional switches occur when signaling pathways and cell–cell communication lead to a new combination of active TFs—or a significant change in the levels of the present active TFs—which then steers the gene regulatory network (GRN) in a new direction, often causing the activation of other TFs (and thereby feedback and feedforward loops), and ultimately leading to the activation of an altered expression profile of that cell or cell type (Davidson, 2001). To gain a deeper understanding of how transcriptional switches are implemented, it is necessary to identify the master regulator(s) that operates the switch and all its direct target genes. For the targets that are TFs themselves, their targets also have to be identified, and so on, until the entire GRN downstream of the master regulator is mapped. Within this network, special attention can be given to focus on regulatory loops, such as feedback loops (e.g., a predicted target is a TF that itself has its regulator as target) and feedforward loops (e.g., a predicted target is a TF that itself has shared targets with its regulator). The master regulator that controls the switch is usually a TF for which the loss-of-function (LOF) causes the loss of that particular cell type, and/or the ectopic gain-of-function (GOF) induces an excess of that cell type. For example, in *Drosophila*, ectopic induction of the gene *eyeless* causes ectopic formation of eyes, and its loss causes the absence of the eye (Halder *et al.*, 1995). Even though in every condition many CRMs are held inactive by the chromatin (Roy *et al.*, 2010), this block is dynamic and can be reversed by the appropriate TFs by recruiting demethylases or histone modifiers (Tapscott, 2005).

The main approaches for the identification of direct target genes of a particular TF are: (i) gene expression profiling using microarrays or RNA-Seq with RNA extracted from the cell type under study, for example before and after the transcriptional switch, or in wild-type versus TF-GOF or

TF-LOF conditions, to identify genes induced or repressed downstream (direct or indirect) of the TF; (ii) chromatin immunoprecipitation (ChIP) using an antibody against a TF, followed by tiling arrays (ChIP-chip) or high-throughput sequencing (ChIP-Seq), to identify all the genomic locations the TF is bound to in that cell type; (iii) computational predictions of all genomic locations that harbor recognition sequences for the TF. All three approaches have their own advantages and limitations, yet they are complementary and are often used in combination. Here, I review a number of strategies with a focus on the computational predictions, how they can be used alone, and how in combination with microarray data and with ChIP data. Besides these three main approaches, a number of auxiliary approaches can be used, both experimental and computational. For example, ChIP against p300, a factor binding to most active CRMs can reduce the search space of transcription factor binding site (TFBS) predictions. Likewise, genome-wide measurements of chromatin activity based on DNaseI hypersensitivity or the presence of particular histone modifications can "filter" the genome and deliver candidate CRMs that are active in the cell type under study (Kaplan *et al.*, 2011). Other auxiliary approaches are the use of existing knowledge to guide TF-target predictions, such as functional annotation (e.g., gene ontology), gene expression patterns, protein–protein interactions, and mutant phenotypes. We refer to several reviews for general principles, earlier work, and comparative studies (Bulyk, 2003; GuhaThakurta, 2006; Haeussler and Joly, 2011; Siggia, 2005; Su *et al.*, 2010; Taverner *et al.*, 2004; van Loo and Marynen, 2009; Vingron *et al.*, 2009; Wasserman and Sandelin, 2004; Wyrick and Young, 2002). Here, we describe both classical and newer strategies of regulatory bioinformatics, with a focus on methods to identify TF-target CRMs and target genes in a Metazoan genome, including the integration of additional data such as gene expression and chromatin-binding data.

## 2. Genome-Wide Binding Site and CRM Predictions Using PWMs

### 2.1. Single binding site detection

A position weight matrix (PWM) is a model for the binding specificity of a TF and can be used to scan a sequence for the presence of DNA words that are significantly more similar to the PWM than to the background (Stormo, 2000) (Fig. 5.1A). PWMs can be constructed from experimentally derived binding sites reported in the literature (i.e., by literature curation (Griffith *et al.*, 2008)), or can be determined experimentally by bacterial-1-hybrid (Noyes *et al.*, 2008), protein-binding-microarrays (Berger *et al.*, 2006), SELEX (Tuerk and Gold, 1990), MITOMI (Maerkl and Quake, 2007), or ChIP-chip/Seq (see review by Stormo and Zhao, 2010). Curated and

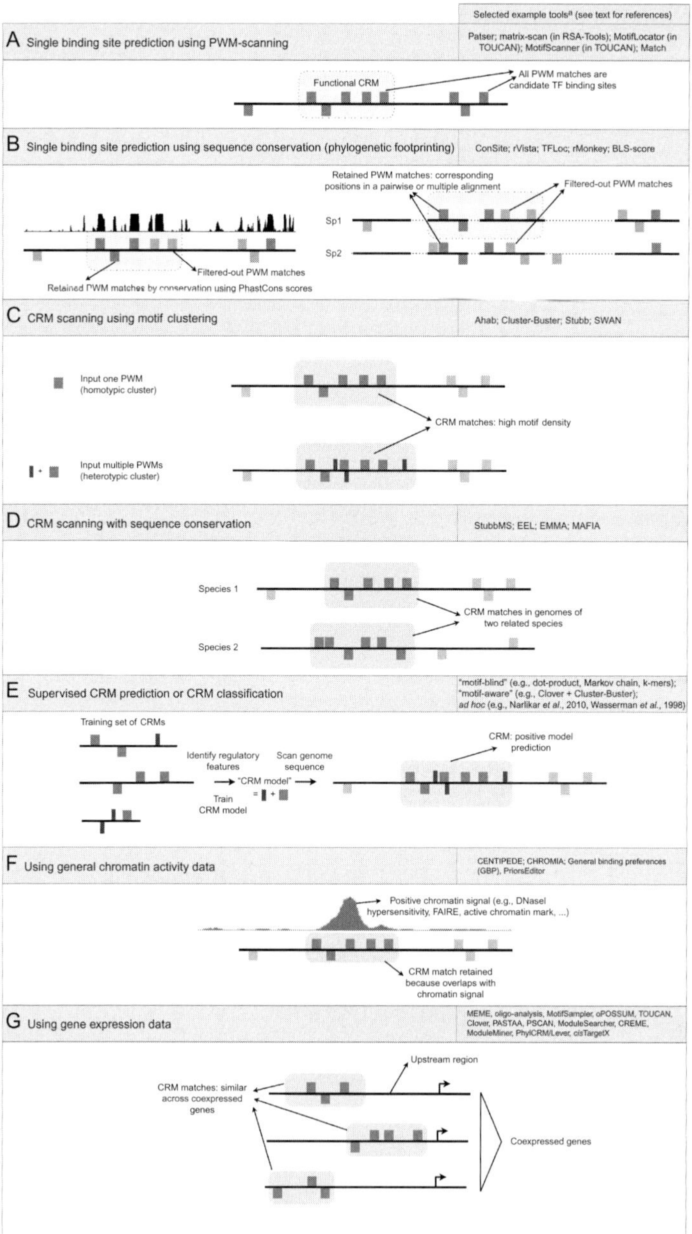


**Figure 5.1** Schematic overview of computational strategies to predict TF binding sites and *cis*-regulatory modules. (A) Single binding site prediction using PWM scanning. A sequence is scored with a position weight matrix (PWM), retaining matches with a normalized score greater than an arbitrary threshold or with a significant $p$-value. Sensitivity and specificity can be balanced by adjusting the threshold, and are usually

experimentally determined PWMs are available from open databases such as JASPAR (Portales-Casamar *et al.*, 2010), UNIPROBE (Robasky and Bulyk, 2010), or the FlyFactorSurvey database (Zhu *et al.*, 2011) and from commercial databases such as TRANSFAC®. In theory, a candidate TF binding site can be considered in a genomic sequence when the log-likelihood ratio of the score of the PWM model over the score of the background model, for a given subsequence, is positive or when the normalized score is greater than an arbitrary threshold (e.g., 90% of the maximal

---

better than consensus matches with regular expressions, although according to the futility theorem (Wasserman and Sandelin, 2004), this approach yields an excessive amount of false positives. (B) Single binding site prediction using sequence conservation (phylogenetic footprinting). Shown here is a filtering of PWM matches (retained matches are in bold) by their conservation across multiple species using (i) the phastCons score (black continuous track on top); and (ii) using corresponding matches (i.e., orthologous positions) in a pairwise alignment between two genomes of related species. Alternative strategies not shown here allow motif movement in the alignment, or match PWMs only to conserved words (see main text for details). (C) CRM scanning. Prediction of a cluster of motifs for one PWM (top) or multiple PWMs (bottom). CRM matches are genomic regions with a significantly high motif density compared to background sequences, measured by counting PWM matches in a sliding window or measured by the log-likelihood ratio as calculated by a hidden Markov model. (D) CRM scanning with sequence conservation. Similar as in (C) but now using multiple genomes. Shown here is network-level conservation (or CRM preservation), that requires that the orthologous genomic regions all have positive CRM matches, without requiring that the individual motifs are located at orthologous positions in the alignment. This can be achieved by CRM scanning in each genome independently or by HMM implementations that incorporate conserved and nonconserved states. (E) Supervised CRM prediction or CRM classification A "training set" of CRMs (left) is used to identify regulatory features shared among similar CRMs, such as overrepresented motifs, PWMs, or k-mers; or "foreground" Markov chains, compared to a background set (e.g., random genomic regions or randomly selected CRMs). The optimized CRM model can consist of PWMs for known TFs ("motif-aware," e.g., Wasserman and Fickett, 1998) or can be "motif-blind" (e.g., Kantorovitz *et al.*, 2009) or a mixture of both (e.g., Narlikar *et al.*, 2010). (F) Using general chromatin activity data. The gray track represents a positive chromatin signal measured in the biological condition under study, such as DNaseI hypersensitivity, FAIRE, active histone modification marks (e.g., H3K27Ac), or general CRM coactivators such as p300. The integration with PWM and CRM scanning can be done using a HMM (e.g., CHROMIA (Won *et al.*, 2010)) or using other methods such as classification (e.g., GBP (Ernst *et al.*, 2010)). (G) Using gene expression data. When gene expression data obtained in the biological condition under study is available, motif and CRM prediction can be performed in the promoters of a set of coexpressed genes (a gene signature), although in higher eukaryotes the *cis*-regulatory space extends much further than only the proximal promoter. Recent techniques integrate CRM scanning, sequence conservation, and enrichment statistics of the gene signature along a whole-genome ranking to identify candidate motifs and CRMs that may underlie the observed coexpression pattern (Aerts *et al.*, 2010; Warner *et al.*, 2008). [a] I apologize to researchers whose work or software tools are not cited.

PWM score). Several standalone implementations for *PWM scanning* are available, such as Patser (Hertz and Stormo, 1999), Match (Kel *et al.*, 2003), MotifLocator (Aerts *et al.*, 2003b), MotifScanner (Aerts *et al.*, 2003b), Matrix-Scan (Thomas-Chollier *et al.*, 2008), and others. Also, several software tools, either web applications or standalone tools, provide front ends to these methods, such as RSAT (Turatsinze *et al.*, 2008) and TOUCAN (Aerts *et al.*, 2005). Even though PWM scanning methods have improved, for example using higher order background models (Marchal *et al.*, 2003) and *p*-value estimation (Hertz and Stormo, 1999; Thomas-Chollier *et al.*, 2008), and even though alternative models to PWMs have been proposed (e.g., hidden Markov models (HMMs); Marinescu *et al.*, 2005), in practice any single-site detection method yields an excessive number of false positive predictions, simply because the actual TF binding sites are too short and too variable. Consequently, nearly every gene in the genome will have a match to the PWM of nearly every TF. This excess in false positives is called the futility theorem (Wasserman and Sandelin, 2004). There are some exceptions where direct PWM scanning delivers an acceptable accuracy, for example in very compact genomes with small intergenic search spaces (e.g., bacteria or yeast), but PWM scanning, by itself, is not successful in Metazoa. Nevertheless, PWM scanning is often useful as a starting point to generate a preliminary list of candidate-binding sites, that can be further filtered using measures of sequence conservation, binding site clustering, binding site combinations, or binding site overrepresentation across a set of putatively coregulated sequences.

## 2.2. Single binding site detection with sequence conservation

TF binding sites are often under purifying selection as shown by cross-species comparison of ChIP-derived or curated TF binding sites (Birney *et al.*, 2007; Eisen, 2007; He *et al.*, 2011b; Roy *et al.*, 2010; Sandmann *et al.*, 2006; Zeitlinger *et al.*, 2007). Consequently, PWM predictions can be filtered by phylogenetic footprinting (Tagle *et al.*, 1988; Wasserman *et al.*, 2000), retaining a PWM match only if the TF binding site is conserved (Fig. 5.1B). Note that conserved binding sites are often the strongest sites (Håndstad *et al.*, 2011) and weak sites may be missed (although a PWM approach will favor strong sites anyway, when stringent thresholds are applied). Also, species-specific sites or sites under positive selection (He *et al.*, 2011a) will also be missed. Deciding whether a site is conserved or not can be done arbitrarily, for example requiring that the orthologous sequences of the site, as determined by pairwise or multiple alignments, need to be the same across a number of species, or that the corresponding position in the alignment is also a positive hit for the PWM. Examples of available software tools that perform simple PWM scanning across species are ConSite (Sandelin *et al.*, 2004) and rVista (Loots and Ovcharenko, 2004), which are mostly used to predict conserved PWM instances in a promoter of a particular gene under study. A similar approach is used by the program TFLOC (Transcription Factor binding site LOCater) that identifies PWM matches on *multiz* multiple alignments (Fujita *et al.*, 2010). Candidate-binding sites are scored by the sum of normalized log-odds scores of each species, and a threshold is determined empirically from the distribution of maximal scores across all 5-kb upstream regions. Hence, genome-wide scores and conservation are used to determine the highest confidence sites. The predictions of TFLOC for all TRANSFAC PWMs are available as a track in the UCSC Genome Browser (Fujita *et al.*, 2010). A reverse strategy can also be used, starting with the detection of sequence conservation, potentially followed by matching conserved DNA words with PWMs or consensus sequences, as done, for example, in EvoPrinter (Odenwald *et al.*, 2005) using multiple alignments; in FootPrinter (Blanchette and Tompa, 2003) using parsimony scores in the phylogenetic tree; and in rMonkey (Moses *et al.*, 2006) using an evolutionary model. Additional confidence in conserved motifs can be gained by comparing the conservation of a particular motif or k-mer across the genome with random expectations. This strategy is used in combination with the branch length score (BLS) of the motif along the phylogenetic tree throughout several applications from the Kellis laboratory (Kellis *et al.*, 2003; Stark *et al.*, 2007; Xie *et al.*, 2005). In these studies, the BLS score is combined with statistics based on genome-wide occurrences of conserved sites compared to random motif shuffles. When a particular consensus site or PWM has significantly more *conserved* matches in the genome than expected by chance (e.g., using shuffled motifs), genome-wide motif prediction for those motifs is feasible, although the search space is usually still limited to proximal regions to achieve acceptable false positive rates (Kheradpour *et al.*, 2007). The target gene predictions for highly conserved motifs for *Drosophila*, detected using the BLS and confidence scores and then matched with known consensus sites or PWMs, can be found online (Kheradpour and Stark). When a new PWM is determined for a TF, it can be matched with the k-mers of this motif collection to check whether it has favorable conservation properties; or alternatively the BLS scoring procedure, including PWM shuffles, has to be applied to the PWM under study.

Importantly, the turnover of TFBS can be high, hence even a "functionally conserved" site may actually not be found at the same position in the alignment (Ludwig *et al.*, 2000). Yet, they often reside within conserved regions (see further below). To also capture "moved" sites within the same CRM, some approaches allow for motif movement, such as the BLS implementation and—by definition—the network-level conservation approach. Tavazoie and colleagues used such an approach to identify conserved motifs on a genomic scale, requiring a motif to be present in orthologous promoters, but not necessarily aligned (Elemento and Tavazoie, 2005; Pritsker *et al.*, 2004). This kind of binding site conservation, being present in the same region but not necessarily aligned, is also called binding site "preservation" (Berman *et al.*, 2004). For a further overview of conservation and divergence of CRMs, we refer to a recent review (Meireles-Filho and Stark, 2009). In conclusion, the use of binding site conservation or preservation may be useful for proximal promoters, yet additional cues are often necessary to further increase the specificity of TF binding site and target gene predictions.

## 2.3. CRM detection using motif clustering

A regulatory element driving a particular expression pattern of a target gene (e.g., promoter, enhancer) does usually not consist of a single transcription factor binding site. Rather, transcriptional regulation is controlled by CRMs that harbor a local density of TF binding sites (Davidson, 2001). CRMs range in size between a few hundred bp to several kb and often contain multiple TFBSs for several distinct TFs. A classical example is the *even skipped* (*eve*) stripe 2 enhancer in *Drosophila*, containing 6 Kr, 5 Bcd, 3 Gt, and 3 Hb binding sites within 500bp (Small *et al.*, 1992). Another well-characterized example in *Drosophila* is the *eve* muscle and heart enhancer (MHE), containing 6 Mad, 4 Ets, 4 Tin, 2 Twi, and 1 dTCF binding sites (Halfon *et al.*, 2000). Well-studied examples of CRMs in mammalian genomes are the human β-globin locus control region (binding sites for [GATA1, EKLF, NF-E2, SOX6, BCL11A]), which implements a transcriptional switch by driving globin expression specifically in erythroid cells (Levings and Bungert, 2002), or the mouse CRM of the *Ckm* gene (binding sites for [Srf, Ap2, Trex, Sp1, Mef2, and Mdf] driving *Ckm* expression specifically in muscle cells (Davidson, 2001; Sternberg *et al.*, 1988). Such clusters of binding sites for distinct TFs is called a heterotypic CRM, while a cluster of sites for a single TF is called a homotypic CRM. However, this distinction is merely useful to describe the computational methods, either detecting heterotypic or homotypic motif clusters, because in Metazoa, most CRMs identified to date are heterotypic (Davidson, 2001).

The majority of CRMs adhere to the so-called "billboard model" with a free architecture, meaning that the cluster of TF binding sites has no restrictions concerning the order of binding sites, the strand, and the spacing between them, as long as the TFs are not sterically hindered (Kulkarni and Arnosti, 2003). Therefore, the computational detection of a local density of TF binding sites within a confined genomic region allows identifying CRMs (Fig. 5.1C). A straightforward computational strategy to detect TFBS clusters is the sliding window approach, in which a window of genomic sequence is considered as a positive hit when the number of predicted PWM matches reaches an arbitrary threshold. This strategy can be relatively easily implemented by a python or perl script that parses the output of a PWM scanner (see above). This approach has been successfully applied to identify CRMs involved in several developmental processes in *Drosophila*. Markstein *et al.* (2002) identified direct target CRMs for dorsal (dl) requiring at least three matches to the consensus sequence in a window of 400bp. Rebeiz *et al.* (2002) found Su(H) targets by searching for homotypic clusters of the Su(H) consensus site in 100–5000bp windows. Halfon *et al.* (2002) searched for heterotypic clusters of dTcf, Mad, Tin, Twi, and Pnt binding sites in windows of 500bp having at least two matches for each PWM (one for dTcf). Berman *et al.* (2002) identified target CRMs of Cad, Bcd, Kr, Kni, and Hb by retaining all genomic windows of 700bp with at least 15 PWM matches in total. All these studies validated a number of predicted CRMs by *in vivo* enhancer–reporter assays.

A more sophisticated strategy to identify motif clusters, which does not require arbitrary thresholds on the number of PWM matches and the window size, is the use of HMMs (Eddy, 2004). For CRM prediction, the hidden states are the intra- and intercluster background states, and the actual PWMs, and a CRM is predicted when the likelihood that the CRM sequence is generated by this HMM is greater than the likelihood of being generated by the background model only (Wu and Xie, 2010). Since the first implementation of a HMM for CRM finding in 1997 by Crowley *et al.* (1997), multiple implementations have become available, such as Ahab (Rajewsky *et al.*, 2002), Cluster-Buster (Frith *et al.*, 2003), and SWAN (Kim *et al.*, 2010). Their usage is relatively straightforward as they require the PWMs as input, together with the sequence files of the genomic regions under study. As whole-genome HMM-scoring can become CPU-intensive, reduced time complexity is important. Cluster-Buster achieved this through a linear-time heuristic and can score the entire *Drosophila* genome with one PWM in 52s (at 2.27GHz) and the entire human genome in 941s. This allows to score the entire genome, and related genomes with hundreds of PWMs (i.e., entire PWM libraries such as JASPAR) on a high-performance computing cluster (Aerts *et al.*, 2010; see further below). Ahab was applied by Schroeder *et al.* (2004) to detect clusters of sites for Bcd, Hb, Cad, TorRE, Stat92E, Kr, Kni, Gt, and Tll around a set of 29 preselected genes involved in blastoderm segmentation. Of the 16 predicted CRMs, 13 showed the correct reporter gene expression.

Importantly, none of the above methods uses restrictions on TF binding site arrangements, in agreement with the billboard model. Nevertheless, some exceptions to the billboard model exist, such as the human interferon-β enhancer, which contains a well-defined architecture of binding sites for ATF-2, c-Jun, IRF-3A, IRF-7B, IRF-3C, p50, and RelA (Panne *et al.*, 2007). This type of CRMs adhere to the "enhanceosome" model. Intriguingly, a Rel family factor in *Drosophila*, Dorsal, also binds to Enhanceosome-compliant CRMs in a cooperative manner with Twist, with particular restrictions to the site orientation and the distance between the Dorsal and Twist sites (Papatsenko and Levine, 2007). To identify CRMs with architectural restrictions, to my knowledge, no particular scanners are available.

However, to some extent, enhancer element locator (EEL) incorporates information about the position of PWM matches (see Section 4), and restrictions on TFBS positions can be easily applied in a postanalysis on the output of the prediction of free CRMs with a HMM.

## 2.4. CRM detection using motif clustering and comparative genomics

Motif clustering improves the specificity of CRM predictions compared to single PWM scanning. Nevertheless, when applied to the whole noncoding genome, this strategy still results in too low sensitivity and specificity (Aerts *et al.*, 2007). To improve the accuracy, cross-species comparisons of motif clusters can be used, in a variety of flavors (Fig. 5.1D). A straightforward and often used option is to exploit the presence of conserved noncoding sequences (CNSs) in vertebrate genomes, which are conserved sequence "islands" in the noncoding genome and very often are proxies for regulatory regions (Pennacchio *et al.*, 2006; Yuh *et al.*, 1998). CNSs are collected from pairwise or multiple alignments (e.g., from the UCSC Genome Browser, Ensembl, or the VISTA Genome Browser) of the entire noncoding genome, or of large upstream sequences; then this collection can be scored for motif clusters. Note that, this is not always an effective approach: for vertebrate genomes, many tested predictions based on sequence conservation alone appear to be false positives; and for invertebrate genomes, due to compact genomes and high degrees of conservation, the CNSs do not appear as islands and they are more difficult to demarcate, at least when using the currently available genomes for 12 *Drosophila* species (Peterson *et al.*, 2009). More detailed analyses of the conservation of the individual motifs within the cluster has also been applied, for example, by Blanchette *et al.* (2006) who identify both homotypic and heterotypic CRMs in the human genome for all PWMs in TRANSFAC. Their method favors aligned PWM matches in all three species utilized (human–mouse–rat), and then detects clusters by the statistical significance of the density of conserved hits in a CRM compared to the rest of the genome. Predictions based on this method are available from the PReMod database (Ferretti *et al.*, 2007). The EEL applies a variant technique by performing the actual alignment between genomic sequences using the PWM matches (Hallikas *et al.*, 2006).

An elegant bioinformatics solution to combine motif clusters with sequence conservation is to extend the HMMs to detect motif clusters (see above) to include conserved/nonconserved hidden states, as done for example in STUBBMS (Sinha *et al.*, 2004b), EMMA (He *et al.*, 2009), and MAFIA (Majoros and Ohler, 2010). Sinha *et al.* (2004b) have shown that STUBB on multiple species (STUBBMS) yields better prediction performances than STUBB on a single genome, and have made genome-wide predictions of conserved CRMs using STUBBMS available online through a tool called

Genome Surveyor (Kazemian *et al.*, 2011). Importantly, the billboard property of CRMs has profound implications for the detection of purifying selection on CRMs, because a CRM can be functionally conserved when enough binding sites for the correct input TFs are locally clustered, regardless of the actual sequence alignment of the sites. This freedom is allowed in methods using network-level conservation or CRM "preservation." In *Drosophila*, Berman *et al.* (2004) found that preservation of heterotypic motif clusters between *D. melanogaster* and *D. pseudoobscura* allows to distinguish positive from negative CRMs. In a benchmark study across 34 TFs in *Drosophila*, it was shown that conserved homotypic motif clusters significantly improve prediction accuracies, with a slight advantage for network-level conservation approaches compared to motif-level alignment-based approaches (Aerts *et al.*, 2007). This study also showed that, starting with known PWMs as input, the best CRM prediction accuracies are still not good enough, for most TFs, to be applied to the whole genome, yielding too low specificity. Therefore, usually additional data are required to optimize the predictions, such as chromatin-binding data, gene expression data, or more complex enhancer models trained on similar CRMs.

## 3. Learning from Experimentally Validated CRMs

So far, we have discussed computational predictions of TF binding sites using one or more PWMs as input. However, sometimes the PWM or PWM combination is not known in advance (Ivan *et al.*, 2008). Moreover, even if some PWMs are known *a priori*, the CRMs we are searching for may contain particular characteristics, like other cofactor motifs or particular nucleotide compositions that could contribute to making the correct distinction between bona fide target CRMs and false positive predictions. This strategy can be applied if several CRMs are already known that contain binding sites for the TF under study, which can be used for feature discovery (Fig. 5.1E). Alternatively, it can also be applied when no actual CRMs but simply a number of individual TF binding sites are known, by using the genomic sequence of the binding sites together with flanking sequence (e.g., 100 bp or more on both sides). A hallmark study of CRM classification was performed by Wasserman and Fickett (1998), predicting target CRMs for the TF combination [MEF2, MYF, SP1, SRF, TEF]. They used a set of 29 experimentally determined CRMs that drive skeletal muscle-specific expression and trained a classifier on the PWM scores for these five PWMs on the positive training set using logistic regression. Applying the classifier to large genomic sequences identified a number of new CRMs located near muscle-specific genes. A similar approach was taken by the same authors to identify target CRMs of [HNF1, HNF3, HNF4, C/EBP] driving liver-specific expression (Krivan and Wasserman, 2001). These studies have inspired much of the later CRM classification work. In most of the later work, this approach has been extended to not only include *a priori* known PWMs but to learn the regulatory features from the positive training set by *pattern discovery*. Nth-order Markov chains trained on the positive set (capturing higher order nucleotide compositions shared among the positive sequences), or significantly overrepresented k-mers or PWMs identified in the positive set are then used as features in the predictive model. Interestingly, when libraries of PWMs from known TFs are used for feature discovery, the candidate input TFs that target the CRMs can be discovered, and newly predicted CRMs can be considered as candidate targets of these TFs. For example, in a recent study by Narlikar *et al.* (2010), 77 experimentally determined heart enhancers were used as positive set, and feature discovery identified 30 PWMs from TRANSFAC/JASPAR, five *de novo* motifs (using the PRIORITY algorithm), and two Markov features using a LASSO model for feature discovery. Eighteen of the 26 candidate TFs in the model were previously known to be active in the heart. To discover new CRMs involved in heart development, the model was applied to all human–mouse CNSs (see above). Positively classified CNSs were shown to be preferentially located near heart-related genes and *in vivo* validation of a subset of predicted CRMs in zebrafish showed that 16 of the 26 tested CRMs drove reporter expression during heart development. Kantorovitz *et al.* (2009) applied a similar strategy to 33 sets of similar CRMs from *Drosophila*, derived from the REDFly database (Gallo *et al.*, 2010) and to eight sets of similar CRMs from mouse, derived from the VISTA Enhancer Browser (Visel *et al.*, 2007). Regulatory feature selection was performed by several methods and included Markov chain features, Dot product-based scores, and overrepresented k-mers. This method is called "motif-blind" CRM detection because it only learns sequence characteristics and does not consider the identity of putative input TFs. By comparison with a model based on overrepresented PWMs (i.e., "motif-aware"), they showed that the motif-blind procedure often outperforms the motif-aware strategy. Finally, these authors also showed that cross-species scoring of the model outperforms the single-species version. Several newly predicted CRMs were successfully validated, both in *Drosophila* and mouse, by enhancer–reporter assays. Additional methods for CRM classification exist, including more advanced classifiers such as support vector machines (SVMs) (e.g., KIRMES (Schultheiss *et al.*, 2009)).

In conclusion, CRM classification can be a very powerful approach to identify novel CRMs generating particular expression patterns, although it depends largely on the quality of the training set, and is mostly useful when the goal is not to find direct TF-specific target CRMs but rather tissue-specific or process-specific CRMs.

## 4. CRM Analyses Using Chromatin Binding Data

The revolution in next-generation sequencing provides great opportunities to sequence biochemically purified genomic fragments, most notably sequences that are bound *in vivo* by a particular TF and that can be "pulled-down" by ChIP using an antibody against that TF (Park, 2009) or can be selected by "DamID" (van Steensel and Henikoff, 2000). Numerous studies have shown that this procedure indeed results in *bona fide* targets of the selected TF and that represents a straightforward way to identify regulatory interactions, provided that a specific antibody can be generated and that enough cells of the same type can be harvested. Besides multiple individual examples, several large-scale studies have been performed. In *Drosophila*, the modENCODE consortium ChIP'ed more than 30 TFs across multiple developmental stages during embryogenesis, as well as in multiple cell lines (Nègre *et al.*, 2011; Roy *et al.*, 2010); the BDTNP project ChIP'ed 21 TFs (most in duplicate using a different antibody) during blastoderm segmentation (MacArthur *et al.*, 2009); and Dr. Furlong's lab ChIP'ed five TFs at different time points during embryonic mesoderm development (Jakobsen *et al.*, 2007; Sandmann *et al.*, 2006, 2007). Interestingly, as more and more TF binding sites are identified by ChIP under the same condition, combinatorial codes can be deciphered. Zinzen *et al.* (2009) trained such codes from the five TF temporal ChIP data using a SVM that was shown to be predictive for the particular cell type in which the respective CRM is active. Note, however, that all *in vivo* ChIP experiments have been performed on the entire organism or cell line, which is only possible if the TF under study is active in a large enough fraction of the isolated cells. To achieve cell-type specific signals, new ChIP protocols and amplification techniques will be needed (Shankaranarayanan *et al.*, 2011).

Even though, in theory, ChIP yields all the direct target sites in the particular condition studied, in practice, the ChIP peaks usually need to be further filtered to remove nonspecific regions (e.g., regions bound indirectly by the TF, through a cofactor) and nonfunctional regions (e.g., regions bound coincidentally by the TF without any consequence on the expression of a nearby gene). To this end, *cis*-regulatory sequence analysis can provide additional confidence of direct TF targets (Gordan *et al.*, 2009). A straightforward strategy is to perform *de novo motif discovery* (D'haeseleer, 2006) on the set of TF-bound sequences to identify the recognition sequence (i.e., consensus motif or PWM) of the TF under study, and consequently to select a subset of "direct" target sequences of the TF, namely those with a significant match (or matches) to the PWM. Note that although, in principle, *de novo* motif discovery tools (e.g., MEME, oligo-analysis, MotifSampler, etc.) can be applied to the set of *candidate* target CRMs derived from ChIP, usually they are computationally not efficient when applied to very large sequence sets with thousands of sequences. Recently, several new algorithms have appeared that are optimized for very large sequence sets, such as MEME-ChIP (Machanick and Bailey, 2011), DREME (Bailey, 2011), and Peak-Motifs (Thomas-Chollier *et al.*, 2011). Further, this motif discovery step can be optimized by applying motif clustering and motif conservation (see above), for example retaining only ChIP peaks that contain a cluster of conserved PWM matches. A second useful application of motif discovery on ChIP peak sets is to identify cofactor motifs and to gain more insight in the *cis*-regulatory code of the target CRMs. For example, Sandmann *et al.* (2006) applied motif discovery on Mef2 ChIP peaks and identified specific cofactor motifs at different time points during development, such as Twist in early Mef2-bound CRMs.

Unfortunately, it is not expected that ChIP experiments with this level of detail will be feasible, at least not on the short term, for the more than thousand TFs in a Metazoan genome, under each and every condition (i.e., all species, cell types, developmental times, environmental conditions, and genetic backgrounds). Therefore, alternative approaches need to be developed. One promising direction is the high-throughput measurements of active and repressive histone marks and of chromatin accessibility using DNaseI hypersensitivity (Sabo *et al.*, 2006) or FAIRE (Giresi and Lieb, 2009). These approaches yield genome-wide candidate regulatory regions, independent of the specific TFs that are bound. Alternatively, ChIP against general cofactors that bind to many CRMs, such as p300 for distal CRMs (Blow *et al.*, 2010) or RNAPII for proximal CRMs (Birney *et al.*, 2007), have been used to determine all active CRMs in a specific condition. These approaches will allow to focus the binding site and CRM predictions on the (usually small) fraction of active regulatory elements (Thomas-Chollier *et al.*, 2011) (Fig. 5.1F). A similar principle has been used to increase PWM-based predictions dramatically, for example by Ernst *et al.* (2010) who generated a general binding preferences (GBP) for each nucleotide in the human genome, bacsed on 29 chromatin features (e.g., histone methylation data, DNAseI data, CTCF binding, but also phastCons scores for conservation and location relative to a transcription start site (TSS)) and 14 genome-wide ChIP-derived TF binding data. These authors found that PWM × GBP scores are more accurate to predict functional TF binding sites than PWM-only scores. Moreover, the PWM × GBP predictions outperformed state-of-the-art methods that do not use chromatin data, but only motif clustering and sequence conservation, such as PreMod and TFLoc predictions (see above). A comparison of the PWM × GBP predictions on independent E2F ChIP-chip data corroborated the improved prediction accuracy, although the maximum area under the curve was around 0.75, leaving still considerable room for improvement, both in sensitivity and in specificity. A recent tool that allows incorporating chromatin features as positional priors during motif discovery is PriorsEditor (Klepper and

Drabløs, 2010). Two other examples of methods that predict TF-specific target CRMs using chromatin-binding data are CENTIPEDE (Pique-Regi *et al.*, 2010) and CHROMIA (Won *et al.*, 2010). CENTIPEDE uses a Bayesian mixture model to score candidate PWM matches for their likelihood to be functional (or "bound"), given DNAseI data and histone modification data from lymphoblast cell lines, as well as evolutionary sequence conservation. CHROMIA applies a HMM using not only PWM scores but also histone modification data—obtained from mouse embryonic stem (mES) cells—as input. CHROMIA was used to predict TF binding sites in the genome for 13 TFs involved in mES. Again, it was shown that by incorporating chromatin data, the performance of TF binding site predictions improved significantly compared to methods that are unaware of chromatin data, such as STUBBMS, EEL, and Cluster-Buster (see above), but are still far from perfect, with a positive predictive value (PPV) in distal enhancers of only 4–31%. Note that, importantly, these methods combine multiple chromatin data sets that have all been constructed from the same cell type. Although the GBP study alludes to the general applicability, even to completely different conditions, developmental time points, and so on, this remains to be seen in future validations of these methods, and data, by independent researchers.

## 5. *cis*-Regulatory Analysis of Gene Signatures

A functional *cis*-regulatory element can be defined as a site that is bound by a TF under particular circumstances, and this binding plays a significant role in regulating the transcription initiation rate of the TF-target gene, which is usually located in *cis* to this TF-target site. All methods described thus far do not take a putative functional consequence of TF binding on the target expression into account. Given the availability and the ease of determination of genome-wide expression data by microarrays or RNA-Seq, CRM, and TF binding site predictions are often linked to expression changes and can greatly improve the identification of functional targets. Gene expression profiles can be obtained from specific tissue, cell types, or even individual cells (Islam *et al.*, 2011) under wild-type and TF mutant, knockdown, or overexpression conditions in model organisms. Thus, this approach potentially provides high-resolution insight into the GRN that specifies or operates a particular cell type, provided that direct TF-target relationships can be identified from the expression data by applying *cis*-regulatory sequence analysis.

Gene expression data have been used since the early days of microarrays to infer direct regulatory interactions (Tavazoie *et al.*, 1999). Gene expression data can sometimes be directly correlated ("explained") by predicted *cis*-regulatory elements, for which regression methods have been used (Bussemaker *et al.*, 2001). However, the most commonly used strategy is to start from a set of coexpressed genes—a *gene signature*—and to identify shared regulatory motifs across this set that may underlie the observed coexpression (Fig. 5.1G). By linking a putative TF to this motif, many direct TF-target interactions can be inferred. This approach has been useful in bacteria and yeast, where most regulatory motifs are concentrated in a few hundred base pairs upstream of the translational start site. When enough coregulated promoters are present in the input sequences derived from the set of coexpressed genes, then simple motif overrepresentation methods often identify the correct motif (van Helden *et al.*, 1998). Two categories of methods can be applied in this situation, namely *de novo* motif discovery methods or PWM overrepresentation methods. *De novo* motif discovery identifies overrepresented DNA words or dyads present in the input (foreground) set compared to a control (background) set, either as consensus sequence or as PWM (D'haeseleer, 2006). The former is done by word-counting methods (e.g., oligo-analysis (Defrance *et al.*, 2008)) and the latter by Gibbs sampling (e.g., MotifSampler (Thijs *et al.*, 2002)) or Expectation-Maximization (e.g., MEME (Bailey *et al.*, 2009)). PWM enrichment methods, on the other hand, compare the number of PWM matches in the foreground set to that in the background set, for example, using a hypergeometric distribution (e.g., oPOSSUM (Ho Sui *et al.*, 2007), TOUCAN (Aerts *et al.*, 2003b)) or using shuffles of the motif and/or sequence set (e.g., Clover (Frith *et al.*, 2004)).

However, the application of the same methods to higher eukaryotes has not been fruitful, not even to small sequence search spaces (e.g., promoters; Tompa *et al.*, 2005). An additional complication arises in higher eukaryotes when larger candidate sequences are included outside the proximal promoter (e.g., 50–100 kb upstream, introns, downstream) that further decrease the signal-to-noise ratios. To solve these problems, several approaches have been tested that incorporate some of the data or techniques that we already discussed above for PWM-based target prediction, namely (1) using sequence conservation in the motif discovery method (e.g., PhyME (Sinha *et al.*, 2004a), PhyloGibbs (Siddharthan *et al.*, 2005)); (2) using clusters of motifs and motif *combinations* to identify overrepresented heterotypic CRMs that are similar across the gene set (e.g., *cis*Module (Zhou and Wong, 2004)); (3) combining motif combinations with sequence conservation and PWM libraries (e.g., CREME (Sharan *et al.*, 2003), ModuleSearcher (Aerts *et al.*, 2003a)).

An interesting, more recent, direction for PWM overrepresentation is to replace the background sequence set by the entire genome, and to assess the enrichment of a motif within the foreground set compared to the rest of the genome. Technically, this can be achieved by ranking all genes in the genome by a motif-based scoring of their candidate regulatory regions, for example all proximal promoters (e.g., PASTAA (Roider *et al.*, 2009), PSCAN (Zambelli *et al.*, 2009)). This approach provides a lot of freedom on how the motif-based genome-wide gene ranking is generated. Particularly, the CRM prediction methods described above, that start from a PWM or PWM combination as input and use motif clustering and sequence conservation on the whole genome, can be combined with a gene signature. This has been used by PhylCRM/Lever (Warner *et al.*, 2008), ModuleMiner (Van Loo *et al.*, 2008), and *cis*TargetX (Aerts *et al.*, 2010). These methods thus integrate many of the layers described thus far and provide acceptable accuracies of motif discovery and CRM prediction when a gene signature, but no training CRMs, is available.

## 6. Conclusions and Further Challenges

As more and more high-throughput data becomes available, and as more and more CRMs are characterized (both in terms of *cis*-regulatory logic and in terms of regulatory activity), new strategies for data integration are required to make optimal use of the existing knowledge when searching for additional CRMs. Particularly, the combination of a CRM model, trained on previously known CRMs, with chromatin-mapping data, gene expression data, and genome-wide predictions should be the next step forward in mapping the entire *cis*-targetome of a TF. Nevertheless, multiple challenges remain to be solved in *cis*-regulatory sequence analysis to further increase the applicability of computational strategies.

An unsolved problem in the currently available methods arises when motifs or PWMs are predicted to be involved in the *cis*-regulation of a gene signature or of a set of coregulated CRMs, which still require the association of a candidate TF to the identified motif. For this problem, publicly available functional and expression annotation could be useful to prioritize a list of candidate TFs that could be the binding factors to these motifs based on the protein family information. For example, when an E-box motif is found, then all bHLH factors can be considered as candidates, and only those with matching expression profiles, protein–protein interactions, genetic interactions, GO annotation, or mutant phenotypes could be selected for follow-up studies (Aerts *et al.*, 2006, 2009). Predicted regulators could be matched with the predicted motifs based on TF-family characteristics. Another problem in CRM prediction arises when a CRM is identified through genome-wide searches, for which the target gene is not known. Not seldom a CRM regulates a target gene that is located many kilobases away, and even other genes can be present between the CRM and the gene it regulates. A related problem occurs when motifs and CRMs are predicted on gene signatures, where for each gene in the set an arbitrarily chosen search space is defined. This search space is usually a sequence upstream of the TSS (e.g., 1–2kb, 10kb, or even up to 50kb), sometimes also sequence downstream of the TSS is included, or all introns, or even sequences downstream of the 3′ end of the gene. However, whatever the parameters are, choices have to be made and usually they limit the search space to sequences in the immediate neighborhood of the genes, without allowing very distal regions or sequences across other genes. Ultimately, genome-wide CRM prediction methods will have to overcome this caveat. One direction that may be promising in this regard is the use of insulator positions. Insulators have been identified along the genome, in particular cell types, both in human/mouse and in *Drosophila*, using ChIP-chip against insulator-bound proteins such as CTCF (Martin *et al.*, 2011; Nègre *et al.*, 2010). These insulator positions appear to be rather stable across cell types and conditions, hence can be used to delineate the genomic region, and all candidate target genes contained in it, where a predicted CRM can act upon.

Finally, I believe that the community is in need for high-level software tools that can integrate many heterogeneous data sets and algorithms for regulatory biology. Much progress has been made in genomic data analysis by the online genome browsers (Ensembl, UCSC Genome Browser), the standalone genome browsers (e.g., IGV (Robinson *et al.*, 2011)), next-generation sequence analysis tools, and integrative tools like Galaxy (Giardine *et al.*, 2005). For the analysis of GRNs, software is also available, such as Cytoscape (Smoot *et al.*, 2010) and BioTapestry (Longabaugh *et al.*, 2005). It will be a substantial challenge to build software tools that integrate the motif and CRM prediction methods with genome annotation data, gene expression data, and network tools to efficiently analyze the regulatory genome.

In conclusion, I believe that computational predictions of *cis*-regulatory elements will be exceedingly important to achieve (1) the genome-wide annotation of *cis*-regulatory elements; (2) the mapping of GRNs; (3) the interpretation of *cis*-regulatory logic and prediction of gene expression profiles; and (4) the prediction and interpretation of changes in the regulatory genome, in the context of evolution or disease. Achieving a systems level understanding of the complete regulome underlying a biological process is a key challenge in today's genome biology, given the crucial role that GRNs play as a developmental and evolutionary driving force, and perhaps more importantly, because changes in these networks such as mutations in TFs and in *cis*-regulatory sequences are major inducers of human diseases.

## ACKNOWLEDGMENTS

I thank Carl Herrmann for helpful suggestions on the manuscript and Zeynep Kalender Atak for help with the figure. I apologize to researchers whose work or software tools are not cited. S.A.'s research is funded by FWO grant G.0704.11N, K.U. Leuven research fund grants CREA/10/014 and PF/10/016, HFSP grant RGY0070/2011, and Foundation Against Cancer grant 2010-154.

## REFERENCES

Aerts, S., Van Loo, P., Thijs, G., Moreau, Y., and De Moor, B. (2003a). Computational detection of cis-regulatory modules. *Bioinformatics* **19**(Suppl. 2), ii5–ii14.

Aerts, S., Thijs, G., Coessens, B., Staes, M., Moreau, Y., and de Moor, B. (2003b). Toucan: Deciphering the cis-regulatory logic of coregulated genes. *Nucleic Acids Res.* **31,** 1753–1764.

Aerts, S., van Loo, P., Thijs, G., Mayer, H., de Martin, R., Moreau, Y., and de Moor, B. (2005). TOUCAN 2: The all-inclusive open source workbench for regulatory sequence analysis. *Nucleic Acids Res.* **33,** W393–W396.

Aerts, S., *et al.* (2006). Gene prioritization through genomic data fusion. *Nat. Biotechnol.* **24,** 537–544.

Aerts, S., van Helden, J., Sand, O., and Hassan, B. A. (2007). Fine-tuning enhancer models to predict transcriptional targets across multiple genomes. *PLoS One* **2,** e1115.

Aerts, S., Vilain, S., Hu, S., Tranchevent, L., Yan, J., Moreau, Y., Quan, X., and Hassan, B. (2009). Integrating computational biology and forward genetics in Drosophila. *PLoS Genet.* **5**(1), e1000.

Aerts, S., Quan, X.-J., Claeys, A., Naval Sanchez, M., Tate, P., Yan, J., and Hassan, B. A. (2010). Robust target gene discovery through transcriptome perturbations and genome-wide enhancer predictions in Drosophila uncovers a regulatory basis for sensory specification. *PLoS Biol.* **8,** e1000435.

Bailey, T. L. (2011). DREME: Motif discovery in transcription factor ChIP-seq data. *Bioinformatics* **27,** 1653–1659.

Bailey, T. L., Boden, M., Buske, F. A., Frith, M., Grant, C. E., Clementi, L., Ren, J., Li, W. W., and Noble, W. S. (2009). MEME SUITE: Tools for motif discovery and searching. *Nucleic Acids Res.* **37,** W202–W208.

Berger, M. F., Philippakis, A. A., Qureshi, A. M., He, F. S., Estep, P. W., and Bulyk, M. L. (2006). Compact, universal DNA microarrays to comprehensively determine transcription-factor binding site specificities. *Nat. Biotechnol.* **24,** 1429–1435.

Berman, B. P., Nibu, Y., Pfeiffer, B. D., Tomancak, P., Celniker, S. E., Levine, M., Rubin, G. M., and Eisen, M. B. (2002). Exploiting transcription factor binding site clustering to identify cis-regulatory modules involved in pattern formation in the Drosophila genome developmental. *Proc. Natl. Acad. Sci. USA* 22;**99**(2), 757–762.

Berman, P. B., Pfeiffer, D. B., Laverty, R. T., Salzberg, L. S., Rubin, M. G., Eisen, B. M., and Celniker, E. S. (2004). Computational identification of developmental enhancers: Conservation and function of transcription factor binding-site clusters in Drosophila. *Genome Biol.* **5,** R61.

Birney, E., *et al.* (2007). Identification and analysis of functional elements in 1% of the human genome by the ENCODE pilot project. *Nature* **447,** 799–816.

Blanchette, M., and Tompa, M. (2003). FootPrinter: A program designed for phylogenetic footprinting. *Nucleic Acids Res.* **31,** 3840–3842.

Blanchette, M., *et al.* (2006). Genome-wide computational prediction of transcriptional regulatory modules reveals new insights into human gene expression. *Genome Res.* **16,** 656–668.

Blow, M. J., *et al.* (2010). ChIP-Seq identification of weakly conserved heart enhancers. *Nat. Genet.* **42,** 806–810.

Bulyk, M. L. (2003). Computational prediction of transcription-factor binding site locations. *Genome Biol.* **5,** 201.

Bussemaker, H. J., Li, H., and Siggia, E. D. (2001). Regulatory element detection using correlation with expression. *Nat. Genet.* **27,** 167–171.

Crowley, E. M., Roeder, K., and Bina, M. (1997). A statistical model for locating regulatory regions in genomic DNA. *J Mol Biol.* 1997;**268**(1), 8–14.

D'haeseleer, P. (2006). How does DNA sequence motif discovery work? *Nat. Biotechnol.* **24,** 959–961.

Davidson, E. H. (2001). Genomic regulatory systems. Academic Press, San Diego, USA.

Defrance, M., Janky, R., Sand, O., and Helden, J. V. (2008). Using RSAT oligo-analysis and dyad-analysis tools to discover regulatory signals in nucleic sequences. *Nat Protoc.* **3**(10), 1589–1603.

Eddy, S. (2004). What is a hidden Markov model? *Nat. Biotechnol.* **22,** 1315–1316.

Eisen, M. (2007). Evolution of genes and genomes in the genus *Drosophila. Nature;* **450** (7167), 203–218.

Elemento, O., and Tavazoie, S. (2005). Fast and systematic genome-wide discovery of conserved regulatory elements using a non-alignment based approach. *Genome Biol.* **6,** R18.

Ernst, J., Plasterer, H. L., Simon, I., and Bar-Joseph, Z. (2010). Integrating multiple evidence sources to predict transcription factor binding in the human genome. *Genome Res.* **20,** 526–536.

Ferretti, V., Poitras, C., Bergeron, D., Coulombe, B., Robert, F., and Blanchette, M. (2007). PReMod: A database of genome-wide mammalian cis-regulatory module predictions. *Nucleic Acids Res.* **35,** D122–D126.

Frith, C. M., Li, C. M., and Weng, Z. (2003). Cluster-Buster: Finding dense clusters of motifs in DNA sequences. *Nucleic Acids Res.* **31,** 3666–3668.

Frith, M. C., Fu, Y., Yu, L., Chen, J. F., Hansen, U., and Weng, Z. (2004). Detection of functional DNA motifs via statistical over-representation. *Nucleic Acids Res.* **32,** 1372–1381.

Fujita, P. A., *et al.* (2010). The UCSC Genome Browser database: Update 2011. *Nucleic Acids Res.* **39,** D876–D882.

Gallo, S. M., Gerrard, D. T., Miner, D., Simich, M., Soye, B., Des Bergman, C. M., and Halfon, M. S. (2010). REDfly v3.0: Toward a comprehensive database of transcriptional regulatory elements in Drosophila. *Nucleic Acids Res.* **39**(Database issue), D118–D123.

Giardine, B., *et al.* (2005). Galaxy: A platform for interactive large-scale genome analysis. *Genome Res.* **15,** 1451–1455.

Giresi, P. G., and Lieb, J. D. (2009). Isolation of active regulatory elements from eukaryotic chromatin using FAIRE (Formaldehyde Assisted Isolation of Regulatory Elements). *Methods (San Diego, Calif.)* **48,** 233–239.

Gordan, R., Hartemink, A. J., and Bulyk, M. L. (2009). Distinguishing direct versus indirect transcription factor-DNA interactions. *Genome Res.* **19**(11), 2090–2100.

Griffith, O. L., *et al.* (2008). ORegAnno: An open-access community-driven resource for regulatory annotation. *Nucleic Acids Res.* **36,** D107–D113.

GuhaThakurta, D. (2006). Computational identification of transcriptional regulatory elements in DNA sequence. *Nucleic Acids Res.* **34,** 3585–3598.

Haeussler, M., and Joly, J.-S. (2011). When needles look like hay: How to find tissue-specific enhancers in model organism genomes. *Dev. Biol.* **350,** 239–254.

Halder, G., Callaerts, P., and Gehring, W. J. (1995). Induction of ectopic eyes by targeted expression of the eyeless gene in Drosophila. *Science (New York, N.Y.)* **267,** 1788–1792.

Halfon, M. S., Carmena, A., Gisselbrecht, S., Sackerson, C. M., Jiménez, F., Baylies, M. K., and Michelson, A. M. (2000). Ras pathway specificity is determined by the integration of multiple signal-activated and tissue-restricted transcription factors. *Cell* **103,** 63–74.

Halfon, M., Grad, Y., Church, G., and Michelson, A. (2002). Computation-based discovery of related transcriptional regulatory modules and motifs using an experimentally validated combinatorial model. *Genome Res.* **12,** 1019–1028.

Hallikas, O., Palin, K., Sinjushina, N., Rautiainen, R., Partanen, J., Ukkonen, E., and Taipale, J. (2006). Genome-wide prediction of mammalian enhancers based on analysis of transcription-factor binding affinity. *Cell* **124,** 47–59.

Håndstad, T., Rye, M. B., Drabløs, F., and Sætrom, P. (2011). A ChIP-Seq benchmark shows that sequence conservation mainly improves detection of strong transcription factor binding sites. *PLoS One* **6,** e18430.

He, X., Ling, X., and Sinha, S. (2009). Alignment and prediction of cis-regulatory modules based on a probabilistic model of evolution. *PLoS Comput Biol.* **3,** e1000299.

He, B. Z., Holloway, A. K., Maerkl, S. J., and Kreitman, M. (2011a). Does positive selection drive transcription factor binding site turnover? A test with Drosophila cis-regulatory modules. *PLoS Genet.* **7,** e1002053.

He, Q., Bardet, A. F., Patton, B., Purvis, J., Johnston, J., Paulson, A., Gogol, M., Stark, A., and Zeitlinger, J. (2011b). High conservation of transcription factor binding and evidence for combinatorial regulation across six Drosophila species. *Nat. Genet.* **43,** 414–420.

Hertz, G., and Stormo, G. (1999). Identifying DNA and protein patterns with statistically significant alignments of multiple sequences. *Bioinformatics* **15,** 563–577.

Ho Sui, S. J., Fulton, D. L., Arenillas, D. J., Kwon, A. T., and Wasserman, W. W. (2007). oPOSSUM: Integrated tools for analysis of regulatory motif over-representation. *Nucleic Acids Res.* **35,** W245–W252.

Islam, S., Kjällquist, U., Moliner, A., Zajac, P., Fan, J.-B., Lönnerberg, P., and Linnarsson, S. (2011). Characterization of the single-cell transcriptional landscape by highly multiplex RNA-seq. *Genome Res.* **21**(7), 1160–1167.

Ivan, A., Halfon, M. S., and Sinha, S. (2008). Computational discovery of cis-regulatory modules in Drosophila without prior knowledge of motifs. *Genome Biol.* **9,** R22.

Jakobsen, J. S., Braun, M., Astorga, J., Gustafson, E. H., Sandmann, T., Karzynski, M., Carlsson, P., and Furlong, E. E. (2007). Temporal ChIP-on-chip reveals Biniou as a universal regulator of the visceral muscle transcriptional network. *Genes Dev.* **21,** 2448–2460.

Kantorovitz, M. R., Kazemian, M., Kinston, S., Miranda-Saavedra, D., Zhu, Q., Robinson, G. E., Göttgens, B., Halfon, M. S., and Sinha, S. (2009). Motif-blind, genome-wide discovery of cis-regulatory modules in Drosophila and mouse. *Dev. Cell* **17,** 568–579.

Kaplan, T., Li, X.-Y., Sabo, P. J., Thomas, S., Stamatoyannopoulos, J. A., Biggin, M. D., and Eisen, M. B. (2011). Quantitative models of the mechanisms that control genome-wide patterns of transcription factor binding during early Drosophila development. *PLoS Genet.* **7,** e1001290.

Kazemian, M., Brodsky, M. H., and Sinha, S. (2011). Genome surveyor 2.0: Cis-regulatory analysis in Drosophila. *Nucleic Acids Res.* **39**(Web Server issue), W79–85.

Kel, A. E., Gössling, E., Reuter, I., Cheremushkin, E., Kel-Margoulis, O. V., and Wingender, E. (2003). MATCH: A tool for searching transcription factor binding sites in DNA sequences. *Nucleic Acids Res.* **31,** 3576–3579.

Kellis, M., Patterson, N., Endrizzi, M., Birren, B., and Lander, E. S. (2003). Sequencing and comparison of yeast species to identify genes and regulatory elements. *Nature* **423,** 241–254.

Kheradpour, P., and Stark, A., http://compbio.mit.edu/fly/motif-instances/.

Kheradpour, P., Stark, A., Roy, S., and Kellis, M. (2007). Reliable prediction of regulator targets using 12 Drosophila genomes. *Genome Res.* **17,** 1919–1931.

Kim, J., *et al.* (2010). Functional characterization of transcription factor motifs using cross-species comparison across large evolutionary distances. *PLoS Comput. Biol.* **6,** e1000652.

Klepper, K., and Drabløs, F. (2010). PriorsEditor: A tool for the creation and use of positional priors in motif discovery. *Bioinformatics (Oxford, England)* **26,** 2195–2197.

Krivan, W., and Wasserman, W. W. (2001). A predictive model for regulatory sequences directing liver-specific transcription. *Genome Res.* **11,** 1559–1566.

Kulkarni, M. M., and Arnosti, D. N. (2003). Information display by transcriptional enhancers. *Development* **130,** 6569–6575.

Levings, P. P., and Bungert, J. (2002). The human beta-globin locus control region. *Eur. J. Biochem./FEBS* **269,** 1589–1599.

Longabaugh, W. J., Davidson, E. H., and Bolouri, H. (2005). Computational representation of developmental genetic regulatory networks. *Dev. Biol.* **283,** 1–16.

Loots, G. G., and Ovcharenko, I. (2004). rVISTA 2.0: Evolutionary analysis of transcription factor binding sites. *Nucleic Acids Res.* **32,** W217–W221.

Ludwig, M. Z., Bergman, C., Patel, N. H., and Kreitman, M. (2000). Evidence for stabilizing selection in a eukaryotic enhancer element. *Nature* **403,** 564–567.

MacArthur, S., *et al.* (2009). Developmental roles of 21 Drosophila transcription factors are determined by quantitative differences in binding to an overlapping set of thousands of genomic regions. *Genome Biol.* **10,** R80.

Machanick, P., and Bailey, T. L. (2011). MEME-ChIP: Motif analysis of large DNA datasets. *Bioinformatics* **27,** 1696–1697.

Maerkl, S. J., and Quake, S. R. (2007). A systems approach to measuring the binding energy landscapes of transcription factors. *Science* **315,** 233–237.

Majoros, W. H., and Ohler, U. (2010). Modeling the evolution of regulatory elements by simultaneous detection and alignment with phylogenetic pair HMMs. *PLoS Comput. Biol.* **6,** e1001037.

Marchal, K., Thijs, G., de Keersmaecker, S., Monsieurs, P., de Moor, B., and Vanderleyden, J. (2003). Genome-specific higher-order background models to improve motif detection. *Trends Microbiol.* **11,** 61–66.

Marinescu, V. D., Kohane, I. S., and Riva, A. (2005). MAPPER: A search engine for the computational identification of putative transcription factor binding sites in multiple genomes. *BMC Bioinformatics* **6,** 79.

Markstein, M., Markstein, P., Markstein, V., and Levine, M. S. (2002). Genome-wide analysis of clustered Dorsal binding sites identifies putative target genes in the Drosophila embryo. *Proc. Natl. Acad. Sci. USA* **99,** 763–768.

Martin, D., *et al.* (2011). Genome-wide CTCF distribution in vertebrates defines equivalent sites that aid the identification of disease-associated genes. *Nat. Struct. Mol. Biol.* **18,** 708–714.

Meireles-filho, A. C. A., and Stark, A. (2009). Comparative genomics of gene regulation—Conservation and divergence of cis-regulatory information. *Curr. Opin. Genet. Dev.* **19** (6), 565–570.

Moses, A. M., Pollard, D. A., Nix, D. A., Iyer, V. N., Li, X. Y., Biggin, M. D., and Eisen, M. B. (2006). Large-scale turnover of functional transcription factor binding sites in Drosophila. *PLoS Comput. Biol.* **2,** e130.

Narlikar, L., Sakabe, N. J., Blanski, A. A., Arimura, F. E., Westlund, J. M., Nobrega, M. A., and Ovcharenko, I. (2010). Genome-wide discovery of human heart enhancers. *Genome Res.* **20,** 381–392.

Nègre, N., *et al.* (2010). A comprehensive map of insulator elements for the Drosophila genome. *PLoS Genet.* **6,** e1000814.

Nègre, N., *et al.* (2011). A cis-regulatory map of the Drosophila genome. *Nature* **471,** 527–531.

Noyes, M. B., Meng, X., Wakabayashi, A., Sinha, S., Brodsky, M. H., and Wolfe, S. A. (2008). A systematic characterization of factors that regulate Drosophila segmentation via a bacterial one-hybrid system. *Nucleic Acids Res.* **36**(8), 2547–2560.

Odenwald, W. F., Rasband, W., Kuzin, A., and Brody, T. (2005). EVOPRINTER, a multigenomic comparative tool for rapid identification of functionally important DNA. *Proc. Natl. Acad. Sci. USA* **102,** 14700–14705.

Panne, D., Maniatis, T., and Harrison, S. C. (2007). An atomic model of the interferon-beta enhanceosome. *Cell* **129,** 1111–1123.

Papatsenko, D., and Levine, M. (2007). A rationale for the enhanceosome and other evolutionarily constrained enhancers. *Cur. Biol.* **17,** R955–R957.

Park, P. J. (2009). ChIP-seq: Advantages and challenges of a maturing technology. *Nature reviews* **10,** 669–680.

Pennacchio, L. A., *et al.* (2006). In vivo enhancer analysis of human conserved non-coding sequences. *Nature* **444,** 499–502.

Peterson, B. K., Hare, E. E., Iyer, V. N., Storage, S., Conner, L., Papaj, D. R., Kurashima, R., Jang, E., and Eisen, M. B. (2009). Big genomes facilitate the comparative identification of regulatory elements. *PLoS One* **4,** e4688.

Pique-Regi, R., Degner, J. F., Pai, A. A., Gaffney, D. J., Gilad, Y., and Pritchard, J. K. (2010). Accurate inference of transcription factor binding from DNA sequence and chromatin accessibility data. *Genome Res.* **21,** 447–455.

Portales-Casamar, E., *et al.* (2010). JASPAR 2010: The greatly expanded open-access database of transcription factor binding profiles. *Nucleic Acids Res.* **38,** D105–D110.

Pritsker, M., Liu, Y., Beer, M., and Tavazoie, S. (2004). Whole-genome discovery of transcription factor binding sites by network-level conservation. *Genome Res.* **14,** 99–108.

Rajewsky, N., Vergassola, M., Gaul, U., and Siggia, E. (2002). Computational detection of genomic cis-regulatory modules applied to body patterning in the early Drosophila embryo. *BMC Bioinformatics* **3,** 30.

Rebeiz, M., Reeves, L. N., and Posakony, W. J. (2002). SCORE: A computational approach to the identification of cis-regulatory modules and target genes in whole-genome sequence data. Site clustering. *Proc. Natl. Acad. Sci. USA* **99,** 9888–9893.

Robasky, K., and Bulyk, M. L. (2010). UniPROBE, update 2011: Expanded content and search tools in the online database of protein-binding microarray data on protein-DNA interactions. *Nucleic Acids Res.* **39**(Database issue), D124–D128.

Robinson, J. T., Thorvaldsdóttir, H., Winckler, W., Guttman, M., Lander, E. S., Getz, G., and Mesirov, J. P. (2011). Integrative genomics viewer. *Nat. Biotechnol.* **29,** 24–26.

Roider, H. G., Manke, T., O'Keeffe, S., Vingron, M., and Haas, S. A. (2009). PASTAA: Identifying transcription factors associated with sets of co-regulated genes. *Bioinformatics* **25,** 435–442.

Roy, S., *et al.* (2010). Identification of functional elements and regulatory circuits by Drosophila modENCODE. *Science (New York, N.Y.)* **330,** 1787–1797.

Sabo, P. J., *et al.* (2006). Genome-scale mapping of DNase I sensitivity in vivo using tiling DNA microarrays. *Nat. Methods* **3,** 511–518.

Sandelin, A., Wasserman, W. W., and Lenhard, B. (2004). ConSite: Web-based prediction of regulatory elements using cross-species comparison. *Nucleic Acids Res.* **32,** W249–W252.

Sandmann, T., Jensen, L., Jakobsen, J., Karzynski, M., Eichenlaub, M., Bork, P., and Furlong, E. (2006). A temporal map of transcription factor activity: mef2 directly regulates target genes at all stages of muscle development. *Dev. Cell* **10,** 797–807.

Sandmann, T., Girardot, C., Brehme, M., Tongprasit, W., Stolc, V., and Furlong, E. (2007). A core transcriptional network for early mesoderm development in Drosophila melanogaster. *Genes Dev.* **21,** 436–449.

Schroeder, D. M., Pearce, M., Fak, J., Fan, H., Unnerstall, U., Emberly, E., Rajewsky, N., Siggia, D. E., and Gaul, U. (2004). Transcriptional control in the segmentation gene network of Drosophila. *PLoS Biol.* **2,** E271.

Schultheiss, S. J., Busch, W., Lohmann, J. U., Kohlbacher, O., and Rätsch, G. (2009). KIRMES: Kernel-based identification of regulatory modules in euchromatic sequences. *Bioinformatics* **25,** 2126–2133.

Shankaranarayanan, P., Mendoza-Parra, M.-A., Walia, M., Wang, L., Li, N., Trindade, L. M., and Gronemeyer, H. (2011). Single-tube linear DNA amplification (LinDA) for robust ChIP-seq. *Nat. Methods* **8**(7), 565–567.

Sharan, R., Ovcharenko, I., Ben-Hur, A., and Karp, R. M. (2003). CREME: A framework for identifying cis-regulatory modules in human-mouse conserved segments. *Bioinformatics* **19**(Suppl. 1), i283–i291.

Siddharthan, R., Siggia, D. E., and van Nimwegen, E. (2005). PhyloGibbs: A Gibbs sampling motif finder that incorporates phylogeny. *PLoS Comput. Biol.* **1,** e67.

Siggia, E. D. (2005). Computational methods for transcriptional regulation. *Curr. Opin. Genet. Dev.* **15,** 214–221.

Sinha, S., Blanchette, M., and Tompa, M. (2004a). PhyME: A probabilistic algorithm for finding motifs in sets of orthologous sequences. *BMC Bioinformatics* **5,** 170.

Sinha, S., Schroeder, M., Unnerstall, U., Gaul, U., and Siggia, E. (2004b). Cross-species comparison significantly improves genome-wide prediction of cis-regulatory modules in Drosophila. *BMC Bioinformatics* **5,** 129.

Small, S., Blair, A., and Levine, M. (1992). Regulation of even-skipped stripe 2 in the Drosophila embryo. *EMBO J.* **11,** 4047–4057.

Smoot, M. E., Ono, K., Ruscheinski, J., Wang, P.-L., and Ideker, T. (2010). Cytoscape 2.8: New features for data integration and network visualization. *Bioinformatics* **27,** 431–432.

Stark, A., *et al.* (2007). Discovery of functional elements in 12 Drosophila genomes using evolutionary signatures. *Nature* **450,** 219–232.

Sternberg, E. A., Spizz, G., Perry, W. M., Vizard, D., Weil, T., and Olson, E. N. (1988). Identification of upstream and intragenic regulatory elements that confer cell-type-restricted and differentiation-specific expression on the muscle creatine kinase gene. *Mol. Cell. Biol.* **8,** 2896–2909.

Stormo, G. D. (2000). DNA binding sites: Representation and discovery. *Bioinformatics* **16,** 16–23.

Stormo, G. D., and Zhao, Y. (2010). Determining the specificity of protein-DNA interactions. *Nat. Rev. Genet.* **11,** 751–760.

Su, J., Teichmann, S. A., and Down, T. A. (2010). Assessing computational methods of cis-regulatory module prediction. *PLoS Comput. Biol.* **6,** e1001020.

Tagle, D. A., Koop, B. F., Goodman, M., Slightom, J. L., Hess, D. L., and Jones, R. T. (1988). Embryonic epsilon and gamma globin genes of a prosimian primate (Galago crassicaudatus). Nucleotide and amino acid sequences, developmental regulation and phylogenetic footprints. *J. Mol. Biol.* **203,** 439–455.

Tapscott, S. J. (2005). The circuitry of a master switch: Myod and the regulation of skeletal muscle gene transcription. *Development* **132,** 2685–2695.

Tavazoie, S., Hughes, J. D., Campbell, M. J., Cho, R. J., and Church, G. M. (1999). Systematic determination of genetic network architecture. *Nat. Genet.* **22,** 281–285.

Taverner, N. V., Smith, J. C., and Wardle, F. C. (2004). Identifying transcriptional targets. *Genome Biol.* **5,** 210.

Thijs, G., Marchal, K., Lescot, M., Rombauts, S., de Moor, B., Rouze, P., and Moreau, Y. (2002). A Gibbs sampling method to detect overrepresented motifs in the upstream regions of coexpressed genes. *J. Comput. Biol.* **9,** 447–464.

Thomas-Chollier, M., Sand, O., Turatsinze, J.-V., Janky, R., Defrance, M., Vervisch, E., Brohée, S., and van Helden, J. (2008). RSAT: Regulatory sequence analysis tools. *Nucleic Acids Res.* **36,** W119–W127.

Thomas-Chollier, M., Herrmann, C., Defrance, M., Sand, O., Thieffry, D., and van Helden, J. (2011). RSAT peak-motifs: Motif analysis in full-size ChIP-seq datasets. Nucleic Acids Res, *in press*. doi:10.1093/nar/gkr1104, 9.

Tompa, M., *et al.* (2005). Assessing computational tools for the discovery of transcription factor binding sites. *Nat. Biotechnol.* **23,** 137–144.

Tuerk, C., and Gold, L. (1990). Systematic evolution of ligands by exponential enrichment: RNA ligands to bacteriophage T4 DNA polymerase. *Science* **249,** 505–510.

Turatsinze, J. V., Thomas-chollier, M., Defrance, M., and Helden, J. V. (2008). Using RSAT to scan genome sequences for transcription factor binding sites and cis-regulatory modules. *Nat. Protoc.* **3**(10), 1578–1588 Online.

van Helden, J., Andre, B., and Collado-Vides, J. (1998). Extracting regulatory sites from the upstream region of yeast genes by computational analysis of oligonucleotide frequencies. *J. Mol. Biol.* **281,** 827–842.

van Loo, P., and Marynen, P. (2009). Computational methods for the detection of cis-regulatory modules. *Brief. Bioinform.* **10,** 509–524.

van Loo, P., Aerts, S., Thienpont, B., de Moor, B., Moreau, Y., and Marynen, P. (2008). ModuleMiner—Improved computational detection of cis-regulatory modules: Are there different modes of gene regulation in embryonic development and adult tissues? *Genome Biol.* **9,** R66.

van Steensel, B., and Henikoff, S. (2000). Identification of in vivo DNA targets of chromatin proteins using tethered dam methyltransferase. *Nat. Biotechnol.* **18,** 424–428.

Vingron, M., Brazma, A., Coulson, R., Helden, J. V., Manke, T., Palin, K., Sand, O., and Ukkonen, E. (2009). Integrating sequence, evolution and functional genomics in regulatory genomics. *Genome Biol.* **10**(1), 202.

Visel, A., Minovitsky, S., Dubchak, I., and Pennacchio, L. A. (2007). VISTA Enhancer Browser—A database of tissue-specific human enhancers. *Nucleic Acids Res.* **35,** D88–D92.

Warner, J. B., Philippakis, A. A., Jaeger, S. A., He, F. S., Lin, J., and Bulyk, M. L. (2008). Systematic identification of mammalian regulatory motifs' target genes and functions. *Nat. Methods* **5,** 347–353.

Wasserman, W. W., and Fickett, J. W. (1998). Identification of regulatory regions which confer muscle-specific gene expression. *J. Mol. Biol.* **278,** 167–181.

Wasserman, W. W., and Sandelin, A. (2004). Applied bioinformatics for the identification of regulatory elements. *Nat. Rev. Genet.* **5,** 276–287.

Wasserman, W. W., Palumbo, M., Thompson, W., Fickett, J. W., and Lawrence, C. E. (2000). Human-mouse genome comparisons to locate regulatory sites. *Nat. Genet.* **26,** 225–228.

Won, K. J., Ren, B., and Wang, W. (2010). Genome-wide prediction of transcription factor binding sites using an integrated model. *Genome Biol. 22;* **11**(1), R7.

Wu, J., and Xie, J. (2010). Statistical methods in molecular biology. *Methods Mol. Biol.* **620,** 405–416.

Wyrick, J. J., and Young, R. A. (2002). Deciphering gene expression regulatory networks. *Curr. Opin. Genet. Dev.* **12,** 130–136.

Xie, X., Lu, J., Kulbokas, E., Golub, T., Mootha, V., Lindblad-Toh, K., Lander, E., and Kellis, M. (2005). Systematic discovery of regulatory motifs in human promoters and 3′ UTRs by comparison of several mammals. *Nature* **434,** 338–345.

Yuh, C. H., Bolouri, H., and Davidson, E. H. (1998). Genomic cis-regulatory logic: Experimental and computational analysis of a sea urchin gene. *Science* **279,** 1896–1902.

Zambelli, F., Pesole, G., and Pavesi, G. (2009). Pscan: Finding over-represented transcription factor binding site motifs in sequences from co-regulated or co-expressed genes. *Nucleic Acids Res.* **37,** W247–W252.

Zeitlinger, J., Zinzen, R., Stark, A., Kellis, M., Zhang, H., Young, R., and Levine, M. (2007). Whole-genome ChIP-chip analysis of Dorsal, Twist, and Snail suggests integration of diverse patterning processes in the Drosophila embryo. *Genes Dev.* **21,** 385–390.

Zhou, Q., and Wong, W. H. (2004). CisModule: De novo discovery of cis-regulatory modules by hierarchical mixture modeling. *Proc. Natl. Acad. Sci. USA* **101,** 12114–12119.

Zhu, L. J., *et al.* (2011). FlyFactorSurvey: A database of Drosophila transcription factor binding specificities determined using the bacterial one-hybrid system. *Nucleic Acids Res.* **39,** D111–D117.

Zinzen, R. P., Girardot, C., Gagneur, J., Braun, M., and Furlong, E. E. (2009). Combinatorial binding predicts spatio-temporal cis-regulatory activity. *Nature* **462,** 65–70.

CHAPTER SIX

# Transcriptional Enhancers in Ascidian Development

Wei Wang *and* Lionel Christiaen

## Contents



## Abstract

The study of *cis*-regulatory DNAs that control developmental gene expression is integral to the modeling of comprehensive genomic regulatory networks for embryogenesis. Ascidian embryos provide a unique opportunity for the analysis of *cis*-regulatory DNAs with cellular resolution in the context of a simple but typical chordate body plan. Here, we review landmark studies that have laid the foundations for the study of transcriptional enhancers, among other *cis*-regulatory DNAs, and their roles in ascidian development. The studies using ascidians of the *Ciona* genus have capitalized on a unique electroporation technique that permits the simultaneous transfection of hundreds of fertilized

Department of Biology, Center for Developmental Genetics, New York University, New York, New York, USA

*Current Topics in Developmental Biology*, Volume 98
ISSN 0070-2153, DOI: 10.1016/B978-0-12-386499-4.00006-9

eggs, which develop rapidly and express transgenes with little mosaicism. Current studies using the ascidian embryo benefit from extensively annotated genomic resources to characterize transcript models in silico. The search for functional noncoding sequences can be guided by bioinformatic analyses combining evolutionary conservation, gene coexpression, and combinations of overrepresented short-sequence motifs. The power of the transient transfection assays has allowed thorough dissection of numerous *cis*-regulatory modules, which provided insights into the functional constraints that shape enhancer architecture and diversification. Future studies will benefit from pioneering stable transgenic lines and the analysis of chromatin states. Whole genome expression, functional and DNA binding data are being integrated into comprehensive genomic regulatory network models of early ascidian cell specification with a single-cell resolution that is unique among chordate model systems.

## 1. Introduction

A central task in current developmental biology studies is to understand how tissue-specific gene expression is regulated during embryogenesis. Of utmost interest is the exquisite control of complex spatiotemporal patterns of gene expression that determine the phenomenology of development. Decades of studies using model organisms have led to a paradigm whereby gene regulatory networks control the transcription of virtually all the genes that are relevant for the development of specific cell types, tissues, and organs (Davidson *et al.*, 2002; Stathopoulos and Levine, 2005; with the possible exception of some housekeeping genes that may be expressed ubiquitously and constitutively, independently of tissue-specific gene networks). A key feature of metazoan transcriptional networks is that each gene integrates multiple inputs that define its own expression profile. These combinatorial effects are essential for the generation of increasing complexity during development (e.g., the number of different cell types). Sequence-specific DNA binding transcription factors convey these multiple inputs, often in response to extracellular signals. Therefore, a gene's transcriptional response is hardwired in sequences present in *cis*, which are bound by the upstream *trans*-acting factors and have been referred to as *cis*-regulatory systems (Arnone and Davidson, 1997). The *cis*-regulatory system of a gene controls all aspects of its spatiotemporal transcriptional pattern. Numerous developmental genes have complex pleiotropic patterns of expression, which rely on multiple *cis*-regulatory modules (CRMs) that control transcription in coherent spatiotemporal units (e.g., the different stripes of even-skipped expression in the early *Drosophila* embryo are controlled by separate enhancer modules; Fujioka *et al.*, 1999). Individual CRMs integrate the multiple inputs that define their unique transcriptional outputs by functioning as DNA-binding platforms for several distinct sequence-specific transcription factors

(Arnone and Davidson, 1997; Levine, 2010; Zinzen *et al.,* 2009). Detailed studies of CRMs, such as transcriptional enhancers, require wild type and mutant reporter constructs, some form of transgenesis to place these constructs in the context of a developing embryo, and ways to analyze the transcriptional activity of the reporter constructs with defined spatiotemporal resolution. In this regard, ascidian embryos provide several unique advantages for the study of *cis*-regulatory DNAs in a simple chordate embryo.

Ascidians, or sea squirts, are members of the clade Tunicata, the closest living relatives to the vertebrates (Delsuc *et al.,* 2006). As such, these basal chordates share several developmental and anatomical features with the vertebrates: the tadpole larvae of the ascidian *Ciona intestinalis* possess a simplified yet characteristic chordate body plan composed of only ~2600 cells, which constitute the epidermis, central nervous system, notochord, and tail muscles along with the rudiments of the adult heart, digestive organs, and gonads (Satoh, 1994). Following stereotyped early division patterns and morphogenetic movements, *Ciona* embryos develop into tailbud embryo (e.g., Fig. 6.2C) after ~10h and into swimming tadpoles within ~18h at 18°C. Ascidian embryos have fixed and well-defined early embryonic lineages for each of the above tissues (Nishida, 1987; Nishida and Satoh, 1983, 1985). These characteristics of ascidian embryos have permitted the analysis of *cis*-regulatory DNAs in a simple chordate with cellular resolution (e.g., Fig. 6.2).

The study of *cis*-regulatory DNAs in ascidian embryos began using microinjection to study the regulation of muscle *actin* genes (Hikosaka *et al.,* 1994; Kusakabe *et al.,* 1995) and was sparked by the application of a simple electroporation method that was used to generate transient transgenic embryos and identify a minimal enhancer for *Brachyury* expression in the notochord (Corbo *et al.*, 1997). The notochord is arguably the most prominent chordate synapomorphy, and *Brachyury/T* orthologs are conserved markers and determinants of notochord fate specification and development in the early ascidian embryo (Chiba *et al.,* 2009; Takahashi *et al.,* 1999; Yasuo and Satoh, 1993). Understanding the regulation of *Brachyury* in ascidian embryos thus held the promises of shedding light on conserved mechanisms for notochord development and evolution in chordates. The key innovation that allowed rapid progress in studying the *Ci-Bra* (*C. intestinalis Brachyury*) enhancer was the development of simultaneous plasmid DNA electroporation in hundreds of fertilized eggs (Corbo *et al.*, 1997). This technique greatly facilitates the generation of "transient" transgenic embryos, which display limited mosaicism because, even though the DNA is not integrated in the genome, it is distributed among only the few hundred cells that constitute the early ascidian embryo (Christiaen *et al.*, 2009b; Corbo *et al.,* 2001). Using electroporation of reporter gene constructs, researchers have characterized hundreds of *cis*-regulatory DNAs to various extends. Most of the documented *cis*-regulatory elements identified in ascidians can be found on the ANISEED (Tassy *et al.,* 2010) and/or DBTGR (http://dbtgr.hgc.jp/; Sierro *et al.,* 2006) databases.

Here, we will review the current methods for analyzing *cis*-regulatory DNAs in ascidians, with a special focus on *C. intestinalis* which has been more extensively used than other species.

## 2. Identifying the Landmarks of Transcriptional Units

In order to study the *cis*-regulatory DNA of any given gene expressed during development, one must first identify the key landmarks of the transcript's structure on the genomic DNA sequence (e.g., Fig. 6.1). Ideally,

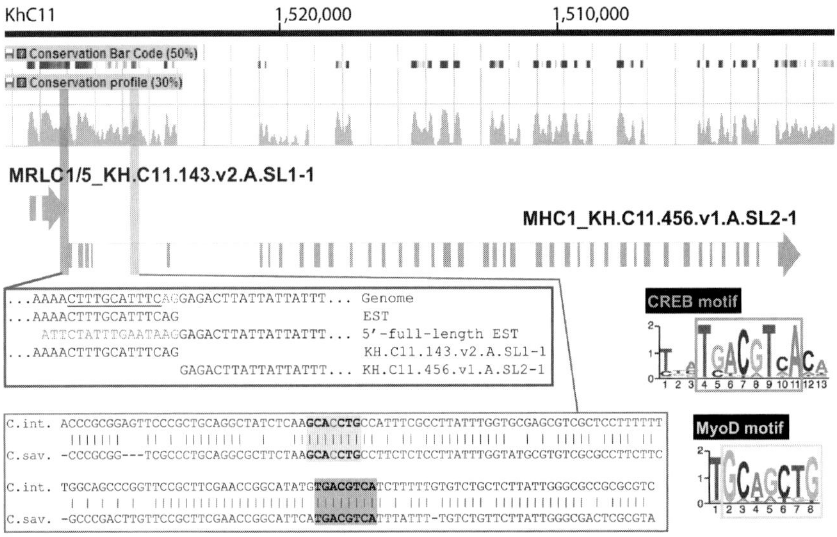


**Figure 6.1** Genomic organization of the MRLC1/5-MHC1 operon. Top bar: 30-kbp region extending from positions 1,500,000 to 1,530,000 of the KhC11 scaffold and encoding the KH transcript models for *myosin regulatory light chain 1/5* and *myosin heavy chain 1*. Green boxes indicate exons, green solid arrows indicate the last exons. The last exon of *MRLC1/5* abuts the first exon of *MHC1* (blue box). This is typical of an operon. The junction contains a splice acceptor site (red AG) preceded by a polypyrimidine tract (underlined) and the 5′ full-length EST, for *MHC1* contains a 5′ SL leader sequence (red) indicative of SL *trans*-splicing. Conservation profile and bars are from the corresponding region on the ANISEED genome browser. Note that only fragments of the whole region could be aligned. A conserved noncoding region is located in the fourth intron of MHC1 (red box). This sequence contains short, conserved blocks almost identical to the over-represented motifs identified upstream of muscle genes by Kusakabe *et al.* (2004) and Johnson *et al.* (2005). These motifs correspond to possible binding sites for CREB (orange box) and MRF/MyoD (yellow box). The figure was modified in part from Satou *et al.* (2008) and from Kusakabe *et al.* (2004). (For interpretation of the references to color in this figure legend, the reader is referred to the Web version of this chapter.)

these include the transcriptional start site (TSS), the translation start site (ATG), exon–intron boundaries, untranslated regions (5′ UTR and 3′ UTR), and possible alternative isoforms. Except for the identification of TSSs, several of these tasks are greatly facilitated for both *Ciona* species by the wealth of sequence data obtained over the past 10 years. The first version of the *C. intestinalis* genome was published in 2002 (Dehal *et al.*, 2002), shortly followed by the genome of *Ciona savignyi* (http://www.broad.mit.edu/annotation/ciona; Hill *et al.*, 2008; Small *et al.*, 2007; Vinson *et al.*, 2005). These two reference sequences served as templates for virtually every subsequent *cis*-regulatory analysis in Ciona.

## 2.1. Transcript models in Ciona

Comprehensive EST datasets obtained by the accumulation of cDNA sequences through several rounds of large-scale sequencing projects permitted the identification and annotation of transcribed regions (Satou *et al.*, 2002a, 2008). Most importantly, these invaluable resources are readily accessible via web-based databases, including the popular ANISEED (http://www.aniseed.cnrs.fr/; Tassy *et al.*, 2010) and Ghost databases (http://ghost.zool.kyoto-u.ac.jp/cgi-bin/gbrowse/kh/; Satou *et al.*, 2005); a comprehensive list of all current databases specifically dedicated to tunicates can be found on the Tunicate Portal, http://www.tunicate-portal.org/). The ANISEED and Ghost databases provide access to the genomic sequence annotated by mapping transcripts models, which have been defined using extensive expression data (Matsumoto *et al.*, 2010; Satou *et al.*, 2002c, 2006, 2008). Currently, the 115.2-Mbp genome of *C. intestinalis* is predicted to contain 15,254 gene loci, which encode 24,025 transcripts (KH transcript models, e.g., Fig.6.1; Satou *et al.*, 2008).

## 2.2. Operons and SL *trans*-splicing

The excess of transcripts models compared to the number of predicted genes can be explained by the existence of alternative transcript isoforms, but 19% (2909/15,254) of the *C. intestinalis* genes are grouped in 1310 operons, which can generate 4248 distinct mRNAs, including alternative isoforms (Satou *et al.*, 2008). The existence of operons must be taken into consideration when designing strategies for *cis*-regulatory DNA analyses since genes within a given operon will presumably be transcribed from the same basal promoter, even though the *Ciona* genes organized in operons do not seem to be strictly coregulated (Zeller, 2010).

Polycistronic pre-mRNAs derived from operons are thought to be resolved by spliced leader (SL) RNA *trans*-splicing (Satou *et al.*, 2006, 2008; e.g., Fig. 6.1). In tunicates, SL *trans*-splicing had been described for the ascidians *C. intestinalis* (Vandenberghe *et al.*, 2001), *C. savignyi*

(Yeats *et al.*, 2010), and *Halocynthia roretzi* (Yuasa *et al.*, 2002) and for the appendicularian *Oikopleura dioica* (Ganot *et al.*, 2004). During SL *trans*-splicing in *C. intestinalis*, a 16-nucleotide leader sequence is added to a splice acceptor site in lieu of the outron within the 5′ UTR of a pre-mRNA (Fig. 6.1; Vandenberghe *et al.*, 2001). The 16-nucleotide 5′ leader sequence originates from common 46 nucleotide donors, the SL RNAs, which are present in multiple copies in the *Ciona* genome (Vandenberghe *et al.*, 2001; Yeats *et al.*, 2010). Recent genome-wide high-throughput sequence analyses have provided quantitative insights into SL *trans*-splicing usage during gene expression in *C. intestinalis* (Matsumoto *et al.*, 2010; Satou *et al.*, 2006). In the latest study, the authors used random-primed SL PCR to create a library enriched with cDNAs derived from SL *trans*-spliced mRNAs and obtained nearly 250,000 sequence reads using the 454 technology (Matsumoto *et al.*, 2010). These data indicate that, in *C. intestinalis* tailbud embryos, ~58% of the total complement of genes (8790/15,254) are expressed and undergo SL *trans*-splicing, which is consistent with previous estimates that about 50% or more of the expressed genes are *trans*-spliced in *Ciona* (Satou *et al.*, 2006). This data clearly indicates that the majority of transcripts in *Ciona* are SL *trans*-spliced. The proportion of genes undergoing SL *trans*-splicing is notably higher than the proportion of genes present in operons (~58% vs. ~20%), which indicates that the resolution of polycistronic RNAs is only one of the possible functions of SL *trans*-splicing in tunicates. Other proposed roles of SL *trans*-splicing include the "sanitization" of 5′ UTRs from possibly detrimental short-open reading frames and other out-of-frame initiation codons (Hastings, 2005).

A natural consequence of widespread SL *trans*-splicing for *cis*-regulatory DNA analyses in tunicates is that prediction of the TSS from cDNA 5′ end sequence data is hindered by the replacement of the initial outron with the 5′ leader oligonucleotide. Khare *et al.* recently used two methods that circumvent this limitation to identify the TSS of the *C. intestinalis Troponin I* gene (Khare *et al.*, 2011). The first method uses modified transgenes where the splice acceptor sites have been mutated to prevent SL *trans*-splicing, followed by 5′ RACE analysis to identify the TSS. The second approach uses sequences of the rare non-SL *trans*-spliced transcripts obtained from high throughput 5′ end sequencing data. Both modified *Troponin I transgenes* and the sequencing data pointed to a TSS located 523bp upstream from the translation initiation codon, while the SL leader is inserted 64 nucleotides upstream of the AUG in the mature *trans*-spliced mRNAs (Vandenberghe *et al.*, 2001; Yeats *et al.*, 2010). A 48-bp region surrounding the TSS contains a TATA box and is required for core promoter activity. Interestingly, the popular basal promoter prediction algorithm NNPP (http://www.fruitfly.org/seq_tools/promoter.html; Reese, 2001) precisely identified the experimentally determined TSS within a 437-bp region in *Troponin I cis*-regulatory DNA.

In summary, the wealth of genomic resources accumulated over the past 10 years and their accessibility via integrated web-interfaced databases greatly facilitate the characterization of virtually all protein-coding transcripts, as a prerequisite to *cis*-regulatory DNA analyses.

## 3. CRMs for Developmental Gene Expression

With 15,254 genes distributed over 115.2 Mbp, the Ciona genome is rather compact (i.e., 1 gene every 7.5 kb on average, compared to 1 gene every 89.6 kb on the human chromosome 16; Martin *et al.,* 2004). As a consequence of this compactness, functional *cis*-regulatory elements could be identified within only 3 to 4 kb of noncoding DNA located upstream of the translation initiation codon (e.g., Corbo *et al.*, 1997; Di Gregorio and Levine, 1999). Taking advantage of the ease of electroporation and the compact *Ciona* genome, Harafuji *et al.* initiated an unbiased genome-wide search for *cis*-regulatory DNAs with enhancer activity (Harafuji *et al.,* 2002). The authors tested the activity of 138 randomly chosen genomic DNA fragments, of 1.7 kb on average, placed upstream of a basal promoter and the reporter gene *lacZ.* This preliminary screen identified 11 fragments that could enhance transcription, including five or more *bona fide* CRMs (Table 6.1; Harafuji *et al.,* 2002). This study pointed to a density of one

**Table 6.1** Summary of systematic enhancer screens performed in *Ciona*

| Filters | Fragments tested | Size (kbp) | Positives | *cis*-regulatory elements | References |
|---|---|---|---|---|---|
| Random | 138 | 1.7 | 11 (8.0%) | 5 (3.6%) | Harafuji *et al.* (2002) |
| *Hox* loci | 222 | ~3 | 29 (13.1%) | 21 (9.5%) | Keys *et al.* (2005) |
| Gene coexpression+ motifs+ conservation | 23 | <2 | 7 (30.4%) | 6* (26.1%) | Johnson *et al.* (2005) |
| Neighboring gene expression+ motifs+ conservation | 23 | <1 | 10 (43.5%) | 10 (43.5%) | Haeussler *et al.* (2010) |
| Motifs+ conservation | 20 | <0.2 | 7 (35%) | 4 (25%) | Khoueiry *et al.* (2010) |

* In this study, one more enhancer was identified but did not drive the expected reporter gene expression in the muscles.

enhancer every 20–40kb in the *Ciona* genome. In a similar study focused on the *Hox* genes, the authors tested the activity of 222 partly overlapping ~3kb fragments that were more densely distributed over >200kb of genomic DNA (Keys *et al.*, 2005). This study identified 21 separate *cis*-regulatory elements. The estimated density of about one enhancer every 10kb suggests that the proportion of false negatives in enhancer screens can be reduced by testing overlapping constructs. These studies indicate that unbiased searches for enhancers in the *Ciona* genome can identify functional elements in about 10% or less of the tested constructs (Table 6.1). In the following sections, we discuss the computational methods that have permitted the use of additional expression and sequence data to guide *cis*-regulatory DNA analyses in ascidians.

### 3.1. Evolutionary conservation of functional *cis*-regulatory elements

Evolutionary conservation (aka phylogenetic footprinting) is a popular guide used to identify functional *cis*-regulatory DNAs from multispecies genome sequence alignments (e.g., Tagle *et al.*, 1988; Pennacchio *et al.*, 2006; Fig. 6.1). As mentioned above, whole genome sequences are available for two related ascidian species, *C. intestinalis* and *C. savignyi*. Early developmental events in these two species are extremely similar and largely interchangeable, in as much as forced hybrids can develop to the tadpole stage (Byrd and Lambert, 2000), but sequence divergence is such that only evolutionarily constrained and presumably functional elements are conserved (Johnson *et al.*, 2004). Pre-computed whole-genome alignments highlighting "islands of conservation" between the two *Ciona* genomes can be explored on browsers and databases such as VISTA (http://pipeline.lbl.gov/cgi-bin/gateway2?bg=cin1&selector=vista; Frazer *et al.*, 2004; Fig. 6.1). Several studies have used evolutionary conservation to identify CRMs controlling the expression of genes encoding transcription factors or signaling molecules (e.g., *Pitx* (Christiaen *et al.*, 2005), *sFRP1/5* (Lamy *et al.*, 2006), *FoxF* (Beh *et al.*, 2007), *ROR* (Auger *et al.*, 2009), *Lef/Tcf* (Squarzoni *et al.*, 2011)). In an attempt to evaluate the relationship between sequence conservation and *cis*-regulatory activity, Johnson *et al.* determined that the functional upstream sequences of eight genes expressed in the muscles, nervous system, notochord, epidermis, and/or endoderm show higher levels of sequence conservation than presumably nonfunctional regions (Johnson *et al.*, 2004). Sequence conservation is not evenly distributed within 300–500bp elements; instead, short blocks of perfectly conserved sequence alternate with divergent regions suggesting that individual transcription factor binding sites (TFBSs) themselves are more likely to be conserved (e.g., Fig. 6.1). In keeping with this hypothesis, a detailed analysis of the *Troponin I* *cis*-regulatory DNA identified short stretches of conserved sequences that concentrate the enhancer activity of the *cis*-regulatory DNA and correspond to candidate binding sites for a possible upstream activator, the myogenic basic helix-loop-helix (bHLH) protein Ci-MRF, a MyoD/Myf ortholog required for muscle development in *Ciona* (Johnson *et al.,* 2004; Meedel *et al.,* 2007; Fig. 6.1).

Another venue to identify candidate functional noncoding DNA computationally exploits the high-allelic polymorphism of *Ciona* genomes (1.2% in a single individual on average; Dehal *et al.,* 2002). Using intraspecies sequence comparisons, Boffelli *et al.* found that noncoding regions with low-mutation rates tend to act as tissues-specific enhancers (Boffelli *et al.,* 2004).

## 3.2. TFBS compositions define CRMs

The identification of functional TFBSs within transcriptional enhancers is essential for *cis*-regulatory DNA analyses. The short sequences corresponding to TFBSs have been referred to as sequence motifs (e.g., Haeussler and Joly, 2011). The above-mentioned methods aimed at identifying CRMs without preconception of their motif composition. Several studies have used an approach that relies on gene coexpression, with or without complementary phylogenetic footprinting, in order to identify overrepresented motifs that correlate with tissue-specific gene expression (Haeussler *et al.,* 2010; Johnson *et al.,* 2005; Kusakabe *et al.,* 2004). The rationale is that tissue-specific combinations of DNA-binding transcription factors, the binding sites of which will be overrepresented in functional enhancers, control cohorts of coregulated genes. Following large scale *in situ* hybridization screens (Kusakabe *et al.,* 2002; Satou *et al.*, 2001b, 2002a), Kusakabe and colleagues selected 50 genes expressed specifically in the photoreceptors, central nervous system, or skeletal tail muscles of the *Ciona* embryos and larvae (Kusakabe *et al.,* 2004; see also DBTGR http://dbtgr.hgc.jp/; Sierro *et al.,* 2006). One thousand base pairs of 5′ flanking sequences for each gene were searched for overrepresented motifs within each coexpression group. This approach identified six unique motifs that were overrepresented upstream of muscle-specific genes, including candidate binding sites for a CREB basic leucine zipper (bZip) protein, and MRF transcription factors. Functional analysis of the *myosin light chain 2 (MLC2)* and *myosin regulatory light chain 2 (MRLC2)* upstream sequences showed that the candidate CREB binding sites are required for reporter gene expression in the primary muscles (Kusakabe *et al.,* 2004; see also Fig. 6.1). Johnson *et al.* (2005) used a similar approach to identify motifs overrepresented in the *cis*-regulatory DNAs of muscle differentiation genes. The authors first performed multiple rounds of truncation/transfection of reporter constructs to identify the minimal *cis*-regulatory element(s) for each one of the 20 conserved muscle genes selected. This preliminary analysis restricted the search for overrepresented motifs to 6050 bp of functional *cis*-regulatory sequences out of the initial 23,620bp of total noncoding DNA included in the analysis. Most notably, the authors identified three overrepresented motifs within this dataset, including candidate binding sites for the same CREB and MRF factors (Johnson *et al.*, 2005). In a separate study, these motifs were showed to contribute significantly to enhancer activity in reporter gene assays (Brown *et al.*, 2007). Johnson *et al.* then searched the genome for CRMs containing variable combination of the three motifs. They identified 269 candidate modules containing at least one conserved >20bp element showing >75% identity between *C. intestinalis* and *C. savignyi* structural muscle genes. Almost 30% of the tested elements drove reporter gene in the muscles, indicating that combinations of CREB, MRF, and/or third orphan motif could constitute a *cis*-regulatory signature for enhancer activity in the muscle precursor cells (Table 6.1; Fig. 6.1; Johnson *et al.*, 2004).

The studies by Kusakabe *et al.* and Johnson *et al.* both used the two *Ciona* species in their analysis, which permits an evaluation of the evolutionary conservation of *cis*-regulatory motifs for coexpressed genes. Kusakabe *et al.* "found no apparent similarity in the nucleotide sequences of upstream regions and introns between the orthologs of the two species", while Johnson found twice more conserved elements (CE; >75% identity over >20bp) in functional elements than expected by chance, and these "positive elements" showed 52% identity over 293bp on average, which is also higher than the genome-wide background. This apparent conundrum indicates that cryptic conservation may be difficult to detect using standard alignments methods but could still be relevant for phylogenetic footprinting of *cis*-regulatory elements.

An in-depth analysis of the above muscle enhancers provided further insights into the relationship between CRM architecture, motifs composition, evolutionary constraints, and enhancer activity (Brown *et al.*, 2007). In an experimental and analytical "tour de force", Brown *et al.* estimated the enhancer activity of each one of the 77 individual CREB, MRF, and Tbx6 motifs present in 19 muscle enhancers derived from *C. intestinalis* and *C. savignyi* genes. The quantitative analysis demonstrates that CRMs from different genes were highly variable in terms of composition, arrangement, and enhancer activity of individual motifs. However, orthologous functional motifs proved to be highly constrained, showing >79% identity between the two *Ciona* species compared to <20% of the genome-wide background. These studies underscore the functional and evolutionary importance of individual *cis*-regulatory motifs that compose transcriptional enhancers following rather flexible rules.

The above studies have characterized transcriptional enhancers for restricted coexpression of tissue-specific differentiation genes. These could arguably be subjected to simpler *cis*-regulatory logics than pleiotropic and transiently expressed developmental control genes. For example, single motifs determine neuron-specific expression of differentiation gene batteries in

*Caenorhabditis elegans* (Flames and Hobert, 2009; Wenick and Hobert, 2004), while multiple enhancers determine the complex pattern of even-skipped expression in the early *Drosophila* embryo (Fujioka *et al.,* 1999). Studies using *Ciona* have aimed at identifying *cis*-regulatory rules that defined enhancer activity in transient and complex embryonic territories such as the neurectoderm (Haeussler *et al.,* 2010; Khoueiry *et al.,* 2010). Haeussler *et al.* set out to identify the *cis*-regulatory elements required to respond to a conserved selector gene for the anterior neurectoderm, the Paired-class K50 homeodomain protein Otx. Analysis of a prototypical enhancer that governs *Pitx* gene expression in the anterior neural boundary (ANB), a subset of the anterior neurectoderm, indicated that a minimal functional enhancer needs at least two copies of three different motifs, including the Otx-binding GATTA pentamer, within 125-bp windows (Christiaen *et al.,* 2005; Haeussler *et al.,* 2010). The authors then compiled all genes with known expression in any part of the anterior neurectoderm and searched for duplicate instances of any of the 512 possible pentamers within 125 bp of conserved noncoding DNA. This analysis identified pairs of GATTA motifs as the only significant *cis*-regulatory signature for anterior neurectodermal genes. This signature was found in 53 conserved noncoding elements, and 10 of the 23 tested fragments drove reporter gene expression in parts of the anterior neurectoderm (Table 6.1; Haeussler *et al.,* 2010). These data suggest that duplicate GATTA motifs within 125 bp are sufficient to render a *cis*-regulatory element competent to respond to Otx activating inputs. However, GATTA motifs alone were not sufficient to drive reporter gene expression; moreover, detailed analysis of the *Pitx* ANB enhancer and the variety of patterns driven by newly identified anterior neurectodermal enhancers indicate that additional TFBSs within *cis*-regulatory elements are required to determine the exquisite precision of spatiotemporal expression patterns in a combinatorial manner.

In a comparable study, Khoueiry *et al.* (2010) determined that the minimal enhancer for Otx activation in the a- and b-line neural lineages requires two GATA and two ETS binding sites without further identifiable architectural constraints (Bertrand *et al.,* 2003; Khoueiry *et al.,* 2010). A genome-wide search for conserved elements containing the above motifs identified 55 candidate CRMs, 25% (4/20) of which could drive reporter gene expression in the a- and/or b-line neural lineages (Table 6.1; Khoueiry *et al.,* 2010). These active elements had low nucleosome occupancy, as computationally predicted using a dinucleotide signature. The authors then documented a significant inverse correlation between predicted nucleosome occupancy and enhancer activity for 79 additional tissue-specific enhancers identified in *Ciona* (Khoueiry *et al.,* 2010). These data suggest that a low affinity for nucleosomes is required together with the appropriate assembly of TFBSs to determine the enhancer activity of CRMs.

Detailed analyses of the Otx neural enhancer in distant ascidian species emphasized the functional importance of individual motifs and the flexibility of design rules for CRMs (Oda-Ishii *et al.,* 2005). Early a- and b-line neural expression of Otx is conserved in the ascidian *H.* roretzi (Hudson and Lemaire, 2001; Wada *et al.,* 1996). The *Hr-Otx* (*H. roretzi Otx*) a- and b-line neural enhancer does not show clear sequence conservation with the orthologous enhancer from *C. intestinalis* but, strikingly, the enhancer from one species can drive reporter gene expression in the a-neural blastomeres of the other species, and vice versa (Oda-Ishii *et al.,* 2005). Both enhancer modules contained GATA and Ets binding sites, which are critical for *Otx* inducion by FGF signaling (Bertrand *et al.,* 2003), but the TFBSs differ in numbers, distances, and relative orientations in the two orthologous enhancers. The authors also showed that multiple copies of Fox, Lhx, and Tbx binding sites were required for the enhancer activity of A- and B-line specific modules in both species, but each module has a distinct *cis*-regulatory architecture (i.e., composition and organization of motifs). This study further reinforces the notion that simple combinations of individual motifs bear most of the functional constraints that shape enhancer evolution.

## 4. Stable Transgenic Lines and Enhancer Traps Identify Novel *cis*-Regulatory DNAs

In most model systems, the study of *cis*-regulatory DNAs requires the generation of stable transgenic lines. Transgenes usually insert randomly in the genome, with some exceptions in *Drosophila* (Groth *et al.,* 2004) or mice (Belteki *et al.,* 2003). As mentioned above, plasmid DNA is not normally integrated in the *Ciona* genome following electroporation (stable transgenic lines can be obtained by electroporation only if a transposase is coelectroporated and the transgene contains the appropriate sites for integration by transposition, see below and Matsuoka *et al.,* 2005). It is not clear at present what chromatin-like structure self-organizes at the "extra-chromosomal arrays" that are thought to contain multiple copies of plasmid DNA per cell and permit transcription in a tissue-specific manner. Because of this limitation, the influence of chromatin structure on transgene expression cannot be readily investigated using electroporation.

Two distinct methods have been developed that permit the generation of transgenic lines with stable insertion of transgenes in *C. intestinalis* (Sasakura *et al.,* 2003) and *C. savignyi* (Deschet *et al.,* 2003; Matsuoka *et al.,* 2004). One of the two methods uses the Tc1/mariner class transposon Minos and characterization of the first transgenic lines identified an enhancer-trap event, where transgene expression was controlled by the *cis*-regulatory elements responsible for expression of a Musashi homolog in the endostyle and peribranchial bands (Awazu *et al.,* 2004). Since this original line was created using transposon-mediated transgenesis, the authors undertook a large-scale enhancer trap screen by remobilizing the Minos transposon (Awazu *et al.,*

2007). The enhancer trap vector used contains the Ci-TPO promoter upstream of *gfp*, flanked by ITRs for transposition by Minos and I-sceI sites for insertion using meganuclease-mediated transgenesis (Awazu *et al.*, 2007; Deschet *et al.*, 2003). Direct insertion of this construct using transposon-mediated transgenesis yielded six enhancer-trap lines out of 111 F1 families (5.4%). By contrast, remobilization of Minos within a stable line generated 19 lines with novel GFP expression pattern out of 24 F1 families (79%), thus demonstrating the efficacy of transposon remobilization for enhancer trapping in *Ciona*. In this study, the authors microinjected transposase mRNAs in transgenic eggs to remobilize the Minos transposon. An alternative method consisting in transposase expression in transgenic gametes greatly facilitated the generation of "remobilized" enhancer trap lines displaying novel GFP expression patterns (Hozumi *et al.*, 2010; Sasakura *et al.*, 2008).

In the above studies, the vast majority of enhancer trap lines showed expression in juveniles and/or adults but only 2 out of 30 lines showed GFP expression in larvae (Awazu *et al.*, 2007; Sasakura *et al.*, 2008). These observations could be due to the specific Ci-TPO promoter that was used (Sasakura *et al.*, 2008), but they also open the possibility that specific mechanisms silence transgene expression during early development. In keeping with this possibility, Sasakura and colleagues identified maternal gfp/gene silencing (MGS). The molecular mechanism for MGS remains elusive but seems to specifically affect the production and/or stability of *gfp* mRNAs produced from integrated transgenes during early development of eggs derived from a single specific transgenic line (Sasakura *et al.*, 2010).

## 5. Chromatin Studies in Tunicates

The studies indicating nucleosome depletion at active *cis*-regulatory elements or the discovery of MGS point to the roles of chromatin modifications in developmental gene regulation in ascidians. The recent development of methods for generating and culturing transgenic ascidians shall open new research opportunities to investigate the structure and function of chromatin in tunicate development, which remain largely unexplored.

The main efforts to study chromatin structure in ascidians have aimed at characterizing DNA methylation at CpG islands (Simmen *et al.*, 1999; Suzuki *et al.*, 2007; Tweedie *et al.*, 1997). Global estimates using methylation-sensitive restriction enzymes indicate that ~25% of the CpG islands are methylated in the *C. intestinalis* genome (Tweedie *et al.*, 1997). Detailed characterization of the methylation patterns in ~100 kb of genomic DNA indicated that transposable elements are not preferred targets for CpG methylation (Simmen *et al.*, 1999). Instead, this and whole genome studies showed that transcriptional units are preferred sites for DNA methylation, even though they are depleted in CpG islands (Simmen *et al.*, 1999; Suzuki *et al.*, 2007; Tweedie *et al.*, 1997; Wang and Leung, 2008). Suzuki *et al.* also reported that the most highly expressed genes were overrepresented in the unmethylated fraction of genes. Taken together, these studies support the idea that CpG methylation of transcribed genes in *Ciona* prevents spurious transcription initiation with potentially deleterious consequences.

Molecular and cellular analyses of chromatin and nuclear properties have elucidated essential roles for histone and RNA polymerase II covalent modifications as well as nuclear architecture for the developmental gene regulation (e.g., reviewed in Berger, 2007; Brookes and Pombo, 2009; Geyer *et al.*, 2011). A single study investigated the nuclear distribution of histone marks in the endoreplicating epidermal cells of the appendicularian *O. dioica*, but domains of specific histone modifications did not seem to correlate with the subnuclear localization of otherwise active transcriptional units (Spada *et al.*, 2007).

It seems likely that conserved repressive and productive histone and RNA polymerase II modifications will exert similar functions in tunicates as those described in other organisms. For example, in animal embryos, early embryonic germline progenitors are prevented from differentiating inappropriately into somatic cells by global repression of zygotic transcription. This germline-specific transcriptional quiescence is associated with an absence of phosphorylation at the serine 2 of the RNA polymerase II C-terminal domain (CTD-Ser2), which depends on the Cdk9-containing P-TEFb complex and occurs during transcriptional elongation (Ahn *et al.*, 2004; Hanyu-Nakamura *et al.*, 2008; Seydoux and Dunn, 1997). This transcriptional inhibition depends on the species-specific maternal factors PIE-1 in *C. elegans* and Pgc in *Drosophila* (Ghosh and Seydoux, 2008; Hanyu-Nakamura *et al.*, 2008). Recently, posterior end mark (PEM), an ascidian-specific localized factor, has been shown to inhibit early zygotic transcription specifically in the germline precursors of *C. intestinalis* and *H. roretzi* (Kumano *et al.*, 2011; Shirae-Kurabayashi *et al.*, 2011). The *PEM* mRNA is localized to the posterior vegetal cortex in the fertilized egg and is segregated during asymmetric cleavages specifically into the germline precursor blastomeres (reviewed in Sardet *et al.*, 2005; Fig. 6.2A). This asymmetric distribution of the PEM mRNAs and rapid degradation permits accumulation of the protein specifically in the nuclei of germline blastomeres (Kumano *et al.*, 2011; Shirae-Kurabayashi *et al.*, 2011). Evidence was presented that a C-terminal WRPW repressor motif is required for PEM-mediated transcription repression, that *Ciona* PEM-1 can interact with the corepressors Groucho-1 and -2 (Shirae-Kurabayashi *et al.*, 2011), and that *Halocynthia* PEM can interact with P-TEFb/Cdk9 (Kumano *et al.*, 2011). Although distinct maternal factors evolved independently in *Drosophila*, *C. elegans*, and ascidians, they all seem to repress transcription in the germline via regulation of P-TEFb/Cdk9-mediated PolII CTD-Ser2 phosphorylation. In spite of these important breakthroughs, chromatin studies in tunicates are still in their infancy, and it remains to be determined how "epigenetic" regulatory inputs are integrated in tunicate gene regulatory networks during development.

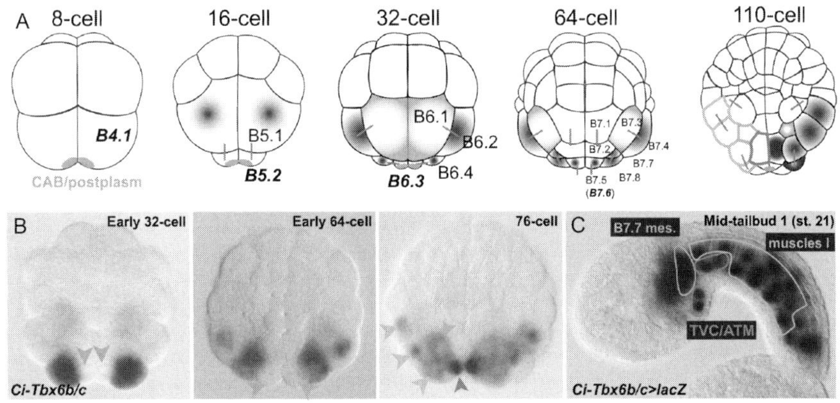


**Figure 6.2** Expression of *Tbx6b/c* and reporter construct in *Ciona intestinalis* embryos. (A) Schematic representations of vegetal views of early stage embryos showing the B-line blastomeres. Short red lines connect sister cells following cleavages. The blue staining represents *Tbx6b/c* transcripts as detected by *in situ* hybridization (panel B, see also Takatori *et al.*, 2004; Yagi *et al.*, 2005). The green ellipses represent the posterior vegetal cortical cytoplasm or postplasm and centrosome attracting body (CAB), where the postplasmic mRNAs localize, including macho-1 and PEM-1 (see text for details). Note that postplasmic mRNAs are segregated into the B7.6 germline precursors by the 64-cell stage (not represented on that drawing). Colored outlines on the left side of 110-cell embryo represent the prospective fate of the blastomeres [orange: mesenchyme, blue: primary muscles, red: trunk ventral cells (TVC: heart and atrial siphon muscles precursors) and anterior tail muscles (ATM)]. (B) Chromogenic *in situ* hybridizations showing *Tbx6b/c* expression at indicated stages. Note the absence of expression in cells carrying the postplasmic RNA PEM-1 (green arrowheads). Colored arrowheads indicate the blastomere outlines in A with the same colors. (C) Lateral view of a mid-tailbud 1 (stage 21) embryo electroporated with a *Tbx6b*>*lacZ* construct (Christiaen *et al.*, 2009a) and stained with X-gal (blue staining). Beta-galactosidase activity is detected in all descendants of the early blastomeres that express *Tbx6b* (outlined with the same colors as in A and B). (See Color Insert.)

# 6. From *cis*-Regulatory Analyses to Developmental Gene Regulatory Network Models

## 6.1. *Cis*-regulatory DNAs can encode the developmental logic

A fundamental property of transcriptional enhancers for metazoan embryogenesis is to encode developmental logic in *cis*-regulatory DNAs. Recent studies of gene regulation in the cardiogenic mesoderm of *Ciona* illustrate some of these basic developmental logics and gene network motifs. In *Ciona*, specification of the cardiogenic mesoderm requires activation of the bHLH transcription factor Mesp specifically in the B7.5 pair of blastomeres (Satou *et al.*, 2004). Mesp activation at the 64-cell stage requires the direct *trans*-activating inputs from Tbx6b/c, which mediate the effects of the general posterior vegetal determinant macho-1 (Davidson *et al.,* 2005; Yagi *et al.,* 2005). However, *Tbx6b/c* genes are expressed in all primary muscle progenitors (B7.4, B7.5, B7.7, and B7.8; Fig. 6.2), but only B7.5 cells turn on *Mesp* because the other Tbx6b-expressing blastomeres lack an essential coactivator of *Mesp*, the LIM homeodomain protein Lhx3, which mediates the effects of β-catenin in the vegetal hemisphere (Christiaen *et al.*, 2009a; Satou *et al.*, 2001a). *Lhx3* is expressed in all endoderm progenitors, which do not express *Mesp* because they lack Tbx6b/c (Christiaen *et al.*, 2009a; Imai *et al.*, 2004; Satou *et al.*, 2001a). Thus, B7.5 blastomeres uniquely activate *Mesp* transcription because they are the only cells to express the two necessary *trans*-activators Tbx6b/c and Lhx3b in the 64-cell embryo. This synergistic developmental logic is recapitulated on the Mesp *cis*-regulatory DNA, where partially overlapping Lhx3 and Tbx6 binding sites are required for reporter gene expression in B7.5 blastomeres (Christiaen *et al.*, 2009a).

By the tailbud stage, the B7.5 blastomeres give birth to two distinct types of cells: anterior tail muscles (ATMs) and the trunk ventral cells (TVCs), which migrate anteriorly and constitute the heart and atrial siphon muscle progenitors (Davidson and Levine, 2003; Hirano and Nishida, 1997; Satou *et al.,* 2004; Stolfi *et al.,* 2010); Fig. 6.2C). Mesp function is required for TVC specification and heart development together with an FGF signal that activates MAP Kinase (MAPK) signaling in the anterior B7.5 granddaughter cells, thus inducing the TVC fate (Davidson *et al.,* 2006; Satou *et al.,* 2004). The developmental logic whereby Mesp provides the competence to respond to an FGF signal by forming heart progenitors is recapitulated on the minimal TVC enhancer of *FoxF*, which is also required for TVC migration and is activated specifically in the TVCs in direct response to the Ets1/2-mediated FGF/MAPK signal and to an unidentified E-box binding factor (Beh *et al.*, 2007).

FoxF- and Ets-mediated FGF signaling control TVC migration in part by upregulating the expression of downstream effector genes such as *RhoDF*, which encodes a small Rho GTPase expressed in the TVCs and notochord cells (Christiaen *et al.,* 2008). The RhoDF TVC enhancer contains both Ets1/2 and FoxF binding sites, which are required for reporter gene activation. Since FoxF is also a direct target of activated Ets1/2, the three genes function in a coherent feed-forward loop, which is a simple network motif commonly found in metazoan GRNs (Davidson, 2009).

## 6.2. Toward comprehensive GRN models of early ascidian development

The in-depth analysis of *cis*-regulatory DNAs controlling developmental gene expression have laid the foundations for system's approaches aiming at deciphering comprehensive GRNs (Davidson, 2006; Davidson *et al.,* 2002; Levine and Davidson, 2005; Stathopoulos and Levine, 2005). Building comprehensive network models requires thorough experimental support for direct and indirect regulatory interactions. Such interactions have been studied in ascidian embryos using *cis*-regulatory analyses complemented by (1) gene expression studies by *in situ* hybridizations, quantitative RT-PCR and expression microarrays; (2) systematic perturbations of the function of regulatory genes using morpholino oligonucleotides (morpholinos or MOs), dominant negatives, and/or overexpression constructs; and (3) whole-genome DNA binding profiling using chromatin immunoprecipitation and tiling arrays (ChIP-chip) (Christiaen *et al.,* 2008; Imai *et al.,* 2004, 2006, 2009; Kubo *et al.,* 2010). These approaches, combined with numerous gene-specific studies, permit the modeling of hypothetical regulatory networks for the development of the nervous system, the notochord, and the muscles to name just a few. Here, for simplicity we will review some of the experimental evidence that support a GRN model for development of the primary muscles in ascidians, which form cell autonomously in the B-line of blastomeres (Deno *et al.,* 1984; Meedel and Whittaker, 1984). The main maternal determinant of primary muscle fate is the zinc finger DNA-binding transcription factor of the Zic family, macho-1 (Nishida and Sawada, 2001; Satou *et al.*, 2002b). Macho-1 proteins activate downstream target genes in the posterior vegetal hemisphere of the early ascidian embryo, including other DNA binding transcription regulators such as Tbx6a/b/c and Snail (Sawada *et al.,* 2005; Yagi *et al.*, 2004a; Figs. 6.2 and 6.3). Tbx6b and -c mediate the effects of macho-1 on muscle gene expression (Kugler *et al.,* 2010; Yagi *et al.,* 2005). In *Ciona* species, another Zic family transcription factor called ZicL can presumably rescue some of the effects of macho-1 loss of function even though their binding preferences differ substantially (Imai *et al.,* 2002; Satou *et al.*, 2002b; Yagi *et al.*, 2004b). ZicL is also required for primary muscle development, but it does not seem to be directly activated by macho-1 in the B lineage (Anno *et al.,* 2006), and ZicL is capable of upregulating *Tbx6* genes in the absence of macho-1 (Yagi *et al.,* 2005).

Comprehensive gene annotation and expression studies using *in situ* hybridization have documented the near-full complement of transcription factors and signaling molecules expressed in or near the muscle precursors (Hino *et al.,* 2003; Imai *et al.,* 2004; Miwata *et al.,* 2006; Satou *et al.*, 2003a,b; Takatori *et al.,* 2004; Wada *et al.,* 2003; Yagi *et al.,* 2003; Yamada *et al.,* 2003). The *Ciona* genome is predicted to encode 669 transcription factors, including 356 zinc finger proteins as well as canonical members of the homeodomain, bHLH, bZIP, Ets, Forkhead/Fox, HMG, T box, and nuclear receptor families. Comprehensive *in situ* hybridization screens have identified 65 transcription factors and 25 signaling molecules that are expressed zygotically in the pregastrula stage embryo (Imai *et al.,* 2004). Twenty-eight of the above transcription factors are expressed in muscle precursors between the 16- and 112-cell stage, including ZicL, Tbx6b/c, Snail, and MRF/MyoD.

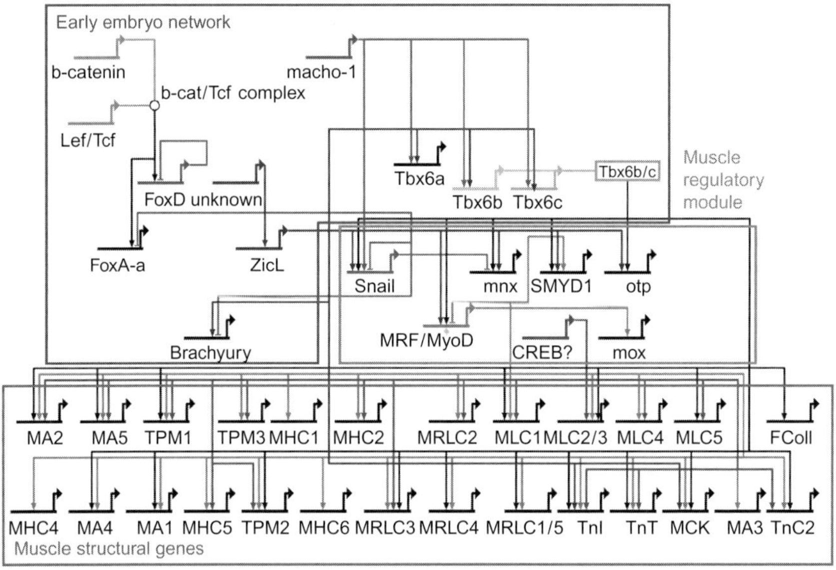


**Figure 6.3** Tentative genomic regulatory network model for the early specification of primary muscle precursors in the B-line blastomeres. This model was built using Biotapestry (Longabaugh *et al.*, 2005) and the regulatory interactions inferred from data presented in the literature including Imai *et al.* (2006), Johnson *et al.* (2005), Kubo *et al.* (2010), Kusakabe *et al.* (2004) and Yagi *et al.* (2004a,b, 2005)). Note that no regulatory connection was predicted between the unknown CREB factor and *myosin heavy chain-1 (MHC-1)* because the published analysis focused on the upstream genomic DNA, but the regulatory element may be located in the fourth intron (Fig. 6.1). (See Color Insert.)

These studies laid the foundation for extensive functional analyses combining gene knock-down by microinjection of morpholinos, whole-mount *in situ* hybridization, and quantitative RT-PCR analyses (Imai *et al.*, 2006). In this breath-taking study, the function of 27 patterned regulatory genes was systematically impaired using morpholinos, followed by *in situ* hybridizations and/or quantitative RT-PCR for 76 regulatory genes. More than 3000 combinations of gene expression and morphant background were integrated to build preliminary gene regulatory network models for the main chordate tissues specified in the early ascidian embryos.

Very few of these regulatory interactions had been documented by *cis*-regulatory DNA analyses. Kubo *et al.* combined chromatin immunoprecipitation with microarray technology (ChIP-chip) to examine the *in vivo* DNA-binding profiles of 11 zygotically expressed transcription factors controlling gene expression in the early *Ciona* embryo (Kubo *et al.*, 2010). By combining the analysis of ChIP data with previous expression profiles and functional analyses, the structure of the core networks for early *Ciona* embryogenesis was determined at the *cis*-regulatory level. The current network models suggest that most of the previously established regulatory interactions are in fact direct (Kubo *et al.*, 2010). Regarding the "muscle network", whole-genome DNA-binding data supported the regulatory interactions between ZicL, Tbx6, and MRF/MyoD and corroborated the above-mentioned bioinformatics analyses by showing that both MRF/MyoD and Tbx6b bind directly to the enhancer regions of structural muscle genes, but the predicted CREB-family transcription factor remains elusive (Fig. 6.3).

## 7. Concluding Remarks and Future Directions

In summary, *cis*-regulatory DNA analyses conducted in ascidians have characterized *bona fide* transcriptional enhancers of developmental gene expression with unprecedented cellular resolution in chordates. These studies have repeatedly highlighted the importance of simple combinations of specific motifs in determining the precise spatiotemporal patterns of developmental gene expression. In combination with large scale functional and expression data, *cis*-regulatory analyses are now allowing the elaboration of comprehensive GRN models for the formation of a simplified version of the chordate body plan.

In future studies, the use of stable transgenic lines and the application of state-of-the-art methods for the analysis of histone and RNA polymerase modifications will illuminate the influence of nuclear dynamics on the deployment of tissue-specific GRNs.

## *ACKNOWLEDGMENTS*

Our work is supported by grants 10SDG4310061 from the American Heart Association, R01GM096032 from NIGMS/NIH, R01HL108643 from NHLBI/NIH, by the New York Cardiac Center, and by the New York University College of Arts and Sciences.

## *REFERENCES*

Ahn, S. H., Kim, M., and Buratowski, S. (2004). Phosphorylation of serine 2 within the RNA polymerase II C-terminal domain couples transcription and 3′ end processing. *Mol. Cell* **13**(1), 67–76.

Anno, C., Satou, A., and Fujiwara, S. (2006). Transcriptional regulation of ZicL in the Ciona intestinalis embryo. *Dev. Genes Evol.* **216**(10), 597–605.

Arnone, M. I., and Davidson, E. H. (1997). The hardwiring of development: Organization and function of genomic regulatory systems. *Development* **124**(10), 1851–1864.

Auger, H., Lamy, C., Haeussler, M., Khoueiry, P., Lemaire, P., and Joly, J. S. (2009). Similar regulatory logic in Ciona intestinalis for two Wnt pathway modulators, ROR and SFRP-1/5. *Dev. Biol.* **329**(2), 364–373.

Awazu, S., Sasaki, A., Matsuoka, T., Satoh, N., and Sasakura, Y. (2004). An enhancer trap in the ascidian Ciona intestinalis identifies enhancers of its Musashi orthologous gene. *Dev. Biol.* **275**(2), 459–472.

Awazu, S., Matsuoka, T., Inaba, K., Satoh, N., and Sasakura, Y. (2007). High-throughput enhancer trap by remobilization of transposon Minos in Ciona intestinalis. *Genesis* **45**(5), 307–317.

Beh, J., Shi, W., Levine, M., Davidson, B., and Christiaen, L. (2007). FoxF is essential for FGF-induced migration of heart progenitor cells in the ascidian Ciona intestinalis. *Development* **134**(18), 3297–3305.

Belteki, G., Gertsenstein, M., Ow, D. W., and Nagy, A. (2003). Site-specific cassette exchange and germline transmission with mouse ES cells expressing phiC31 integrase. *Nat. Biotechnol.* **21**(3), 321–324.

Berger, S. L. (2007). The complex language of chromatin regulation during transcription. *Nature* **447**(7143), 407–412.

Bertrand, V., Hudson, C., Caillol, D., Popovici, C., and Lemaire, P. (2003). Neural tissue in ascidian embryos is induced by FGF9/16/20, acting via a combination of maternal GATA and Ets transcription factors. *Cell* **115**(5), 615–627.

Boffelli, D., Weer, C. V., Weng, L., Lewis, K. D., Shoukry, M. I., Pachter, L., Keys, D. N., and Rubin, E. M. (2004). Intraspecies sequence comparisons for annotating genomes. *Genome Res.* **14**(12), 2406–2411.

Brookes, E., and Pombo, A. (2009). Modifications of RNA polymerase II are pivotal in regulating gene expression states. *EMBO Rep.* **10**(11), 1213–1219.

Brown, C. D., Johnson, D. S., and Sidow, A. (2007). Functional architecture and evolution of transcriptional elements that drive gene coexpression. *Science* **317**(5844), 1557–1560.

Byrd, J., and Lambert, C. C. (2000). Mechanism of the block to hybridization and selfing between the sympatric ascidians Ciona intestinalis and Ciona savignyi. *Mol. Reprod. Dev.* **56**(4), 541.

Chiba, S., Jiang, D., Satoh, N., and Smith, W. C. (2009). Brachyury null mutant-induced defects in juvenile ascidian endodermal organs. *Development* **136**(1), 35–39.

Christiaen, L., Bourrat, F., and Joly, J. S. (2005). A modular cis-regulatory system controls isoform-specific pitx expression in ascidian stomodaeum. *Dev. Biol.* **277**(2), 557–566.

Christiaen, L., Davidson, B., Kawashima, T., Powell, W., Nolla, H., Vranizan, K., and Levine, M. (2008). The transcription/migration interface in heart precursors of Ciona intestinalis. *Science* **320**(5881), 1349–1352.

Christiaen, L., Stolfi, A., Davidson, B., and Levine, M. (2009a). Spatio-temporal intersection of Lhx3 and Tbx6 defines the cardiac field through synergistic activation of Mesp. *Dev. Biol.* **328**(2), 552–560.

Christiaen, L., Wagner, E., Shi, W., and Levine, M. (2009b). Electroporation of transgenic DNAs in the sea squirt Ciona. *Cold Spring Harb. Protoc.* **2009**(12), pdb.prot5345.

Corbo, J. C., Levine, M., and Zeller, R. W. (1997). Characterization of a notochord-specific enhancer from the Brachyury promoter region of the ascidian, Ciona intestinalis. *Development* **124**(3), 589–602.

Corbo, J. C., Di Gregorio, A., and Levine, M. (2001). The ascidian as a model organism in developmental and evolutionary biology. *Cell* **106**(5), 535–538.

Davidson, E. (2006). The Regulatory Genome. Academic Press, St. Louis.

Davidson, E. H. (2009). Network design principles from the sea urchin embryo. *Curr. Opin. Genet. Dev.* **19**(6), 535–540.

Davidson, B., and Levine, M. (2003). Evolutionary origins of the vertebrate heart: Specification of the cardiac lineage in Ciona intestinalis. *Proc. Natl. Acad. Sci. USA* **100**(20), 11469–11473.

Davidson, E. H., Rast, J. P., Oliveri, P., Ransick, A., Calestani, C., Yuh, C. H., Minokawa, T., Amore, G., Hinman, V., Arenas-Mena, C., *et al.* (2002). A genomic regulatory network for development. *Science* **295**(5560), 1669–1678.

Davidson, B., Shi, W., and Levine, M. (2005). Uncoupling heart cell specification and migration in the simple chordate Ciona intestinalis. *Development* **132**(21), 4811–4818.

Davidson, B., Shi, W., Beh, J., Christiaen, L., and Levine, M. (2006). FGF signaling delineates the cardiac progenitor field in the simple chordate, Ciona intestinalis. *Genes Dev.* **20**(19), 2728–2738.

Dehal, P., Satou, Y., Campbell, R. K., Chapman, J., Degnan, B., De Tomaso, A., Davidson, B., Di Gregorio, A., Gelpke, M., Goodstein, D. M., *et al.* (2002). The draft genome of Ciona intestinalis: Insights into chordate and vertebrate origins. *Science* **298** (5601), 2157–2167.

Delsuc, F., Brinkmann, H., Chourrout, D., and Philippe, H. (2006). Tunicates and not cephalochordates are the closest living relatives of vertebrates. *Nature* **439**(7079), 965–968.

Deno, T., Nishida, H., and Satoh, N. (1984). Autonomous muscle cell differentiation in partial ascidian embryos according to the newly verified cell lineages. *Dev. Biol.* **104**(2), 322–328.

Deschet, K., Nakatani, Y., and Smith, W. C. (2003). Generation of Ci-Brachyury-GFP stable transgenic lines in the ascidian Ciona savignyi. *Genesis* **35**(4), 248–259.

Di Gregorio, A., and Levine, M. (1999). Regulation of Ci-tropomyosin-like, a Brachyury target gene in the ascidian, Ciona intestinalis. *Development* **126**(24), 5599–5609.

Flames, N., and Hobert, O. (2009). Gene regulatory logic of dopamine neuron differentiation. *Nature* **458**(7240), 885–889.

Frazer, K. A., Pachter, L., Poliakov, A., Rubin, E. M., and Dubchak, I. (2004). VISTA: Computational tools for comparative genomics. *Nucleic Acids Res.* **32**(Web Server issue), W273–W279.

Fujioka, M., Emi-Sarker, Y., Yusibova, G. L., Goto, T., and Jaynes, J. B. (1999). Analysis of an even-skipped rescue transgene reveals both composite and discrete neuronal and early blastoderm enhancers, and multi-stripe positioning by gap gene repressor gradients. *Development* **126**(11), 2527–2538.

Ganot, P., Kallesoe, T., Reinhardt, R., Chourrout, D., and Thompson, E. M. (2004). Spliced-leader RNA trans splicing in a chordate, Oikopleura dioica, with a compact genome. *Mol. Cell. Biol.* **24**(17), 7795–7805.

Geyer, P. K., Vitalini, M. W., and Wallrath, L. L. (2011). Nuclear organization: Taking a position on gene expression. *Curr. Opin. Cell Biol.* **23**(3), 354–359.

Ghosh, D., and Seydoux, G. (2008). Inhibition of transcription by the Caenorhabditis elegans germline protein PIE-1: Genetic evidence for distinct mechanisms targeting initiation and elongation. *Genetics* **178**(1), 235–243.

Groth, A. C., Fish, M., Nusse, R., and Calos, M. P. (2004). Construction of transgenic Drosophila by using the site-specific integrase from phage phiC31. *Genetics* **166**(4), 1775–1782.

Haeussler, M., and Joly, J. S. (2011). When needles look like hay: How to find tissue-specific enhancers in model organism genomes. *Dev. Biol.* **350**(2), 239–254.

Haeussler, M., Jaszczyszyn, Y., Christiaen, L., and Joly, J. S. (2010). A cis-regulatory signature for chordate anterior neuroectodermal genes. *PLoS Genet.* **6**(4), e1000912.

Hanyu-Nakamura, K., Sonobe-Nojima, H., Tanigawa, A., Lasko, P., and Nakamura, A. (2008). Drosophila Pgc protein inhibits P-TEFb recruitment to chromatin in primordial germ cells. *Nature* **451**(7179), 730–733.

Harafuji, N., Keys, D. N., and Levine, M. (2002). Genome-wide identification of tissue-specific enhancers in the Ciona tadpole. *Proc. Natl. Acad. Sci. USA* **99**(10), 6802–6805.

Hastings, K. E. (2005). SL trans-splicing: Easy come or easy go? *Trends Genet.* **21**(4), 240–247.

Hikosaka, A., Kusakabe, T., and Satoh, N. (1994). Short upstream sequences associated with the muscle-specific expression of an actin gene in ascidian embryos. *Dev. Biol.* **166**(2), 763–769.

Hill, M. M., Broman, K. W., Stupka, E., Smith, W. C., Jiang, D., and Sidow, A. (2008). The C. savignyi genetic map and its integration with the reference sequence facilitates insights into chordate genome evolution. *Genome Res.* **18**(8), 1369–1379.

Hino, K., Satou, Y., Yagi, K., and Satoh, N. (2003). A genomewide survey of developmentally relevant genes in Ciona intestinalis. VI. Genes for Wnt, TGFbeta, Hedgehog and JAK/STAT signaling pathways. *Dev. Genes Evol.* **213**(5–6), 264–272.

Hirano, T., and Nishida, H. (1997). Developmental fates of larval tissues after metamorphosis in ascidian Halocynthia roretzi. I. Origin of mesodermal tissues of the juvenile. *Dev. Biol.* **192**(2), 199–210.

Hozumi, A., Kawai, N., Yoshida, R., Ogura, Y., Ohta, N., Satake, H., Satoh, N., and Sasakura, Y. (2010). Efficient transposition of a single Minos transposon copy in the genome of the ascidian Ciona intestinalis with a transgenic line expressing transposase in eggs. *Dev. Dyn.* **239**(4), 1076–1088.

Hudson, C., and Lemaire, P. (2001). Induction of anterior neural fates in the ascidian Ciona intestinalis. *Mech. Dev.* **100**(2), 189–203.

Imai, K. S., Satou, Y., and Satoh, N. (2002). Multiple functions of a Zic-like gene in the differentiation of notochord, central nervous system and muscle in Ciona savignyi embryos. *Development* **129**(11), 2723–2732.

Imai, K. S., Hino, K., Yagi, K., Satoh, N., and Satou, Y. (2004). Gene expression profiles of transcription factors and signaling molecules in the ascidian embryo: Towards a comprehensive understanding of gene networks. *Development* **131**(16), 4047–4058.

Imai, K. S., Levine, M., Satoh, N., and Satou, Y. (2006). Regulatory blueprint for a chordate embryo. *Science* **312**(5777), 1183–1187.

Imai, K. S., Stolfi, A., Levine, M., and Satou, Y. (2009). Gene regulatory networks underlying the compartmentalization of the Ciona central nervous system. *Development* **136**(2), 285–293.

Johnson, D. S., Davidson, B., Brown, C. D., Smith, W. C., and Sidow, A. (2004). Noncoding regulatory sequences of Ciona exhibit strong correspondence between evolutionary constraint and functional importance. *Genome Res.* **14**(12), 2448–2456.

Johnson, D. S., Zhou, Q., Yagi, K., Satoh, N., Wong, W., and Sidow, A. (2005). De novo discovery of a tissue-specific gene regulatory module in a chordate. *Genome Res.* **15**(10), 1315–1324.

Keys, D. N., Lee, B. I., Di Gregorio, A., Harafuji, N., Detter, J. C., Wang, M., Kahsai, O., Ahn, S., Zhang, C., Doyle, S. A., *et al.* (2005). A saturation screen for cis-acting regulatory DNA in the Hox genes of Ciona intestinalis. *Proc. Natl. Acad. Sci. USA* **102**(3), 679–683.

Khare, P., Mortimer, S. I., Cleto, C. L., Okamura, K., Suzuki, Y., Kusakabe, T., Nakai, K., Meedel, T. H., and Hastings, K. E. (2011). Cross-validated methods for promoter/transcription start site mapping in SL trans-spliced genes, established using the Ciona intestinalis troponin I gene. *Nucleic Acids Res.* **39**(7), 2638–2648.

Khoueiry, P., Rothbacher, U., Ohtsuka, Y., Daian, F., Frangulian, E., Roure, A., Dubchak, I., and Lemaire, P. (2010). A cis-regulatory signature in ascidians and flies, independent of transcription factor binding sites. *Curr. Biol.* **20**(9), 792–802.

Kubo, A., Suzuki, N., Yuan, X. Y., Nakai, K., Satoh, N., Imai, K. S., and Satou, Y. (2010). Genomic cis-regulatory networks in the early Ciona intestinalis embryo. *Development* **137** (10), 1613–1623.

Kugler, J. E., Gazdoiu, S., Oda-Ishii, I., Passamaneck, Y. J., Erives, A. J., and Di Gregorio, A. (2010). Temporal regulation of the muscle gene cascade by Macho1 and Tbx6 transcription factors in Ciona intestinalis. *J. Cell Sci.* **123**(Pt 14), 2453–2463.

Kumano, G., Takatori, N., Negishi, T., Takada, T., and Nishida, H. (2011). A maternal factor unique to ascidians silences the germline via binding to P-TEFb and RNAP II regulation. *Curr. Biol.* **21**(15), 1308–1313.

Kusakabe, T., Hikosaka, A., and Satoh, N. (1995). Coexpression and promoter function in two muscle actin gene complexes of different structural organization in the ascidian Halocynthia roretzi. *Dev. Biol.* **169**(2), 461–472.

Kusakabe, T., Yoshida, R., Kawakami, I., Kusakabe, R., Mochizuki, Y., Yamada, L., Shin-i, T., Kohara, Y., Satoh, N., Tsuda, M., *et al.* (2002). Gene expression profiles in tadpole larvae of Ciona intestinalis. *Dev. Biol.* **242**(2), 188–203.

Kusakabe, T., Yoshida, R., Ikeda, Y., and Tsuda, M. (2004). Computational discovery of DNA motifs associated with cell type-specific gene expression in Ciona. *Dev. Biol.* **276** (2), 563–580.

Lamy, C., Rothbacher, U., Caillol, D., and Lemaire, P. (2006). Ci-FoxA-a is the earliest zygotic determinant of the ascidian anterior ectoderm and directly activates Ci-sFRP1/5. *Development* **133**(15), 2835–2844.

Levine, M. (2010). Transcriptional enhancers in animal development and evolution. *Curr. Biol.* **20**(17), R754–R763.

Levine, M., and Davidson, E. H. (2005). Gene regulatory networks for development. *Proc. Natl. Acad. Sci. USA* **102**(14), 4936–4942.

Longabaugh, W. J., Davidson, E. H., and Bolouri, H. (2005). Computational representation of developmental genetic regulatory networks. *Dev. Biol.* **283**(1), 1–16.

Martin, J., Han, C., Gordon, L. A., Terry, A., Prabhakar, S., She, X., Xie, G., Hellsten, U., Chan, Y. M., Altherr, M., *et al.* (2004). The sequence and analysis of duplication-rich human chromosome 16. *Nature* **432**(7020), 988–994.

Matsumoto, J., Dewar, K., Wasserscheid, J., Wiley, G. B., Macmil, S. L., Roe, B. A., Zeller, R. W., Satou, Y., and Hastings, K. E. (2010). High-throughput sequence analysis of Ciona intestinalis SL trans-spliced mRNAs: Alternative expression modes and gene function correlates. *Genome Res.* **20**(5), 636–645.

Matsuoka, T., Awazu, S., Satoh, N., and Sasakura, Y. (2004). Minos transposon causes germline transgenesis of the ascidian Ciona savignyi. *Dev. Growth Differ.* **46**(3), 249–255.

Matsuoka, T., Awazu, S., Shoguchi, E., Satoh, N., and Sasakura, Y. (2005). Germline transgenesis of the ascidian Ciona intestinalis by electroporation. *Genesis* **41**(2), 67–72.

Meedel, T. H., and Whittaker, J. R. (1984). Lineage segregation and developmental autonomy in expression of functional muscle acetylcholinesterase mRNA in the ascidian embryo. *Dev. Biol.* **105**(2), 479–487.

Meedel, T. H., Chang, P., and Yasuo, H. (2007). Muscle development in Ciona intestinalis requires the b-HLH myogenic regulatory factor gene Ci-MRF. *Dev. Biol.* **302**(1), 333–344.

Miwata, K., Chiba, T., Horii, R., Yamada, L., Kubo, A., Miyamura, D., Satoh, N., and Satou, Y. (2006). Systematic analysis of embryonic expression profiles of zinc finger genes in Ciona intestinalis. *Dev. Biol.* **292**(2), 546–554.

Nishida, H. (1987). Cell lineage analysis in ascidian embryos by intracellular injection of a tracer enzyme. III. Up to the tissue restricted stage. *Dev. Biol.* **121**(2), 526–541.

Nishida, H., and Satoh, N. (1983). Cell lineage analysis in ascidian embryos by intracellular injection of a tracer enzyme. I. Up to the eight-cell stage. *Dev. Biol.* **99**(2), 382–394.

Nishida, H., and Satoh, N. (1985). Cell lineage analysis in ascidian embryos by intracellular injection of a tracer enzyme. II. The 16- and 32-cell stages. *Dev. Biol.* **110**(2), 440–454.

Nishida, H., and Sawada, K. (2001). macho-1 encodes a localized mRNA in ascidian eggs that specifies muscle fate during embryogenesis. *Nature* **409**(6821), 724–729.

Oda-Ishii, I., Bertrand, V., Matsuo, I., Lemaire, P., and Saiga, H. (2005). Making very similar embryos with divergent genomes: Conservation of regulatory mechanisms of Otx between the ascidians Halocynthia roretzi and Ciona intestinalis. *Development* **132**(7), 1663–1674.

Pennacchio, L. A., Ahituv, N., Moses, A. M., Prabhakar, S., Nobrega, M. A., Shoukry, M., Minovitsky, S., Dubchak, I., Holt, A., Lewis, K. D., *et al.* (2006). In vivo enhancer analysis of human conserved non-coding sequences. *Nature* **444**(7118), 499–502.

Reese, M. G. (2001). Application of a time-delay neural network to promoter annotation in the Drosophila melanogaster genome. *Comput. Chem.* **26**(1), 51–56.

Sardet, C., Dru, P., and Prodon, F. (2005). Maternal determinants and mRNAs in the cortex of ascidian oocytes, zygotes and embryos. *Biol. Cell* **97**(1), 35–49.

Sasakura, Y., Awazu, S., Chiba, S., and Satoh, N. (2003). Germ-line transgenesis of the Tc1/mariner superfamily transposon Minos in Ciona intestinalis. *Proc. Natl. Acad. Sci. USA* **100**(13), 7726–7730.

Sasakura, Y., Konno, A., Mizuno, K., Satoh, N., and Inaba, K. (2008). Enhancer detection in the ascidian Ciona intestinalis with transposase-expressing lines of Minos. *Dev. Dyn.* **237**(1), 39–50.

Sasakura, Y., Suzuki, M. M., Hozumi, A., Inaba, K., and Satoh, N. (2010). Maternal factor-mediated epigenetic gene silencing in the ascidian Ciona intestinalis. *Mol. Genet. Genomics* **283**(1), 99–110.

Satoh, N. (1994). Developmental Biology of Ascidians. Cambridge University Press, England.

Satou, Y., Imai, K. S., and Satoh, N. (2001a). Early embryonic expression of a LIM-homeobox gene Cs-lhx3 is downstream of beta-catenin and responsible for the endoderm differentiation in Ciona savignyi embryos. *Development* **128**(18), 3559–3570.

Satou, Y., Takatori, N., Yamada, L., Mochizuki, Y., Hamaguchi, M., Ishikawa, H., Chiba, S., Imai, K., Kano, S., Murakami, S. D., *et al.* (2001b). Gene expression profiles in Ciona intestinalis tailbud embryos. *Development* **128**(15), 2893–2904.

Satou, Y., Takatori, N., Fujiwara, S., Nishikata, T., Saiga, H., Kusakabe, T., Shin-i, T., Kohara, Y., and Satoh, N. (2002a). Ciona intestinalis cDNA projects: Expressed sequence tag analyses and gene expression profiles during embryogenesis. *Gene* **287**(1–2), 83–96.

Satou, Y., Yagi, K., Imai, K. S., Yamada, L., Nishida, H., and Satoh, N. (2002b). macho-1-related genes in Ciona embryos. *Dev. Genes Evol.* **212**(2), 87–92.

Satou, Y., Yamada, L., Mochizuki, Y., Takatori, N., Kawashima, T., Sasaki, A., Hamaguchi, M., Awazu, S., Yagi, K., Sasakura, Y., *et al.* (2002c). A cDNA resource from the basal chordate Ciona intestinalis. *Genesis* **33**(4), 153–154.

Satou, Y., Imai, K. S., Levine, M., Kohara, Y., Rokhsar, D., and Satoh, N. (2003a). A genomewide survey of developmentally relevant genes in Ciona intestinalis. I. Genes for bHLH transcription factors. *Dev. Genes Evol.* **213**(5–6), 213–221.

Satou, Y., Sasakura, Y., Yamada, L., Imai, K. S., Satoh, N., and Degnan, B. (2003b). A genomewide survey of developmentally relevant genes in Ciona intestinalis. V. Genes for receptor tyrosine kinase pathway and Notch signaling pathway. *Dev. Genes Evol.* **213**(5–6), 254–263.

Satou, Y., Imai, K. S., and Satoh, N. (2004). The ascidian Mesp gene specifies heart precursor cells. *Development* **131**(11), 2533–2541.

Satou, Y., Kawashima, T., Shoguchi, E., Nakayama, A., and Satoh, N. (2005). An integrated database of the ascidian, Ciona intestinalis: Towards functional genomics. *Zoolog. Sci.* **22**(8), 837–843.

Satou, Y., Hamaguchi, M., Takeuchi, K., Hastings, K. E., and Satoh, N. (2006). Genomic overview of mRNA 5′-leader trans-splicing in the ascidian Ciona intestinalis. *Nucleic Acids Res.* **34**(11), 3378–3388.

Satou, Y., Mineta, K., Ogasawara, M., Sasakura, Y., Shoguchi, E., Ueno, K., Yamada, L., Matsumoto, J., Wasserscheid, J., Dewar, K., *et al.* (2008). Improved genome assembly and evidence-based global gene model set for the chordate Ciona intestinalis: New insight into intron and operon populations. *Genome Biol.* **9**(10), R152.

Sawada, K., Fukushima, Y., and Nishida, H. (2005). Macho-1 functions as transcriptional activator for muscle formation in embryos of the ascidian Halocynthia roretzi. *Gene Expr. Patterns* **5**(3), 429–437.

Seydoux, G., and Dunn, M. A. (1997). Transcriptionally repressed germ cells lack a subpopulation of phosphorylated RNA polymerase II in early embryos of Caenorhabditis elegans and Drosophila melanogaster. *Development* **124**(11), 2191–2201.

Shirae-Kurabayashi, M., Matsuda, K., and Nakamura, A. (2011). Ci-Pem-1 localizes to the nucleus and represses somatic gene transcription in the germline of Ciona intestinalis embryos. *Development* **138**(14), 2871–2881.

Sierro, N., Kusakabe, T., Park, K. J., Yamashita, R., Kinoshita, K., and Nakai, K. (2006). DBTGR: A database of tunicate promoters and their regulatory elements. *Nucleic Acids Res.* **34**(Database issue), D552–D555.

Simmen, M. W., Leitgeb, S., Charlton, J., Jones, S. J., Harris, B. R., Clark, V. H., and Bird, A. (1999). Nonmethylated transposable elements and methylated genes in a chordate genome. *Science* **283**(5405), 1164–1167.

Small, K. S., Brudno, M., Hill, M. M., and Sidow, A. (2007). A haplome alignment and reference sequence of the highly polymorphic Ciona savignyi genome. *Genome Biol.* **8** (3), R41.

Spada, F., Koch, J., Sadoni, N., Mitchell, N., Ganot, P., De Boni, U., Zink, D., and Thompson, E. M. (2007). Conserved patterns of nuclear compartmentalization are not observed in the chordate Oikopleura. *Biol. Cell* **99**(5), 273–287.

Squarzoni, P., Parveen, F., Zanetti, L., Ristoratore, F., and Spagnuolo, A. (2011). FGF/MAPK/Ets signaling renders pigment cell precursors competent to respond to Wnt signal by directly controlling Ci-Tcf transcription. *Development* **138**(7), 1421–1432.

Stathopoulos, A., and Levine, M. (2005). Genomic regulatory networks and animal development. *Dev. Cell* **9**(4), 449–462.

Stolfi, A., Gainous, T. B., Young, J. J., Mori, A., Levine, M., and Christiaen, L. (2010). Early chordate origins of the vertebrate second heart field. *Science* **329**(5991), 565–568.

Suzuki, M. M., Kerr, A. R., De Sousa, D., and Bird, A. (2007). CpG methylation is targeted to transcription units in an invertebrate genome. *Genome Res.* **17**(5), 625–631.

Tagle, D. A., Koop, B. F., Goodman, M., Slightom, J. L., Hess, D. L., and Jones, R. T. (1988). Embryonic epsilon and gamma globin genes of a prosimian primate (Galago crassicaudatus). Nucleotide and amino acid sequences, developmental regulation and phylogenetic footprints. *J. Mol. Biol.* **203**(2), 439–455.

Takahashi, H., Hotta, K., Erives, A., Di Gregorio, A., Zeller, R. W., Levine, M., and Satoh, N. (1999). Brachyury downstream notochord differentiation in the ascidian embryo. *Genes Dev.* **13**(12), 1519–1523.

Takatori, N., Hotta, K., Mochizuki, Y., Satoh, G., Mitani, Y., Satoh, N., Satou, Y., and Takahashi, H. (2004). T-box genes in the ascidian Ciona intestinalis: Characterization of cDNAs and spatial expression. *Dev. Dyn.* **230**(4), 743–753.

Tassy, O., Dauga, D., Daian, F., Sobral, D., Robin, F., Khoueiry, P., Salgado, D., Fox, V., Caillol, D., Schiappa, R., *et al.* (2010). The ANISEED database: Digital representation, formalization, and elucidation of a chordate developmental program. *Genome Res.* **20** (10), 1459–1468.

Tweedie, S., Charlton, J., Clark, V., and Bird, A. (1997). Methylation of genomes and genes at the invertebrate-vertebrate boundary. *Mol. Cell. Biol.* **17**(3), 1469–1475.

Vandenberghe, A. E., Meedel, T. H., and Hastings, K. E. (2001). mRNA 5′-leader trans-splicing in the chordates. *Genes Dev.* **15**(3), 294–303.

Vinson, J. P., Jaffe, D. B., O'Neill, K., Karlsson, E. K., Stange-Thomann, N., Anderson, S., Mesirov, J. P., Satoh, N., Satou, Y., Nusbaum, C., *et al.* (2005). Assembly of polymorphic genomes: Algorithms and application to Ciona savignyi. *Genome Res.* **15**(8), 1127–1135.

Wada, S., Katsuyama, Y., Sato, Y., Itoh, C., and Saiga, H. (1996). Hroth an orthodenticle-related homeobox gene of the ascidian, Halocynthia roretzi: Its expression and putative roles in the axis formation during embryogenesis. *Mech. Dev.* **60**(1), 59–71.

Wada, S., Tokuoka, M., Shoguchi, E., Kobayashi, K., Di Gregorio, A., Spagnuolo, A., Branno, M., Kohara, Y., Rokhsar, D., Levine, M., *et al.* (2003). A genomewide survey of developmentally relevant genes in Ciona intestinalis. II. Genes for homeobox transcription factors. *Dev. Genes Evol.* **213**(5–6), 222–234.

Wang, Y., and Leung, F. C. (2008). GC content increased at CpG flanking positions of fish genes compared with sea squirt orthologs as a mechanism for reducing impact of DNA methylation. *PLoS One* **3**(11), e3612.

Wenick, A. S., and Hobert, O. (2004). Genomic cis-regulatory architecture and trans-acting regulators of a single interneuron-specific gene battery in C. elegans. *Dev. Cell* **6**(6), 757–770.

Yagi, K., Satou, Y., Mazet, F., Shimeld, S. M., Degnan, B., Rokhsar, D., Levine, M., Kohara, Y., and Satoh, N. (2003). A genomewide survey of developmentally relevant genes in Ciona intestinalis. III. Genes for Fox, ETS, nuclear receptors and NFkappaB. *Dev. Genes Evol.* **213**(5–6), 235–244.

Yagi, K., Satoh, N., and Satou, Y. (2004a). Identification of downstream genes of the ascidian muscle determinant gene Ci-macho1. *Dev. Biol.* **274**(2), 478–489.

Yagi, K., Satou, Y., and Satoh, N. (2004b). A zinc finger transcription factor, ZicL, is a direct activator of Brachyury in the notochord specification of Ciona intestinalis. *Development* **131**(6), 1279–1288.

Yagi, K., Takatori, N., Satou, Y., and Satoh, N. (2005). Ci-Tbx6b and Ci-Tbx6c are key mediators of the maternal effect gene Ci-macho1 in muscle cell differentiation in Ciona intestinalis embryos. *Dev. Biol.* **282**(2), 535–549.

Yamada, L., Kobayashi, K., Degnan, B., Satoh, N., and Satou, Y. (2003). A genomewide survey of developmentally relevant genes in Ciona intestinalis. IV. Genes for HMG transcriptional regulators, bZip and GATA/Gli/Zic/Snail. *Dev. Genes Evol.* **213**(5–6), 245–253.

Yasuo, H., and Satoh, N. (1993). Function of vertebrate T gene. *Nature* **364**(6438), 582–583.

Yeats, B., Matsumoto, J., Mortimer, S. I., Shoguchi, E., Satoh, N., and Hastings, K. E. (2010). SL RNA genes of the ascidian tunicates Ciona intestinalis and Ciona savignyi. *Zoolog. Sci.* **27**(2), 171–180.

Yuasa, H. J., Kawamura, K., Yamamoto, H., and Takagi, T. (2002). The structural organization of ascidian Halocynthia roretzi troponin I genes. *J. Biochem.* **132**(1), 135–141.

Zeller, R. W. (2010). Computational analysis of Ciona intestinalis operons. *Integr. Comp. Biol.* **50**(1), 75–85.

Zinzen, R. P., Girardot, C., Gagneur, J., Braun, M., and Furlong, E. E. (2009). Combinatorial binding predicts spatio-temporal cis-regulatory activity. *Nature* **462**(7269), 65–70.

CHAPTER SEVEN

# A Dynamic Network of Morphogens and Transcription Factors Patterns the Fly Leg

Carlos Estella,*[,1] Roumen Voutev,* *and* Richard S. Mann*

## Contents



## Abstract

Animal appendages require a proximodistal (PD) axis, which forms orthogonally from the two main body axes, anteroposterior and dorsoventral. In this review, we discuss recent advances that begin to provide insights into the molecular mechanisms controlling PD axis formation in the *Drosophila* leg. In this case, two morphogens, Wingless (Wg) and Decapentaplegic (Dpp), initiate a genetic

* Department of Biochemistry and Molecular Biophysics, Columbia University, New York, New York, USA
[1] Current Address: Fundación Instituto Valenciano de Infertilidad (FIVI), Valencia University, and Instituto Universitario IVI/INCLIVA, Valencia, Spain

*Current Topics in Developmental Biology*, Volume 98
ISSN 0070-2153, DOI: 10.1016/B978-0-12-386499-4.00007-0

cascade that, together with growth of the leg imaginal disc, establishes the PD axis. The analysis of *cis*-regulatory modules (CRMs) that control the expression of genes at different positions along the PD axis has been particularly valuable in dissecting this complex process. From these experiments, it appears that only one concentration of Wg and Dpp are required to initiate PD axis formation by inducing the expression of *Distal-less* (*Dll*), a homeodomain-encoding gene that is required for leg development. Once *Dll* is turned on, it activates the medially expressed gene *dachshund* (*dac*). Cross-regulation between *Dll* and *dac*, together with cell proliferation in the growing leg imaginal disc, results in the formation of a rudimentary PD axis. Wg and Dpp also initiate the expression of ligands for the EGFR pathway, which in turn induces the expression of a series of target genes that pattern the distal-most portion of the leg.

## 1. Introduction

Animal appendages are external projections from the body wall that are used for very diverse functions including locomotion, grooming, and feeding. In the thorax of diptera, such as the fruit fly *Drosophila melanogaster*, there are dorsal appendages required for flight—a pair of wings in the second thoracic (T2) segment and a pair of halteres in T3—and three pairs of legs used for walking and grooming. The fly leg, the subject of this review, is composed of 10 morphologically unique segments: coxa, trochanter, femur, tibia, tarsal segments 1–5, and the claw. Together, these segments comprise the proximodistal (PD) axis, in which the proximal coxa is closest to the body and the claw is furthest from the body (Fig. 7.1).

Unlike the two other primary body axes (anteroposterior, AP; dorsoventral, DV), for each appendage, the PD axis is established during embryogenesis *de novo*. In contrast, at all stages of development, even in the unfertilized egg, rudimentary AP and DV axes exist. Thus, in this respect, the PD axis is unique among the main body axes. This topic, how so-called

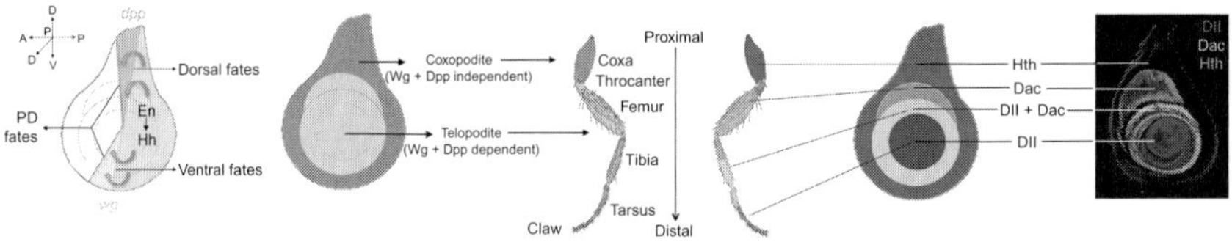


**Figure 7.1** Overview of fly leg development. On the left shows the relationship between En, Hh, *wg*, and *dpp* and the definition of the telopodite (Hh, Wg, and Dpp-dependent domain) and the coxopodite (Hh, Wg, and Dpp-independent domain). On the right shows the relationship between the three primary PD gene expression domains established by Hth, Dac, and Dll. (See Color Insert.)

secondary developmental fields are established from preexisting developmental information, has been debated for decades both from theoretical perspectives and by classical developmental biologists (reviewed by Baker, 2011). Data generated over the past several years have provided novel mechanistic and molecular insights that build upon these earlier studies, providing interesting connections between cell division, secreted morphogens, and the use of dedicated *cis*-regulatory modules (CRMs) for transcriptional regulation of genes expressed along the PD axis. It is the goal of this review to summarize our current understanding of the intimate interplay between these components, orchestrated over developmental time, which establishes, elaborates, and fine-tunes the leg's PD axis.

## 2. The Molecular Players in PD Axis Formation

As for much of the adult fly, fly legs are derived from imaginal discs, elliptical sheets of epithelia that are highly folded by the end of larval development. The fate map of the leg disc is such that cells at its center will give rise to distal-most structures, while cells further away from the center generate more proximal structures (Fig. 7.1). Imaginal discs do not only give rise to appendages: cells at the periphery of the leg disc, for example, generate the ventral portion of the adult body wall, the ventral and sterno pleura. Thus, in the leg imaginal disc, the PD axis—from distal claw to proximal body wall—is displayed as concentric rings within these elliptical epithelia (Fig. 7.1).

Many genes are expressed in rings or subdomains in the leg disc that mark distinct positions along the PD axis (Abu-Shaar and Mann, 1998; Campbell and Tomlinson, 1998; Diaz-Benjumea *et al.*, 1994; Duncan *et al.*, 1998; Emmons *et al.*, 1999; Erkner *et al.*, 1999; previously reviewed by Kojima, 2004; Kojima *et al.*, 2000; Mardon *et al.*, 1994). From this perspective, the problem of how the PD axis is established can be reformulated by asking the simpler question: how are these gene expression domains established? As will be discussed more below, much attention has been focussed on dissecting the regulation of two genes that are broadly expressed in distal and medial domains of the leg disc, *Distal-less* (*Dll*) and *dachshund* (*dac*), respectively (Fig. 7.1). *Dll*, in particular, is a critical player in leg development as it is one of the earliest known markers for the appendage, not just in flies, but also throughout the animal kingdom (Cohen, 1990; Cohen *et al.*, 1989; Panganiban *et al.*, 1997; reviewed in Panganiban and Rubenstein, 2002). Moreover, in flies both *Dll* and *dac* are required for the development of their respective distal and medial domains of the leg (Cohen and Jurgens, 1989; Mardon *et al.*, 1994). *homothorax* (*hth*), encoding a homeodomain transcription factor, and *teashirt* (*tsh*), encoding a zinc-finger transcription factor, are coexpressed in an even more proximal domain (Abu-Shaar and

Mann, 1998; Erkner *et al.*, 1999; Rieckhof *et al.*, 1997; Wu and Cohen, 1999). The domains resulting from the expression of *hth/tsh* (proximal), *dac* (medial), and *Dll* (distal), together with regions that have overlapping expression of these factors, broadly define the PD axis (Fig. 7.1). Other genes, also expressed at specific positions along the PD axis, for example, those expressed in the tarsal segments, are activated later in development in response to EGFR signaling and are required for forming the joints that separate each of the leg segments (see below).

Theoretical modeling and classical limb grafting experiments both led to the idea that the juxtaposition of three different cell types—in particular, posterior, anterior dorsal, and anterior ventral—leads to the induction of new PD axes (reviewed in Baker, 2011; Fig. 7.2). We now have a molecular understanding of this phenomenon, namely, that the juxtaposition of cells expressing Decapentaplegic (Dpp) next to cells expressing Wingless (Wg), two secreted morphogens used widely in animal development, is sufficient within the context of leg development to generate a new PD axis (Campbell *et al.*, 1993; Diaz-Benjumea *et al.*, 1994; Lecuit and Cohen, 1997). In the wild-type leg imaginal disc, *dpp* and *wg* are expressed along the AP compartment boundary in dorsal and ventral cells, respectively, both in response to Hedgehog (Hh) emanating from the posterior compartment (Basler and Struhl, 1994; Fig. 7.2). Accordingly, in the wild-type leg disc, cells expressing Wg and Dpp are adjacent to each other only at the center of the disc, which will give rise to the distal-most portion of the appendage. Thus, the expression patterns of Hh (posterior), Dpp (dorsal-anterior), and Wg (ventral-anterior) account for the three cell types initially postulated by the grafting and theoretical studies (Fig. 7.2).

Significant effort over the past several years has attempted to connect these two sets of molecular players in PD axis formation. At the top of the hierarchy are Wg and Dpp, which together are sufficient for initiating a PD axis in the leg, inducing the correct expression domains of *Dll*, *dac*, and

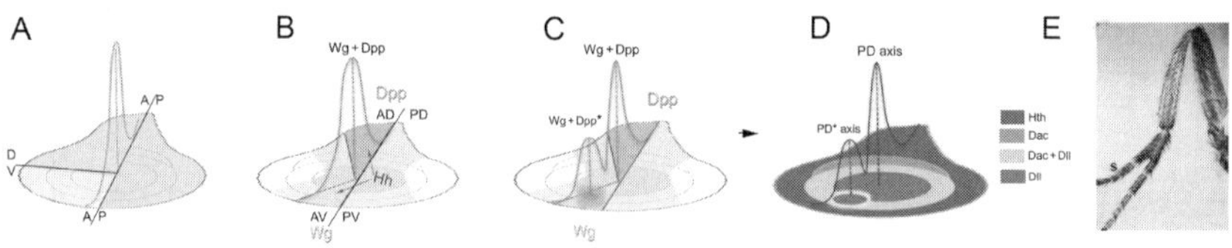


**Figure 7.2** Wg+Dpp initiate the PD axis. (A) Meinhardt's "three-sector" model for induction of the PD axis. (B) Hh, from the P compartment, induces Wg (yellow) in the anterior ventral (AV) domain and Dpp (blue) in the anterior dorsal (AD) domain; note that Wg- and Dpp-expressing cells are only adjacent in the center of the wild-type disc. (C–E) An ectopic source of Dpp (C) in the ventral domain induces an ectopic PD axis visualized in the disc (D) and in the adult appendage (E; from Campbell *et al.*, 1993). (For interpretation of the references to color in this figure legend, the reader is referred to the Web version of this chapter.)

other genes expressed along this axis. How do these morphogens activate these PD genes? Is the activation direct or indirect? What is the relationship between the leg primordia and the dorsal appendage (e.g., wing) primordia? What genes provide the ventral identity of the appendage? Here, we provide an update on efforts to answer these questions and provide an initial view of the molecular architecture underlying PD axis specification.

## 3. The Initial Establishment of the PD Axis is Encoded in the *cis*-Regulatory Architecture of *Dll*

A common principle that has emerged from studying transcriptional regulatory mechanisms in many developmental genes is that dedicated CRMs drive small subsets of complex expression patterns (reviewed in Maeda and Karch, 2011). Dissection of the *Dll* locus revealed a similar level of CRM dedication, but with the additional finding that distinct *Dll* CRMs control *Dll* expression in cells that have different degrees of developmental potential (Fig. 7.3). The discovery of these CRMs has allowed

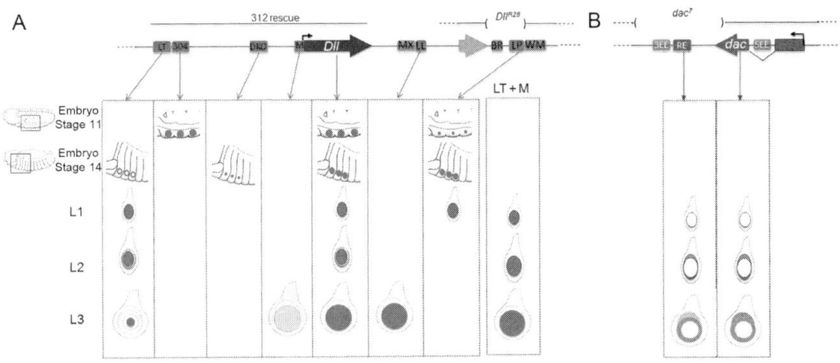


**Figure 7.3** *Dll* and *dac* CRMs. (A,B) Schematic of the *Dll* (A) and *dac* (B) genomic regions showing the positions of identified CRMs (colored boxes) and transcription units (large arrows). The expression patterns driven by individual CRMs is indicated and compared to the intact genes. All CRMs are mentioned in the text except for *DllMX*, *DllWM*, *dac3EE*, and *dac5EE*, which are not active during leg development, or *DllBR*, which is active very late in leg development (Galindo *et al.*, 2011; Pappu *et al.*, 2005). A *Dll* rescue transgene ("312 rescue") and a small *Dll* deficiency ($Dll^{R28}$) both result in a nearly complete PD axis, with defects primarily in the tarsal segments. $dac^7$, which removes *dacRE*, is a deficiency that eliminates both *dac* expression and the medial Dac domain in the leg. (For interpretation of the references to color in this figure legend, the reader is referred to the Web version of this chapter.)

lineage tracing experiments to be carried out, which helped redefine the fate map of the early appendage primordia. Below, we discuss this *cis*-regulatory architecture and its implications for leg development.

## 3.1. *Dll304*

The first sign of appendage formation in the *Drosophila* embryo is the activation of *Dll* at ~6 h after egg laying (AEL; stage 11) in circular domains comprising ~20–30 cells per thoracic hemisegment (Cohen, 1990). At this stage, an early *Dll* regulatory element situated 11 kb 5′of the start of *Dll* transcription, called the *Dll* early enhancer or *Dll304*, is able to drive a pattern similar to *Dll* (Vachon *et al.*, 1992). Genetically, *Dll* and *Dll304* activity at this early stage depend on a positive input from Wg, but not from Dpp (Cohen, 1990; Cohen *et al.*, 1993). In fact, *Dll304* is repressed dorsally and ventrally by the Dpp and EGFR pathways, respectively (Goto and Hayashi, 1997; Kubota *et al.*, 2000). Although a molecular dissection of the Wg, Dpp, or EGFR inputs into this enhancer has not yet been described, direct repressive inputs from the abdominal Hox proteins Ultrabithorax (Ubx) and Abdominal-A (AbdA) have been identified (Castelli-Gair and Akam, 1995; Gebelein *et al.*, 2002, 2004; Mann, 1994; Vachon *et al.*, 1992). By directly repressing *Dll304*, these Hox inputs block *Dll* expression and consequently limb development in the abdominal segments.

In the 1970s, Wieschaus and Gehring used X-ray somatic recombination and gynandromorphs to genetically follow marked cells by lineage analysis into the adult (Gehring *et al.*, 1976; Wieschaus and Gehring, 1976). When induced at the blastoderm stage or earlier, marked clones often included both the T2 leg and wing, suggesting a common embryonic origin to both appendages. This conclusion has been corroborated by using the *Dll304* element to carry out lineage tracing experiments, at first by following β-galactosidase perdurance from *Dll304-lacZ* reporter genes and later by genetic methods (Cohen *et al.*, 1993; McKay *et al.*, 2009). The results show that the cells that express *Dll304* give rise not only to the entire ventral appendage (the entire leg) but also to all parts of the two dorsal appendages (the wing in T2 and haltere in T3; Fig. 7.4). In fact, together the ~20–30 early *Dll304*-expressing cells in each thoracic hemisegment have the potential to give rise to the entire adult thorax.

## 3.2. *DllLT*

The activity of *Dll304* decays within a few hours, but *Dll* expression is maintained in a similarly positioned group of cells in each thoracic hemisegment, suggesting that other regulatory elements must assume control of *Dll* expression. One such element is the late-acting enhancer or "**L**eg **T**rigger" (*DllLT*), situated adjacent to *Dll304* (Cohen *et al.*, 1993; Estella *et al.*, 2008; Vachon *et al.*, 1992; Fig. 7.3). *DllLT* begins to be active at ~8 h and is robustly

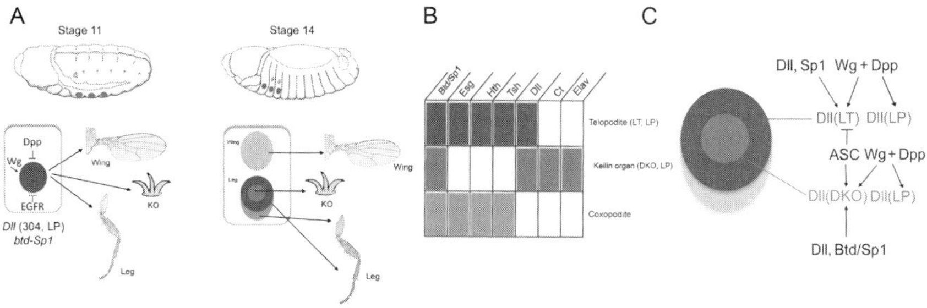


**Figure 7.4** Embryonic appendage fate map. (A) Cells expressing *Dll* at stage 11 via the 304 CRM can give rise to the entire adult thorax, while those expressing *Dll* at stage 14 give rise to either the KO or telopodite, depending whether DKO or LT is driving expression, respectively. (B) Genes expressed in the progenitors to the telopodite, KO, and coxopodite. (C) Fate map and regulatory network defining the activity of *Dll* CRMs. The blue cells, expressing *esg* but not *Dll*, are fated to become coxopodite. (See Color Insert.)

expressed by ~10h AEL. This element is not active in all *Dll*-expressing cells but just in ~15 cells at the periphery of the *Dll* expression domain in each hemisegment (Cohen *et al.*, 1993; McKay *et al.*, 2009; Fig. 7.4). *DllLT*, in contrast to *Dll304*, requires positive inputs from both Wg and Dpp. Moreover, the Wg and Dpp input into *DllLT* is direct, mediated by several binding sites for their respective downstream transcription factors Pangolin (Pan) and Mothers against Dpp (Mad; Estella *et al.*, 2008). Interestingly, *DllLT* also requires *Dll* for its activation, presumably derived from the earlier acting *Dll304* element in the early primordia (Castelli-Gair and Akam, 1995; Estella and Mann, 2010; McKay *et al.*, 2009).

Not only is there a temporal hand off from one *Dll* CRM to another, lineage tracing studies reveal that *DllLT*-expressing cells are dramatically more limited in developmental potential compared to *Dll304*-expressing cells. Specifically, the ~15 *DllLT*-expressing cells per hemisegment only give rise to the mature *Dll* and *dac* expression domains of the leg disc, but not to more proximal regions of the leg nor to any part of the dorsal appendages (McKay *et al.*, 2009; Fig. 7.4). This restricted lineage is interesting for two reasons. For one, the combined Dll+Dac domain coincides with the so-called "telopodite" or "true leg" that was originally described by the American entomologist Robert Snodgrass (Snodgrass, 1935; Fig. 7.1). According to Snodgrass, the telopodite is evolutionarily distinct from the more proximal coxopodite, which evolved as a primitive and unsegmented extension from the body wall. Second, as shown more recently, the telopodite, but not the coxopodite, depends on Hh, Wg, and Dpp signaling (Diaz-Benjumea *et al.*, 1994; Gonzalez-Crespo and Morata, 1996), and *DllLT* directly integrates the Wg and Dpp pathways for its activation (Estella *et al.*, 2008). Thus, the telopodite/coxopodite subdivision concept gained significant molecular support by the observation that a single *Dll* CRM, *DllLT*, is dependent on Wg+Dpp signaling and is active in cells that give rise to the entire telopodite, and only the telopodite.

The telopodite/coxopodite subdivision idea is also supported by genetic analysis of the *Dll* gene, which encodes a homeodomain transcription factor: *Dll* null mutants lack the telopodite but retain the coxopodite (Cohen *et al.*, 1993). There are two interesting follow-up points to be made here. First, even though *Dll* (via the *Dll304* CRM) is expressed earlier, in the progenitors of the entire leg and wing, *Dll* function is only required for the development of the telopodite (Campbell *et al.*, 1993; Cohen *et al.*, 1993). Second, these genetic results imply that *Dll* is required for establishing the Dac domain, even though it is not expressed in part of the Dac domain by the end of larval development. As discussed below, these observations have recently been supported by a molecular analysis of a *dac* CRM, which requires direct Dll input for its activity (Giorgianni and Mann, 2011).

## 3.3. *DllDKO*

*DllLT* is active in only ~15 of the *Dll*-expressing cells in each thoracic hemisegment of stage 14 embryos. The remaining *Dll*-expressing cells have a neural identity as revealed by the expression of genes such as *achaete* (*ac*) and *cut* (*ct*; Bolinger and Boekhoff-Falk, 2005; Cohen and Jurgens, 1989; McKay *et al.*, 2009). A third *Dll* CRM, named *DllDKO* (for **D**istal-less **K**eilin's **O**rgan) was identified ~3kb 5′ from the start of *Dll* transcription and is specifically activated in these *Dll*-expressing, *DllLT*-negative cells (Figs. 7.3 and 7.4). *DllDKO* receives positive input from members of the *achaete–scute* complex (ASC) and *Dll*, thus restricting its activity to the neurogenic cells in the limb primordia (McKay *et al.*, 2009; Fig. 7.4).

At this point in embryogenesis, cells in the leg and wing primordia express the zinc-finger transcription factor *escargot* (*esg*), which is required to maintain diploidy and therefore an imaginal disc fate (Fuse *et al.*, 1996; Hayashi *et al.*, 1993). Notably, there is a gap in *esg* expression in the leg primordia that is filled by the expression of the neuronal genes *ct* and *ac*, suggesting that these cells do not contribute to the leg imaginal disc but instead are fated to form the Keilin's Organ (KO), a larval sensory organ thought to be a vestige of larva legs present in more primitive insects (Bolinger and Boekhoff-Falk, 2005; Keilin, 1915). Lineage tracing experiments confirm that *DllDKO*-expressing cells do not give rise to any adult structures, consistent with the idea that they are dedicated to forming the KO (McKay *et al.*, 2009; Fig. 7.4).

## 3.4. Other *Dll* CRMs

*Dll304*, *DllLT*, and *DllDKO* are all located within a 14-kb region 5′ to the start of *Dll* transcription (Fig. 7.3). Recently, additional *Dll* CRMs 3′ to the *Dll* transcription unit that are also able to produce some aspects of the *Dll* expression pattern in the leg imaginal disc have been identified (Galindo *et al.*, 2011; Fig. 7.3). One element, called *DllLP* for **L**eg **P**rimordium, is active in a subset of *Dll*-expressing cells in stage 10 embryos and remains active until the end of first larval instar when its activity decays. *DllLP*, like *DllLT*, is activated by Wg and Dpp in the embryo, although it is not known if this activation is direct. A second 3′ element, called *DllLL* for **L**eg **L**ate, is only active in mid-third instar larvae and depends, like *DllLT*, on *Dll* for its activity. Although specific deletions do not exist to assess the necessity (or sufficiency) of each individual CRM, the available data suggest that both the 5′ and 3′ enhancers are important for wild-type *Dll* expression and function. While *Dll* null mutants lack the entire telopodite, a *Dll* minigene that includes the 5′ CRMs (including *Dll304* and *DllLT*) but not the 3′ CRMs significantly rescues telopodite development with the exception of the tarsal segments (Galindo *et al.*, 2011; Vachon *et al.*, 1992). This partial rescue, which could be due to inaccurate timing or levels of expression, is very similar to a *Dll* deletion allele that removes the *DllLP* element but leaves the 5′ elements intact (Galindo *et al.*, 2011; Fig. 7.3). Given the complexity of *Dll* regulation during development, it is not surprising that it is governed by multiple CRMs with partially overlapping activities.

The lineage tracing experiments cited above using individual *Dll* CRMs, combined with additional gene expression studies, resulted in a revised fate map of the ventral appendage primordia (Bolinger and Boekhoff-Falk, 2005; McKay *et al.*, 2009; Fig. 7.4). At stage 14, this fate map comprises three domains that correspond to a rudimentary PD axis: (1) an *esg* on, *hth* on, *tsh* on, and *Dll* off domain that is fated to form the coxopodite and the body wall, (2) an *esg* on, *hth* on, *tsh* on, and *Dll* on (*DllLT* on) domain will give rise to the entire telopodite, and (3) an *esg* off, *hth* off, *tsh* off, and *Dll* on (*DllDKO* on) domain fated to generate the KO (Fig. 7.4). This revised fate map is distinct from an earlier version in which the overlap between *hth* and *Dll* was not recognized, likely based on an analogy with third instar discs where the expression of these factors is largely nonoverlapping (Fig. 7.1; Gonzalez-Crespo and Morata, 1996; Gonzalez-Crespo *et al.*, 1998). The overlapping expression of *Dll*, *hth*, and *tsh* in the embryonic telopodite precursors is surprising and is no longer observed by the second instar stage, when these cells begin to divide. Interestingly, the coxopodite precursors (*hth* on, *tsh* on, *Dll* off) begin to proliferate slightly earlier than the telopodite precursors (Bryant and Schneiderman, 1969; McKay *et al.*, 2009). The different timing in proliferation between these two domains may be a consequence of the coexpression of *hth* and *Dll* in the telopodite precursors. Consistently, the forced expression of *hth* in *Dll*-expressing cells blocks cell proliferation and telopodite formation (Azpiazu and Morata, 2002; McKay *et al.*, 2009). Why this temporal asynchrony in the start of proliferation exists between these two domains of the leg is unknown.

## 4. The Role of *Sp1* in Distinguishing Ventral Appendage from Dorsal Appendage Fates

Although the above focus on *Dll* regulation reveals how the initial PD axis and leg fate map are established, several questions remain concerning this early stage of leg development. Some of these questions are answered by two paralogous genes, *buttonhead* (*btd*) and *Sp1* (Estella and Mann, 2010; Estella *et al.*, 2003; Wimmer *et al.*, 1996). Both genes encode Sp family zinc-finger transcription factors that share a similar expression pattern throughout development. Despite their similar expression patterns, the lack of *Sp1* function (but not *btd*) completely abolishes leg formation. Unlike *Dll*, which is required only for telopodite development, *Sp1* is required for the development of both the coxopodite and telopodite. Also noteworthy is that the closest vertebrate homolog of *Sp1* is *Sp8*, which is required for limb development in the mouse, suggesting an evolutionarily conserved role for these transcription factors (Bell *et al.*, 2003; Kawakami *et al.*, 2004; Treichel *et al.*, 2003).

The *Sp1* expression pattern is also consistent with an early role for this transcription factor in leg development. In parallel to *Dll304*, *Sp1* is first activated at stage 10 in the initial appendage primordia. *Sp1* activation requires Wg and is repressed dorsally by Dpp and in the abdomen by Ubx. However, in contrast to *Dll*, *btd* and *Sp1* are both expressed throughout the entire leg primordia, coinciding with *esg* expression (Estella *et al.*, 2003). In embryos without *btd* and *Sp1* function, *Dll* is activated normally but decays as the embryo develops. This is likely because *Dll304* does not require *Sp1/btd* function, but *DllLT* does (Estella and Mann, 2010). Thus, *Sp1* is initially activated in parallel to *Dll* but is required for the maintenance of *Dll* expression and is required for both telopodite and coxopodite fates. *disconnected* (*disco*) has also been proposed to play a role in the maintenance of *Dll* expression (Dey *et al.*, 2009).

In addition to being required for the entire ventral appendage, *Sp1* also appears to be required for suppressing dorsal appendage fates. When clonal analysis was used to generate ventral appendages devoid of both *Sp1* and *btd*, in some cases dramatic homeotic transformations from leg to wing were observed in the adult (Estella and Mann, 2010). Together, these experiments suggest that *Sp1* is a ventral appendage selector gene, and that in its absence dorsal appendage fates become derepressed.

Although many questions remain concerning this early aspect of appendage fate specification, there are several other relevant and interesting observations in the literature. First, as noted above, the dorsal appendage primordia (wing and haltere) and ventral appendage primordia (leg) are derived from the same set of *Dll304*-expressing cells. As the embryo develops, a small number of cells leave this early ventral (*Dll304*-expressing) primordia, migrate dorsally, stop expressing *Dll*, and contribute to the establishment of the dorsal appendage primordia (Cohen *et al.*, 1993; McKay *et al.*, 2009). Second, the balance between the sizes of the ventral and dorsal appendage primordia is sensitive to two opposing signaling pathways: Dpp and EGFR. High levels of Dpp signaling promote dorsal appendage and coxopodite fates, while high levels of EGFR signaling repress dorsal appendage fates and promote ventral appendage fates (Goto and Hayashi, 1997; Kubota *et al.*, 2000). Finally, the T-box transcription factor-encoding genes *Dorsocross1* (*Doc1*), *Doc2*, and *Doc3* are required for establishing the dorsal, but not the ventral, primordia (Hamaguchi *et al.*, 2004; Reim *et al.*, 2003). Although the relationships between these various inputs (*Sp1*, Dpp, EGFR, and *Doc*) are not understood, these observations provide tantalizing hints at a complex process that establishes the fates of these two appendage primordia from a common group of cells.

## 5. Elaboration of the PD Axis: The Role of *brk*

By the end of embryogenesis, a rudimentary PD axis of the leg is apparent in the expression patterns of *Dll* (via *DllLT*, *DllDKO*, and *DllLP*), *hth*, and *tsh*. How are these initial patterns elaborated upon to create the mature PD axis present in the third larval instar stage? The separation between these two time points is huge both in terms of time (96h) and tissue growth (from ~60 cells to ~10,000 cells). Also, at the end of embryogenesis, *dac* has not yet been activated but begins to be expressed in a circular medial domain in the second instar. Nevertheless, despite these dramatic changes, we are beginning to understand how the late embryonic patterns of gene expression evolve and are eventually stabilized during this phase of development.

As discussed above, many observations in the literature support the idea that the juxtaposition of Wg- and Dpp-expressing cells in the center of the leg disc leads to the formation of the PD axis, including the proper domains of *Dll* and *dac* expression. The activation of these genes by Wg and Dpp does not rely on a third signal, arguing that these two signals are both necessary and sufficient to induce the leg's PD axis (Lecuit and Cohen, 1997). How can the obligate synergy between Wg and Dpp be explained at the molecular level? One answer is that, in the leg disc as in other places during fly development, Dpp functions mainly by repressing the transcriptional repressor, *brinker* (*brk*; Campbell and Tomlinson, 1999; Jazwinska *et al.*, 1999; Minami *et al.*, 1999). In contrast, Wg is an obligate activator of *Dll* and *dac* (Lecuit and Cohen, 1997). In the absence of Dpp, *brk* expands throughout the leg disc and *Dll* and *dac* fail to be activated. Importantly, however, in *brk* and *dpp* double mutants, *Dll* and *dac* expression domains are rescued, in patterns reminiscent to those in a wild-type leg (Estella and Mann, 2008). Thus, in the absence of *brk*, Dpp is not required for either *Dll* or *dac* activation. Further, by uncovering a role for Brk repression in *dac* and *Dll* regulation, they provided an explanation why both Wg and Dpp signals are required for PD axis formation: in the early disc, *DllLT* and *Dll* are only expressed in cells that (1) do not have Brk and (2) are exposed to high Wg levels. It was also suggested that different ratios of Wg (a positive input) and Brk (a negative input) determined whether *dac* or *Dll* would be activated (Estella and Mann, 2008). According to this idea, *Dll* is activated by high levels of Wg and very low or no Brk. In contrast, *dac* is activated by lower levels of Wg and is less sensitive to Brk repression than *Dll*.

## 6. Elaboration of the PD Axis: The Role of a Transcription Factor Cascade and Cross-regulation

In addition to positing that Wg and Dpp are critical for initiating the PD axis, Lecuit and Cohen proposed a gradient model to account for the PD axis expression patterns of *dac* and *Dll*. According to this model, the expression of *dac* and *Dll* along the PD axis depends on the levels of Wg and Dpp a cell perceives: high concentrations of both Wg and Dpp activate *Dll* and repress *dac* in the center of the leg disc; intermediate levels activate *dac* but not *Dll* in medial regions of the disc; and low levels of these morphogens fail to activate either gene (Fig. 7.5A). Two main observations supported this model (Lecuit and Cohen, 1997): (1) ectopic expression of *wg* in the dorsal half of the leg (where Dpp levels are high) recapitulated the wild-type nested pattern of PD gene expression, with *Dll* expressed closest to the Wg source and *dac* further away and (2) mutant clones for *mad* or *disheveled* (*dsh*; essential components of the Dpp and Wg pathways, respectively) in the center of the leg disc derepressed *dac*. In addition, based on their expression patterns, the highest levels of Wg and Dpp would be located in the center of the leg disc where the expression of these two morphogens meet, and the levels would gradually diminish in cells closer to the periphery of the disc (Fig. 7.2).

Although very attractive, the gradient model is difficult to envision at the molecular level, where the inputs from the Wg and Dpp signaling pathways must converge onto the CRMs of *Dll* and *dac* (Fig. 7.5A). The model is also difficult to reconcile with the observation that *brk dpp* mutant discs,

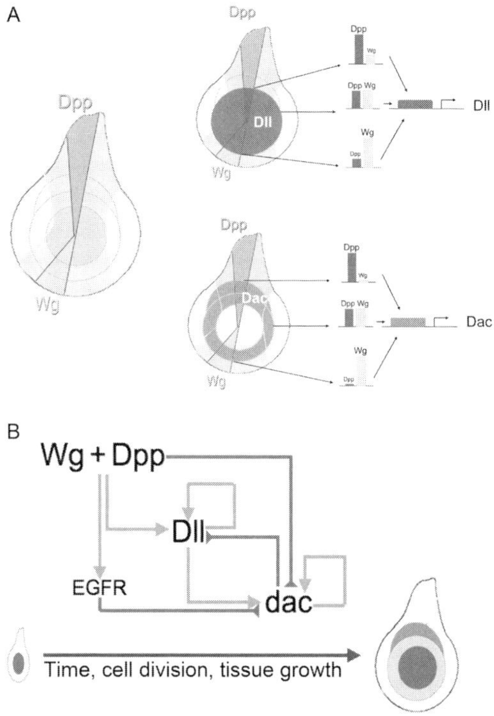


**Figure 7.5** Gradient versus cascade models. (A) The gradient model, highlighting that, depending on a cell's position in the disc, *Dll* and *dac* CRMs must interpret very different ratios of Dpp:Wg signaling. (B) The cascade model, in which Wg+Dpp are only required to initiate PD axis formation by activating *Dll* and ligands for the EGFR pathway. Dll in turn activates *dac*, and both *Dll* and *dac* maintain their expression in a Wg+Dpp-independent manner. EGFR activity maintains *dac* repression, while Wg+Dpp repress *dac* in the center of the leg disc early in leg development. (See Color Insert.)

which do not have a Dpp gradient, have a PD axis (Estella and Mann, 2008). The dissection of *Dll* and *dac* CRMs has provided insights into how these genes respond to Wg and Dpp signaling. As described above, *DllLT* is first activated in the progenitors of the telopodite cells in late embryogenesis. Although *DllLT* remains active during larval development, in third instar discs, it is restricted to cells at the center of the disc, close to where the Wg and Dpp domains touch. This is a small subset of the overall *Dll* domain at this time (Fig. 7.3). Moreover, in contrast to *Dll*, *DllLT* continuously depends on Wg and Dpp signals and it integrates these inputs directly by the binding of the transcription factors Mad, Brk, and Pan (Estella *et al.*, 2008). Although Wg and Dpp meet in other tissues such as the wing disc,

*DllLT* activity is restricted to the leg imaginal disc by *Sp1* and *Dll*, although it is not yet known if this regulation is direct (Estella and Mann, 2010; McKay *et al.*, 2009).

As noted above, the dependency of *Dll* on Wg and Dpp inputs is only transient: by the second instar, these signals are no longer required for *Dll* expression in the leg disc. It has been suggested that the Wg- and Dpp-independent expression phase or "maintenance" is achieved in combination with another *Dll cis*-regulatory element that includes the *Dll* transcription start site. On its own, this element, named *DllM* for **M**aintenance, is weakly active in *Dll*-expressing cells. But when placed in *cis* with *DllLT*, it produces an accurate and robust *Dll* expression pattern in the leg disc (Fig. 7.3; Estella *et al.*, 2008). Moreover, *DllLT+M*, like *Dll*, is able to maintain its expression in the absence of Wg and Dpp inputs, in part by a positive autoregulatory feedback loop. The *M* element contains Dll binding sites that are required for maintenance. Interestingly, the *M* element is also able to produce an accurate Dll pattern when placed close to other *Dll* CRMs that, on their own, do not drive a Dll-like expression pattern (Estella *et al.*, 2008). Although it is very likely that additional CRMs are involved (Estella *et al.*, 2008; Galindo *et al.*, 2011; Vachon *et al.*, 1992), the "trigger-maintenance" mechanism provides a molecular explanation for how Wg and Dpp activate *Dll* and how *Dll* expression is maintained as the disc grows, despite widely varying levels of Wg+Dpp signaling in Dll domain. Importantly, this mechanism does not require *Dll* CRMs to integrate gradients of Dpp and Wg inputs.

The molecular dissection of a *dac* CRM (*dacRE* for **R**ing **E**nhancer), which recapitulates most of the medial expression pattern of this gene (Fig. 7.3B), suggests that *dac* also does not need to interpret Wg and Dpp gradients for its activation in the medial leg domain (Giorgianni and Mann, 2011). Eliminating most of the putative Pan- and Mad-binding sites has no or very little effect on *dacRE* activity in third instar leg discs, suggesting that this element is not integrating intermediate levels of Wg and Dpp. Instead of being activated by Wg and Dpp, *dacRE* is directly activated by Dll, consistent with lineage tracing experiments showing that the Dac domain is derived from *DllLT*-expressing cells (McKay *et al.*, 2009). Moreover, *dacRE* repression in the distal tip of the leg by Wg and Dpp is transient and is maintained by other transcription factors expressed later in development (see below). In summary, instead of using a gradient mechanism, these results suggest that this phase of PD axis formation depends on a genetic cascade, in which Wg+Dpp activate *Dll*, and Dll activates *dac* (Fig. 7.5B).

If gradients of Wg and Dpp signaling are not required for *Dll* and *dac* activation, how are these two genes differentially expressed along the PD axis of the leg? One plausible scenario is as follows (Fig. 7.6): During the first and second instar leg disc, high levels of Wg and Dpp activate *Dll* expression, in part via *DllLT*, and repress *dac* via the *dacRE* element. As the disc grows, *dac* is activated by Dll in cells where the levels of Wg and Dpp

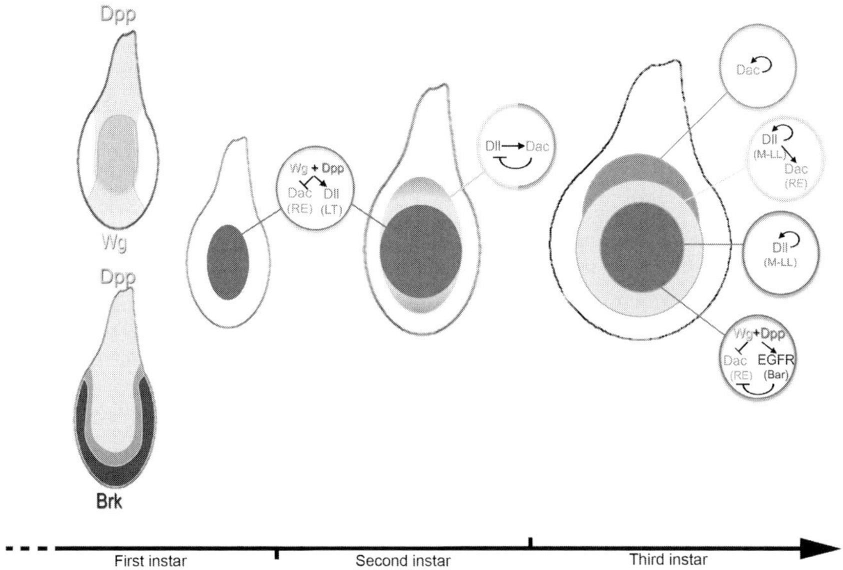


**Figure 7.6** Summary of the dynamic network establishing the PD axis. The relationship between Dpp, Wg, and Brk triggers the expression of *Dll* and represses *dac* in the center of the disc. As the disc grows, Dll activates *dac* in cells that escape repression by Wg+Dpp. These initial domains are likely maintained by a combination of autoregulation, cross-regulation, and transcriptional memory systems. (See Color Insert.)

signaling are low enough to escape repression. As cells continue to divide, some *Dll*- and *dac*-expressing cells will end up removed from peak Wg and Dpp levels and lose *Dll* expression, generating the Dac-only domain. In this manner, the three primary PD domains of gene expression, Dll only, Dll+ Dac, and Dac only, have been specified. At about the same time, the expression of these genes starts to become independent of the activating signals and may be maintained, in part, by an autoregulatory mechanism (via the *M* element for *Dll*) or by a transcriptional memory mechanism (Kim *et al.*, 2008). One key aspect to this model is that Wg+Dpp are only required to initiate the PD cascade: by initially activating *Dll* (and repressing *dac*) in the center of the imaginal disc, Wg+Dpp trigger the cascade but are no longer essential after this step (Figs. 7.5B and 7.6). Although there are many aspects to this model that need confirmation, such a mechanism, coupled to tissue growth, is sufficient to account for the generation of stable domains of gene expression along the PD axis.

Finally, we highlight the role of cross-regulation and additional repressive interactions for the maintenance of PD domains. As noted above, the coxopodite domain is defined by the expression of *hth* and *tsh* in the body wall and coxa (Abu-Shaar and Mann, 1998; Erkner *et al.*, 1999; Gonzalez-Crespo and Morata, 1996; Rieckhof *et al.*, 1997; Wu and Cohen, 1999, 2000). There is evidence that the combined input from Wg and Dpp signaling represses *hth* and *tsh* expression in the telopodite cells via activation of *Dll* and *dac* (Abu-Shaar and Mann, 1998; Gonzalez-Crespo *et al.*, 1998) and by a *Dll*-independent mechanism (Wu and Cohen, 1999). This *Dll*-independent repression mechanism may be mediated via the *elbow–no ocelli* gene complex, which depends on Wg and Dpp signaling for expression and delimits the appendage field in the leg imaginal disc (Weihe *et al.*, 2004). In addition, repression of *dac* by the homeodomain proteins Bar (BarH1/BarH2) is one way in which *dac* repression is maintained in distal leg cells (Giorgianni and Mann, 2011; Kojima *et al.*, 2000). Additional cross-regulation between PD genes is also observed among other transcription factors expressed in the tarsal segments, downstream of EGFR signaling (see below).

## 7. Patterning the DV Axis

If gradients of Wg and Dpp are not used to establish the PD axis, what purpose might they serve? The posterior expression of the homeodomain transcription factors *engrailed* (*en*) and *invected* (*inv*) divides the leg into anterior and posterior compartments, which have distinct cell lineages (Morata and Lawrence, 1975). In contrast to strict lineage restrictions along the AP axis, the distinction between dorsal and ventral fates is controlled by the secreted molecules Wg and Dpp in a nonlineage-dependent manner. As noted above, in the leg imaginal disc, Hh signals from the posterior compartment to anterior compartment cells to activate the expression of *wg* and *dpp* in the ventral and dorsal halves of the disc, respectively (Basler and Struhl, 1994). The anterior dorsal and anterior ventral expression of these two genes in the leg imaginal disc is maintained by a mutual antagonism that prevents the expression and/or signaling of these pathways in the other half of the disc. Dpp specifies dorsal fates and represses ventral ones, whereas Wg specifies ventral fates and represses dorsal ones (Brook and Cohen, 1996; Jiang and Struhl, 1996; Johnston and Schubiger, 1996; Morimura *et al.*, 1996; Penton and Hoffmann, 1996; Struhl and Basler, 1993; Theisen *et al.*, 1996; Wilder and Perrimon, 1995). Hypomorphic mutations in *wg* result in the mirror image duplication of the dorsal leg pattern at the expense of the ventral pattern. Analogously, mutations in *dpp* have the opposite effects (Brook and Cohen, 1996; Jiang and Struhl, 1996; Theisen *et al.*, 1996). Thus, although gradients of Wg and Dpp signaling may not be required for specifying distinct PD axis fates, they appear to play a critical role in establishing positional information along the DV axis.

In the tip of the leg, all cells are likely to perceive high levels of Wg and Dpp, so understanding how a cell is able to discriminate between these signals to promote dorsal or ventral fates is an important question. One possibility is that there are dorsal and/or ventral selector genes activated downstream of these signals. Recently, Svendsen *et al.* (2009) characterized the expression and function of two redundant Tbx20 transcription factors, *H15* and *midline* (*mid*), two genes that can fulfil this role (Svendsen *et al.*, 2009). These genes are expressed in the ventral half of the leg disc, in a broader domain than *wg* that coincides with the domain deleted in *wg* mutants. In *H15* and *mid* mutants, ventral leg structures are transformed to the corresponding dorsal ones, without affecting *wg* or *dpp* expression, probably due to the low levels of Dpp signaling found in the ventral domain (Blackman *et al.*, 1991). Moreover, ectopic *mid* and *H15* expression is able to induce ventral fates in dorsal cells. Although Dpp-dependent dorsal fates have been suggested to be mediated by the dorsal-specific T-box gene *optomotor-blind* (*omb*; Grimm and Pflugfelder, 1996; Maves and Schubiger, 1998), *omb* is not derepressed in *mid* mutant clones (Svendsen *et al.*, 2009). One possibility is that the DV "ground state" is dorsal, but this is argued against by the observation that lateral structures are formed in legs with reduced expression of both *wg* and *dpp* (Held *et al.*, 1994). Thus, other factors in addition to *omb* are probably required to define dorsal fates.

## 8. EGFR Signaling Patterns the Tarsus

While Wg and Dpp play an important role in initiating the PD axis, by the early third instar, Wg and Dpp are no longer required for the PD axis and the role of further elaborating this axis is handed off to the EGFR signaling pathway (Campbell, 2002; Galindo *et al.*, 2002; Fig. 7.7). Shifts of the temperature-sensitive mutant $Egfr^{[ts]}$ ($Egfr^{[tsla]}/Egfr^{[null]}$) to the restrictive temperature in the beginning of the third larval stage lead to development of legs without pretarsus and one or more tarsal segments depending on the

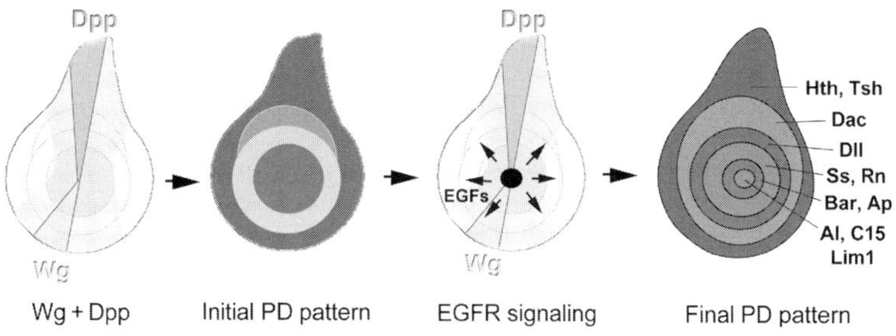


**Figure 7.7** EGFR signaling patterns the tarsal segments. After the initial PD domains are established, EGFR ligands are produced at the center of the disc and activate a series of secondary PD targets in the progenitors to the tarsal segments. (See Color Insert.)

severity of the restrictive temperature. Similar results were obtained when dominant-negative forms or inhibitors of EGFR signaling components were ectopically expressed in leg discs (Campbell, 2002; Galindo *et al.*, 2002, 2005). Genetic experiments suggest that initiation of EGFR signaling in the center of leg discs is dependent on *wg*, *dpp*, and *Dll* (Galindo *et al.*, 2002, 2005). Although it is not known how this occurs at a mechanistic level, it is plausible that in this case graded EGFR activity may be important for setting up distinct PD fates in the tarsus.

EGFR signaling occurs in waves, by the consecutive activation of ligands, proteases, and inhibitors (reviewed in Shilo, 2005). In leg imaginal discs, EGFR signaling is initiated in early third instar larvae by expression of the secreted ligand Vein (Vn) in the central region of leg disc (Campbell, 2002; Galindo *et al.*, 2002), which we refer to as the EGF Organizing Center (EOC). Shortly after *vein* is activated, another component of the EGFR signaling cascade is expressed: the protease Rhomboid (Rho; Campbell, 2002; Galindo *et al.*, 2005). Rho is required for processing of the membrane-bound ligands Spitz (Spi), Keren (Krn), and Gurken (Grk); without Rho these ligands are not secreted (reviewed in Urban, 2006). Although none of the membrane-bound ligands have been detected in the center of leg discs, Vn cannot be the only ligand that plays a role in the EOC because *vn* single mutant leg discs do not phenocopy *Egfr* mutants (Campbell, 2002; Galindo *et al.*, 2005). Only the triple mutant *ru rho vn* produces strong phenotypes that resemble medium *Egfr*$^{[ts]}$ leg mutants (*roughoid* (*ru*) encodes a paralog of *rho*). Small regions of wild-type tissue in the center of otherwise *ru rho vn* mutant leg discs can rescue tarsal formation (Campbell, 2002), providing additional evidence that the EOC serves as a source of secreted EGFR ligands that pattern the tarsus.

*Drosophila* EGFR signaling is a typical Ras–Raf–MAPK (Map kinase) signaling pathway. In the leg imaginal disc, phosphorylated MAPK is detected shortly after expression of *vn* in the EOC (Campbell, 2002). Moreover, the Ets transcription factor PointedP2 (PntP2), a common downstream effector of receptor tyrosine kinase (RTK) signaling in *Drosophila* (Brunner *et al.*, 1994; O'Neill *et al.*, 1994; Scholz *et al.*, 1997), is also expressed in the EOC, and misexpression of a dominant-negative form of *pntP2* (*pntP2*$^{[DN]}$) abolishes tarsal segments 4 and 5 and pretarsus (Galindo *et al.*, 2005). Although the truncation phenotype in this experiment is not as strong as that of *Egfr*$^{[ts]}$, it is possible that PntP2$^{[DN]}$ cannot completely block PntP2 activity or that additional EGFR downstream effectors participate in this process. Interestingly, gene repression by general repressors such as Capicua and Groucho was recently shown to be relieved by EGFR signaling in other developmental contexts (Ajuria *et al.*, 2011; Cinnamon *et al.*, 2008). Similar mechanisms might be involved in leg disc patterning by EGFR.

The expression of several transcription factors required for PD patterning is dramatically changed in *Egfr* mutants, *Egfr* mosaic clones, or ectopic EGFR activation. The results of several studies (Campbell, 2002, 2005;

Galindo *et al.*, 2002, 2005) show that high levels of EGFR signaling are required for activation of *aristaless* (*al*), *C15* (*clawless*), and *Lim1* in small overlapping circular domains in the very center of the leg discs (Fig. 7.7). These homeodomain proteins are required for the specification of the pretarsus (Campbell, 2005; Campbell and Tomlinson, 1998; Kojima *et al.*, 2005; Pueyo and Couso, 2004; Pueyo *et al.*, 2000; Tsuji *et al.*, 2000). Lower levels of EGFR signaling are required for expression of another set of homeodomain-encoding genes, *BarH1/BarH2* and *apterous* (*ap*), in rings surrounding the *al/C15/Lim1* domain. Apterous and BarH1/BarH2 specify tarsal segments 4 and 5 (Kojima *et al.*, 2000; Pueyo and Couso, 2004; Pueyo *et al.*, 2000). *tarsal-less* (*tal*), *spineless* (*ss*), *rotund* (*rn*), and *bric-a-brac* (*bab*) are expressed in concentric rings just proximal to the rings of *BarH1/BarH2* and *ap*. These genes are required for the proper development of tarsal segments 1–5 (Couderc *et al.*, 2002; Duncan *et al.*, 1998; Galindo *et al.*, 2007; Godt *et al.*, 1993; St Pierre *et al.*, 2002). *tal*, *ss*, *rn*, and *bab* are repressed directly or indirectly by EGFR signaling in the EOC since mild loss of EGFR signaling abolishes *al*, *C15*, and *Lim1* expression, while *Bar*, *ap*, *ss*, *rn*, and *bab* expression shifts toward the center of leg discs. Strong loss of EGFR signaling abolishes expression of most tarsal patterning genes while at the same time allows for *dac* expression in the center of leg discs (Campbell, 2002; Galindo *et al.*, 2002). This latter observation is consistent with mutagenesis studies of the *dacRE* element, which suggested that Bar is a direct repressor of *dac* in third instar leg discs (Giorgianni and Mann, 2011). Spineless and the zinc-finger transcription factor Rotund are also involved in delimiting the distal margin of the *dac* expression domain in mid-third instar—*ss; rn* double mutant leg discs show expansion of the *dac* domain distally and the *BarH1/H2* domain proximally until they juxtapose each other (Pueyo and Couso, 2008).

In addition to being targets of EGFR signaling, the tarsal PD patterning genes show complex interactions with each other. For example, Al and C15 form a protein complex and are required for the expression of *Lim1*, while Lim1 together with its cofactor Chip maintains the expression domain of *al* and *C15* by repressing *BarH1/H2* in the very center of leg discs (Campbell, 2005; Kojima *et al.*, 2005; Pueyo and Couso, 2004). BarH1/H2, in turn, keeps *dac* off in the EOC (Giorgianni and Mann, 2011; Kojima *et al.*, 2000) and also activates the polycistronic gene *tal* in a ring between its own expression domain and the *dac* domain (Pueyo and Couso, 2008). Interestingly, all of the above-mentioned EGFR targets in the tarsus encode transcription factors except for *tal*, which encodes four short peptides (Galindo *et al.*, 2007). During embryogenesis and later during leg joint formation, these peptides act posttranslationally to cleave and modulate the activity (repressor vs. activator) of the zinc-finger transcription factor Shavenbaby (Svb; Kondo *et al.*, 2010; Pueyo and Couso, 2011). As a result of Tal function, a cell-nonautonomous signal is released that triggers the expression of *spineless* and *rotund* in cells in between the domains of Dac and

BarH1/H2 (Pueyo and Couso, 2008). *bab* (BTB/POZ pair of genes *bab1/bab2*) is expressed in a similar domain as *rn* and *ss* (Godt *et al.*, 1993), but its early regulation seems to be independent of Tal and Ss. However, *bab1/2* later expression is aberrant in *ss* or *tal* mutants suggesting that it is at least partially regulated by Ss (Chu *et al.*, 2002; Pueyo and Couso, 2008).

## 9. Leg Segmentation and Growth

The process of leg segmentation, or forming the joints that separate each of the leg segments, is one critical downstream consequence of PD gene expression (Rauskolb, 2001). Several genes and pathways required for forming the joints have been defined (Bishop *et al.*, 1999; Chu *et al.*, 2002; Ciechanska *et al.*, 2007; de Celis Ibeas and Bray, 2003; de Celis *et al.*, 1998; Galindo *et al.*, 2005; Greenberg and Hatini, 2009, 2011; Hao *et al.*, 2003; Kerber *et al.*, 2001; Mishra *et al.*, 2001; Pueyo and Couso, 2011; Rauskolb and Irvine, 1999; Shirai *et al.*, 2007). A key step in leg segmentation is the induction of the Notch ligands Delta and Serrate (Bishop *et al.*, 1999; de Celis *et al.*, 1998; Rauskolb and Irvine, 1999), which activate the Notch signal transduction pathway (reviewed by Greenwald, 1998; Kimble and Simpson, 1997). Notch activation at the presumptive borders between leg segments, in turn, induces the expression of several downstream genes, including *d-AP2* and the *odd-skipped* family members, *drumstick* (*drm*), *odd-skipped* (*odd*), *brother of odd with entrails limited* (*bowl*), and *sister of odd and bowl* (*sob*; de Celis Ibeas and Bray, 2003; Hao *et al.*, 2003; Kerber *et al.*, 2001; Rauskolb and Irvine, 1999). Together with the nuclear protein Lines, which antagonizes Bowl, the activity of these genes establishes a feedback mechanism that stabilizes these borders, allowing joint morphogenesis to ensue (Greenberg and Hatini, 2009). In addition, EGFR signaling is induced in the interjoint regions. These waves of EGFR signaling prevent formation of supernumerary joints by antagonizing some of the Notch effector genes (Galindo *et al.*, 2005; Shirai *et al.*, 2007). Other Notch downstream genes, such as *nubbin* and *d-AP2*, play a role in leg growth via mechanisms that have not yet been established (Kerber *et al.*, 2001; Rauskolb and Irvine, 1999).

## 10. Concluding Remarks

The above review reveals that a molecular framework of PD axis formation is now emerging. Yet, many questions remain. For one, the initial stages of dorsal and ventral primordia establishment are not well understood. How, for example, does *Sp1* block dorsal appendage fates? Second, although graded levels of Wg and Dpp activities may not be relevant to elaborating the PD axis, it remains an open question whether a gradient of EGFR signaling is used to turn on its targets in the tarsus. Third, it is not clear how large a role transcriptional memory mechanisms play in maintaining PD gene expression domains once they are initially established, and how the transition from establishment to maintenance occurs. However, now that there are working hypotheses and a molecular framework for how the initial PD pattern is formed, it is likely that these and other questions will soon be answered.

## ACKNOWLEDGMENTS

We are grateful to members of the Mann, Struhl, and Johnston labs for comments, and funding from the NIH (GM058575). R. V. is a CDP Fellow of the Leukemia and Lymphoma Society.

## REFERENCES

Abu-Shaar, M., and Mann, R. S. (1998). Generation of multiple antagonistic domains along the proximodistal axis during Drosophila leg development. *Development* **125,** 3821–3830.

Ajuria, L., Nieva, C., Winkler, C., *et al.* (2011). Capicua DNA-binding sites are general response elements for RTK signaling in Drosophila. *Development* **138,** 915–924.

Azpiazu, N., and Morata, G. (2002). Distinct functions of homothorax in leg development in Drosophila. *Mech. Dev.* **119,** 55–67.

Baker, N. E. (2011). Proximodistal patterning in the Drosophila leg: Models and mutations. *Genetics* **187,** 1003–1010.

Basler, K., and Struhl, G. (1994). Compartment boundaries and the control of Drosophila limb pattern by hedgehog protein. *Nature* **368,** 208–214.

Bell, S. M., Schreiner, C. M., Waclaw, R. R., *et al.* (2003). Sp8 is crucial for limb outgrowth and neuropore closure. *Proc. Natl. Acad. Sci. USA* **100,** 12195–12200.

Bishop, S. A., Klein, T., Arias, A. M., and Couso, J. P. (1999). Composite signalling from Serrate and Delta establishes leg segments in Drosophila through Notch. *Development* **126,** 2993–3003.

Blackman, R. K., Sanicola, M., Raftery, L. A., *et al.* (1991). An extensive 3′ *cis*-regulatory region directs the imaginal disk expression of decapentaplegic, a member of the TGF-beta family in Drosophila. *Development* **111,** 657–666.

Bolinger, R. A., and Boekhoff-Falk, G. (2005). Distal-less functions in subdividing the Drosophila thoracic limb primordium. *Dev. Dyn.* **232,** 801–816.

Brook, W. J., and Cohen, S. M. (1996). Antagonistic interactions between wingless and decapentaplegic responsible for dorsal-ventral pattern in the Drosophila Leg. *Science* **273,** 1373–1377.

Brunner, D., Ducker, K., Oellers, N., *et al.* (1994). The ETS domain protein pointed-P2 is a target of MAP kinase in the sevenless signal transduction pathway. *Nature* **370,** 386–389.

Bryant, P. J., and Schneiderman, H. A. (1969). Cell lineage, growth, and determination in the imaginal leg discs of Drosophila melanogaster. *Dev. Biol.* **20,** 263–290.

Campbell, G. (2002). Distalization of the Drosophila leg by graded EGF-receptor activity. *Nature* **418,** 781–785.

Campbell, G. (2005). Regulation of gene expression in the distal region of the Drosophila leg by the Hox11 homolog, C15. *Dev. Biol.* **278,** 607–618.

Campbell, G., and Tomlinson, A. (1998). The roles of the homeobox genes aristaless and Distal-less in patterning the legs and wings of Drosophila. *Development* **125,** 4483–4493.

Campbell, G., and Tomlinson, A. (1999). Transducing the Dpp morphogen gradient in the wing of Drosophila: Regulation of Dpp targets by brinker. *Cell* **96,** 553–562.

Campbell, G., Weaver, T., and Tomlinson, A. (1993). Axis specification in the developing Drosophila appendage: The role of wingless, decapentaplegic, and the homeobox gene aristaless. *Cell* **74,** 1113–1123.

Castelli-Gair, J., and Akam, M. (1995). How the Hox gene Ultrabithorax specifies two different segments: The significance of spatial and temporal regulation within metameres. *Development* **121,** 2973–2982.

Chu, J., Dong, P. D., and Panganiban, G. (2002). Limb type-specific regulation of bric a brac contributes to morphological diversity. *Development* **129,** 695–704.

Ciechanska, E., Dansereau, D. A., Svendsen, P. C., *et al.* (2007). dAP-2 and defective proventriculus regulate Serrate and Delta expression in the tarsus of Drosophila melanogaster. *Genome* **50,** 693–705.

Cinnamon, E., Helman, A., Ben-Haroush Schyr, R., *et al.* (2008). Multiple RTK pathways downregulate Groucho-mediated repression in Drosophila embryogenesis. *Development* **135,** 829–837.

Cohen, S. M. (1990). Specification of limb development in the Drosophila embryo by positional cues from segmentation genes. *Nature* **343,** 173–177.

Cohen, S. M., and Jurgens, G. (1989). Proximal-distal pattern formation in Drosophila: Cell autonomous requirement for Distal-less gene activity in limb development. *EMBO J.* **8,** 2045–2055.

Cohen, S. M., Bronner, G., Kuttner, F., *et al.* (1989). Distal-less encodes a homoeodomain protein required for limb development in Drosophila. *Nature* **338,** 432–434.

Cohen, B., Simcox, A. A., and Cohen, S. M. (1993). Allocation of the thoracic imaginal primordia in the Drosophila embryo. *Development* **117,** 597–608.

Couderc, J. L., Godt, D., Zollman, S., *et al.* (2002). The bric a brac locus consists of two paralogous genes encoding BTB/POZ domain proteins and acts as a homeotic and morphogenetic regulator of imaginal development in Drosophila. *Development* **129,** 2419–2433.

de Celis Ibeas, J. M., and Bray, S. J. (2003). Bowl is required downstream of Notch for elaboration of distal limb patterning. *Development* **130,** 5943–5952.

de Celis, J. F., Tyler, D. M., de Celis, J., and Bray, S. J. (1998). Notch signalling mediates segmentation of the Drosophila leg. *Development* **125,** 4617–4626.

Dey, B. K., Zhao, X. L., Popo-Ola, E., and Campos, A. R. (2009). Mutual regulation of the Drosophila disconnected (disco) and Distal-less (Dll) genes contributes to proximal-distal patterning of antenna and leg. *Cell Tissue Res.* **338,** 227–240.

Diaz-Benjumea, F. J., Cohen, B., and Cohen, S. M. (1994). Cell interaction between compartments establishes the proximal-distal axis of Drosophila legs. *Nature* **372,** 175–179.

Duncan, D. M., Burgess, E. A., and Duncan, I. (1998). Control of distal antennal identity and tarsal development in Drosophila by spineless-aristapedia, a homolog of the mammalian dioxin receptor. *Genes Dev.* **12,** 1290–1303.

Emmons, R. B., Duncan, D., Estes, P. A., *et al.* (1999). The spineless-aristapedia and tango bHLH-PAS proteins interact to control antennal and tarsal development in Drosophila. *Development* **126,** 3937–3945.

Erkner, A., Gallet, A., Angelats, C., *et al.* (1999). The role of Teashirt in proximal leg development in Drosophila: Ectopic Teashirt expression reveals different cell behaviours in ventral and dorsal domains. *Dev. Biol.* **215,** 221–232.

Estella, C., and Mann, R. S. (2008). Logic of Wg and Dpp induction of distal and medial fates in the Drosophila leg. *Development* **135,** 627–636.

Estella, C., and Mann, R. S. (2010). Non-redundant selector and growth-promoting functions of two sister genes, buttonhead and Sp1, in Drosophila leg development. *PLoS Genet.* **6,** e1001001.

Estella, C., Rieckhof, G., Calleja, M., and Morata, G. (2003). The role of buttonhead and Sp1 in the development of the ventral imaginal discs of Drosophila. *Development* **130,** 5929–5941.

Estella, C., McKay, D. J., and Mann, R. S. (2008). Molecular integration of wingless, decapentaplegic, and autoregulatory inputs into Distalless during Drosophila leg development. *Dev. Cell* **14,** 86–96.

Fuse, N., Hirose, S., and Hayashi, S. (1996). Determination of wing cell fate by the escargot and snail genes in Drosophila. *Development* **122,** 1059–1067.

Galindo, M. I., Bishop, S. A., Greig, S., and Couso, J. P. (2002). Leg patterning driven by proximal-distal interactions and EGFR signaling. *Science* **297,** 256–259.

Galindo, M. I., Bishop, S. A., and Couso, J. P. (2005). Dynamic EGFR-Ras signalling in Drosophila leg development. *Dev. Dyn.* **233,** 1496–1508.

Galindo, M. I., Pueyo, J. I., Fouix, S., *et al.* (2007). Peptides encoded by short ORFs control development and define a new eukaryotic gene family. *PLoS Biol.* **5,** e106.

Galindo, M. I., Fernandez-Garza, D., Phillips, R., and Couso, J. P. (2011). Control of Distal-less expression in the Drosophila appendages by functional 3′ enhancers. *Dev. Biol.* **353,** 396–410.

Gebelein, B., Culi, J., Ryoo, H. D., *et al.* (2002). Specificity of Distalless repression and limb primordia development by abdominal Hox proteins. *Dev. Cell* **3,** 487–498.

Gebelein, B., McKay, D. J., and Mann, R. S. (2004). Direct integration of Hox and segmentation gene inputs during Drosophila development. *Nature* **431,** 653–659.

Gehring, W. J., Wieschaus, E., and Holliger, M. (1976). The use of 'normal' and 'transformed' gynandromorphs in mapping the primordial germ cells and the gonadal mesoderm in Drosophila. *J. Embryol. Exp. Morphol.* **35,** 607–616.

Giorgianni, M. W., and Mann, R. S. (2011). Establishment of medial fates along the proximodistal axis of the Drosophila leg through direct activation of dachshund by Distalless. *Dev. Cell* **20,** 455–468.

Godt, D., Couderc, J. L., Cramton, S. E., and Laski, F. A. (1993). Pattern formation in the limbs of Drosophila: Bric a brac is expressed in both a gradient and a wave-like pattern and is required for specification and proper segmentation of the tarsus. *Development* **119,** 799–812.

Gonzalez-Crespo, S., and Morata, G. (1996). Genetic evidence for the subdivision of the arthropod limb into coxopodite and telopodite. *Development* **122,** 3921–3928.

Gonzalez-Crespo, S., Abu-Shaar, M., Torres, M., *et al.* (1998). Antagonism between extradenticle function and Hedgehog signalling in the developing limb. *Nature* **394,** 196–200.

Goto, S., and Hayashi, S. (1997). Specification of the embryonic limb primordium by graded activity of Decapentaplegic. *Development* **124,** 125–132.

Greenberg, L., and Hatini, V. (2009). Essential roles for lines in mediating leg and antennal proximodistal patterning and generating a stable Notch signaling interface at segment borders. *Dev. Biol.* **330,** 93–104.

Greenberg, L., and Hatini, V. (2011). Systematic expression and loss-of-function analysis defines spatially restricted requirements for Drosophila RhoGEFs and RhoGAPs in leg morphogenesis. *Mech. Dev.* **128,** 5–17.

Greenwald, I. (1998). LIN-12/Notch signaling: Lessons from worms and flies. *Genes Dev.* **12,** 1751–1762.

Grimm, S., and Pflugfelder, G. O. (1996). Control of the gene optomotor-blind in Drosophila wing development by decapentaplegic and wingless. *Science* **271,** 1601–1604.

Hamaguchi, T., Yabe, S., Uchiyama, H., and Murakami, R. (2004). Drosophila Tbx6-related gene, Dorsocross, mediates high levels of Dpp and Scw signal required for the development of amnioserosa and wing disc primordium. *Dev. Biol.* **265,** 355–368.

Hao, I., Green, R. B., Dunaevsky, O., *et al.* (2003). The odd-skipped family of zinc finger genes promotes Drosophila leg segmentation. *Dev. Biol.* **263,** 282–295.

Hayashi, S., Hirose, S., Metcalfe, T., and Shirras, A. D. (1993). Control of imaginal cell development by the escargot gene of Drosophila. *Development* **118,** 105–115.

Held, L. I. J., Heup, M. A., Sappington, J. M., and Peters, S. D. (1994). Interactions of decapentaplegic, wingless, and Distalless in the Drosophila leg. *Rouxs Arch. Dev. Biol.* **203,** 310–319.

Jazwinska, A., Kirov, N., Wieschaus, E., *et al.* (1999). The Drosophila gene brinker reveals a novel mechanism of Dpp target gene regulation. *Cell* **96,** 563–573.

Jiang, J., and Struhl, G. (1996). Complementary and mutually exclusive activities of decapentaplegic and wingless organize axial patterning during Drosophila leg development. *Cell* **86,** 401–409.

Johnston, L. A., and Schubiger, G. (1996). Ectopic expression of wingless in imaginal discs interferes with decapentaplegic expression and alters cell determination. *Development* **122,** 3519–3529.

Kawakami, Y., Esteban, C. R., Matsui, T., *et al.* (2004). Sp8 and Sp9, two closely related buttonhead-like transcription factors, regulate Fgf8 expression and limb outgrowth in vertebrate embryos. *Development* **131,** 4763–4774.

Keilin, D. (1915). Recherches sur les larves de dipteres cyclorhaphes. *Bull. Sci. Fr. Belg.* **49,** 15–198.

Kerber, B., Monge, I., Mueller, M., *et al.* (2001). The AP-2 transcription factor is required for joint formation and cell survival in Drosophila leg development. *Development* **128,** 1231–1238.

Kim, S. N., Jung, K. I., Chung, H. M., *et al.* (2008). The pleiohomeotic gene is required for maintaining expression of genes functioning in ventral appendage formation in Drosophila melanogaster. *Dev. Biol.* **319,** 121–129.

Kimble, J., and Simpson, P. (1997). The LIN-12/Notch signaling pathway and its regulation. *Annu. Rev. Cell Dev. Biol.* **13,** 333–361.

Kojima, T. (2004). The mechanism of Drosophila leg development along the proximodistal axis. *Dev. Growth Differ.* **46,** 115–129.

Kojima, T., Sato, M., and Saigo, K. (2000). Formation and specification of distal leg segments in Drosophila by dual Bar homeobox genes, BarH1 and BarH2. *Development* **127,** 769–778.

Kojima, T., Tsuji, T., and Saigo, K. (2005). A concerted action of a paired-type homeobox gene, aristaless, and a homolog of Hox11/tlx homeobox gene, clawless, is essential for the distal tip development of the Drosophila leg. *Dev. Biol.* **279,** 434–445.

Kondo, T., Plaza, S., Zanet, J., *et al.* (2010). Small peptides switch the transcriptional activity of Shavenbaby during Drosophila embryogenesis. *Science* **329,** 336–339.

Kubota, K., Goto, S., Eto, K., and Hayashi, S. (2000). EGF receptor attenuates Dpp signaling and helps to distinguish the wing and leg cell fates in Drosophila. *Development* **127,** 3769–3776.

Lecuit, T., and Cohen, S. M. (1997). Proximal–distal axis formation in the Drosophila leg. *Nature* **388,** 139–145.

Maeda, R. K., and Karch, F. (2011). Gene expression in time and space: Additive vs hierarchical organization of cis-regulatory regions. *Curr. Opin. Genet. Dev.* **21,** 187–193.

Mann, R. S. (1994). Engrailed-mediated repression of Ultrabithorax is necessary for the parasegment 6 identity in Drosophila. *Development* **120,** 3205–3212.

Mardon, G., Solomon, N. M., and Rubin, G. M. (1994). dachshund encodes a nuclear protein required for normal eye and leg development in Drosophila. *Development* **120,** 3473–3486.

Maves, L., and Schubiger, G. (1998). A molecular basis for transdetermination in Drosophila imaginal discs: Interactions between wingless and decapentaplegic signaling. *Development* **125,** 115–124.

McKay, D. J., Estella, C., and Mann, R. S. (2009). The origins of the Drosophila leg revealed by the cis-regulatory architecture of the Distalless gene. *Development* **136,** 61–71.

Minami, M., Kinoshita, N., Kamoshida, Y., *et al.* (1999). brinker is a target of Dpp in Drosophila that negatively regulates Dpp-dependent genes. *Nature* **398,** 242–246.

Mishra, A., Agrawal, N., Banerjee, S., *et al.* (2001). Spatial regulation of DELTA expression mediates NOTCH signalling for segmentation of Drosophila legs. *Mech. Dev.* **105,** 115–127.

Morata, G., and Lawrence, P. A. (1975). Control of compartment development by the engrailed gene in Drosophila. *Nature* **255,** 614–617.

Morimura, S., Maves, L., Chen, Y., and Hoffmann, F. M. (1996). decapentaplegic overexpression affects Drosophila wing and leg imaginal disc development and wingless expression. *Dev. Biol.* **177,** 136–151.

O'Neill, E. M., Rebay, I., Tjian, R., and Rubin, G. M. (1994). The activities of two Ets-related transcription factors required for Drosophila eye development are modulated by the Ras/MAPK pathway. *Cell* **78,** 137–147.

Panganiban, G., and Rubenstein, J. L. (2002). Developmental functions of the Distal-less/Dlx homeobox genes. *Development* **129,** 4371–4386.

Panganiban, G., Irvine, S. M., Lowe, C., *et al.* (1997). The origin and evolution of animal appendages. *Proc. Natl. Acad. Sci. USA* **94,** 5162–5166.

Pappu, K. S., Ostrin, E. J., Middlebrooks, B. W., *et al.* (2005). Dual regulation and redundant function of two eye-specific enhancers of the Drosophila retinal determination gene dachshund. *Development* **132,** 2895–2905.

Penton, A., and Hoffmann, F. M. (1996). Decapentaplegic restricts the domain of wingless during Drosophila limb patterning. *Nature* **382,** 162–164.

Pueyo, J. I., and Couso, J. P. (2004). Chip-mediated partnerships of the homeodomain proteins Bar and Aristaless with the LIM-HOM proteins Apterous and Lim1 regulate distal leg development. *Development* **131,** 3107–3120.

Pueyo, J. I., and Couso, J. P. (2008). The 11-aminoacid long Tarsal-less peptides trigger a cell signal in Drosophila leg development. *Dev. Biol.* **324,** 192–201.

Pueyo, J. I., and Couso, J. P. (2011). Tarsal-less peptides control Notch signalling through the Shavenbaby transcription factor. *Dev. Biol.* **355,** 183–193.

Pueyo, J. I., Galindo, M. I., Bishop, S. A., and Couso, J. P. (2000). Proximal-distal leg development in Drosophila requires the apterous gene and the Lim1 homologue dlim1. *Development* **127,** 5391–5402.

Rauskolb, C. (2001). The establishment of segmentation in the Drosophila leg. *Development* **128,** 4511–4521.

Rauskolb, C., and Irvine, K. D. (1999). Notch-mediated segmentation and growth control of the Drosophila leg. *Dev. Biol.* **210,** 339–350.

Reim, I., Lee, H. H., and Frasch, M. (2003). The T-box-encoding Dorsocross genes function in amnioserosa development and the patterning of the dorsolateral germ band downstream of Dpp. *Development* **130,** 3187–3204.

Rieckhof, G. E., Casares, F., Ryoo, H. D., *et al.* (1997). Nuclear translocation of extradenticle requires homothorax, which encodes an extradenticle-related homeodomain protein. *Cell* **91,** 171–183.

Scholz, H., Sadlowski, E., Klaes, A., and Klambt, C. (1997). Control of midline glia development in the embryonic Drosophila CNS. *Mech. Dev.* **64,** 137–151.

Shilo, B. Z. (2005). Regulating the dynamics of EGF receptor signaling in space and time. *Development* **132,** 4017–4027.

Shirai, T., Yorimitsu, T., Kiritooshi, N., *et al.* (2007). Notch signaling relieves the joint-suppressive activity of Defective proventriculus in the Drosophila leg. *Dev. Biol.* **312,** 147–156.

Snodgrass, R. E. (1935). Principles of Insect Morphology. McGraw-Hill, New York.

St Pierre, S. E., Galindo, M. I., Couso, J. P., and Thor, S. (2002). Control of Drosophila imaginal disc development by rotund and roughened eye: Differentially expressed transcripts of the same gene encoding functionally distinct zinc finger proteins. *Development* **129,** 1273–1281.

Struhl, G., and Basler, K. (1993). Organizing activity of wingless protein in Drosophila. *Cell* **72,** 527–540.

Svendsen, P. C., Formaz-Preston, A., Leal, S. M., and Brook, W. J. (2009). The Tbx20 homologs midline and H15 specify ventral fate in the Drosophila melanogaster leg. *Development* **136,** 2689–2693.

Theisen, H., Haerry, T. E., O'Connor, M. B., and Marsh, J. L. (1996). Developmental territories created by mutual antagonism between Wingless and Decapentaplegic. *Development* **122,** 3939–3948.

Treichel, D., Schock, F., Jackle, H., *et al.* (2003). mBtd is required to maintain signaling during murine limb development. *Genes Dev.* **17,** 2630–2635.

Tsuji, T., Sato, A., Hiratani, I., *et al.* (2000). Requirements of Lim1, a Drosophila LIM-homeobox gene, for normal leg and antennal development. *Development* **127,** 4315–4323.

Urban, S. (2006). Rhomboid proteins: Conserved membrane proteases with divergent biological functions. *Genes Dev.* **20,** 3054–3068.

Vachon, G., Cohen, B., Pfeifle, C., *et al.* (1992). Homeotic genes of the Bithorax complex repress limb development in the abdomen of the Drosophila embryo through the target gene Distal-less. *Cell* **71,** 437–450.

Weihe, U., Dorfman, R., Wernet, M. F., *et al.* (2004). Proximodistal subdivision of Drosophila legs and wings: The elbow-no ocelli gene complex. *Development* **131,** 767–774.

Wieschaus, E., and Gehring, W. (1976). Clonal analysis of primordial disc cells in the early embryo of Drosophila melanogaster. *Dev. Biol.* **50,** 249–263.

Wilder, E. L., and Perrimon, N. (1995). Dual functions of wingless in the Drosophila leg imaginal disc. *Development* **121,** 477–488.

Wimmer, E. A., Frommer, G., Purnell, B. A., and Jackle, H. (1996). buttonhead and D-Sp1: A novel Drosophila gene pair. *Mech. Dev.* **59,** 53–62.

Wu, J., and Cohen, S. M. (1999). Proximodistal axis formation in the Drosophila leg: Subdivision into proximal and distal domains by Homothorax and Distal-less. *Development* **126,** 109–117.

Wu, J., and Cohen, S. M. (2000). Proximal distal axis formation in the Drosophila leg: Distinct functions of teashirt and homothorax in the proximal leg. *Mech. Dev.* **94,** 47–56.

CHAPTER EIGHT

# Temporal Specification of Neural Stem Cells: Insights from *Drosophila* Neuroblasts

Cédric Maurange*,†

## Contents



* Aix-Marseille Univ, IBDML, Marseille, France
† CNRS, UMR7288, Marseille, France

*Current Topics in Developmental Biology*, Volume 98
ISSN 0070-2153, DOI: 10.1016/B978-0-12-386499-4.00008-2

## Abstract

The mechanisms underlying the temporal specification of neural stem cells (NSCs), a process by which a single progenitor can generate different types of neurons and glia in an invariant order, are still poorly understood in mammals. However, in the past decade, work on *Drosophila* NSCs, called neuroblasts, has identified a series of sequentially expressed transcription factors that lies at the heart of this phenomenon. Here, I highlight some key findings that illuminate the role of these transcription factors during development and the regulatory principles allowing them not only to promote neuronal diversity but also to control the final number of neurons in the different regions of the nervous system. Ultimately, and given recent evidences of evolutionary conservation, cracking the temporal specification code of *Drosophila* neuroblasts may provide new perspectives for the safe manipulation of human NSCs and their therapeutic use.

## Abbreviations

| | |
|---|---|
| Abd-A | abdominal-A |
| Antp | Antennapedia |
| Cas | Castor |
| CNS | central nervous system |
| Col | Collier |
| D | Dichaete |
| Dan | Distal antenna |
| Grh | Grainyhead |
| Hb | Hunchback |
| Kr | Kruppel |
| MB | mushroom body |
| NSC | neural stem cell |
| Pros | Prospero |
| Sqz | Squeeze |
| Svp | Seven-up |
| TTF | temporal transcription factor |
| Ubx | Ultrabithorax |

# 1. Introduction

A fundamental goal in developmental neurobiology is to understand how a small number of seemingly uniform neural stem cells (NSCs), lying in the neural tube of the early vertebrate embryo, can give rise to the tremendous diversity of differentiated neurons and glia found in the adult central nervous system (CNS). In the past 20 years, the identification of various morphogens and their signaling cascades has led to an understanding of how spatial patterning of the neural tube is established, allowing the different NSCs of the tube to adopt different identities and generate unique lineages based on their position along the different axes (Dessaud *et al.*, 2008). But this only resolves half of the issue, as the diversity of neural progeny is also promoted by the ability of each progenitor to change its specification over time. This temporal patterning system ensures that different types of daughter cells are generated, in a stereotypical order, by the same initial progenitor. This property of NSCs is found in most, if not all, regions of the CNS but is particularly evident in the mammalian cerebral cortex (Molyneaux *et al.*, 2007). In the developing cortex, asymmetrically dividing progenitors, called radial glia, located in the ventricular zone, generate different types of neurons. As neurons are born, they migrate toward the surface of the tissue and form six characteristic layers. In this "inside-out" model, the later a neuronal subtype is born, the more superficial the position it occupies. Thus, the layer a neuron occupies reflects the time of its birth. At the end of development, the six layers can be easily visualized on histological sections and are composed of neurons with distinct topologies, expressing characteristic combinations of transcription factors. One of the most striking observations of recent years arose from *in vitro* experiments showing that isolated embryonic cortical NSCs sequentially generate neurons from most cortical layers, in a timing that parallels what is observed during embryonic development (Gaspard *et al.*, 2008; Naka *et al.*, 2008; Shen *et al.*, 2006). These investigations have thus led to the conclusion that an NSC-encoded timing mechanism underlies the sequential appearance of the different types of neurons. However, in contrast to the case of spatial specification, where many molecular players and their mode of action have been characterized, very little is known about how temporal specification of NSCs is orchestrated in mammals.

As in vertebrates, NSCs in insects are multipotent and divide asymmetrically to generate different types of progeny in a stereotypical manner (Doe, 2008; Pearson and Doe, 2004). In this context, *Drosophila* has become over the past 15 years a valuable model system to investigate the mechanisms by which a single NSC generates a wide repertoire of neural fates. In particular, several studies have highlighted the role of a series of sequentially expressed transcription factors in the temporal specification of *Drosophila* NSCs.

The objectives of this review are to summarize the current knowledge concerning this temporal specification system in *Drosophila*, highlight a set of regulatory mechanisms, and outline key questions that remain to be solved.

## 2. *Drosophila* Neuroblasts: Emergence of a Neural Stem Cell Model

*Drosophila* neural progenitors, or neuroblasts, have been extensively studied since the late 1980s. Because neuroblasts are multipotent and generate the differentiated cells that will constitute the adult CNS, they are considered to be the *Drosophila* equivalent of mammalian NSCs (Doe, 2008). Key discoveries in neuroblasts have greatly contributed to the current state-of-the-art in NSC biology. For instance, we now understand better how NSCs are determined from the neuroectoderm, how spatial determinants influence the identity of the newly formed NSC, how asymmetric division allows differentiation of the progeny while maintaining NSC self-renewal, and finally how progeny fate is specified (Brand and Livesey, 2011; Knoblich, 2010; Skeath and Thor, 2003).

*Drosophila* neuroblasts typically undergo several rounds of asymmetric division to self-renew while budding off different types of intermediate progenitors, called ganglion mother cells (GMCs). Generally (but with exceptions), a GMC divides once to give rise to two differentiated neural cells, either neurons or glia. The specific identity of each neuron and glia can be determined by assessing its morphology or the combination of differentiation markers it expresses.

The mechanisms regulating the spatiotemporal specification of neuroblasts have been well studied in the nerve cord that can be considered to be the *Drosophila* spinal cord. The nerve cord extends over three thoracic and eight abdominal segments. Each hemisegment initially contains the same invariant array of 30 neuroblasts that delaminate from the neuroectoderm during early embryogenesis (Doe, 1992; Technau *et al.*, 2006). Each of the 30 neuroblasts inside an embryonic hemisegment can be identified both by its localization (Bate, 1976; Doe and Goodman, 1985) and the unique combination of spatial markers it expresses (Broadus *et al.*, 1995; Doe, 1992). Based on this spatial information, a grid can be designed in which each neuroblast is assigned a unique coordinate according to its row and column number (NB1-1, NB1-2,..., NB2-1, NB2-2, etc.). In addition, lineage-tracing experiments by injecting a fluorescent dye in neuroblasts have permitted characterization of the neuronal and glial composition of each embryonic lineage that varies from 2 to 40 progeny depending on the neuroblast (Bossing *et al.*, 1996; Schmid *et al.*, 1999, 1997). It was established that single neuroblasts can give rise to motoneurons, interneurons, and glia. Therefore, during this first neurogenic period, neuroblasts generate the progeny that will form the nervous system required for larval life. At the end of embryogenesis, some neuroblasts, mostly abdominal, undergo apoptosis, while the remaining enter a period of quiescence. Mitotic activity in the surviving neuroblasts is restored toward the end of the first larval stage, allowing most neuroblasts to generate about 100 additional progeny (Truman and Bate, 1988). This second neurogenic period contributes to building the adult CNS. During the metamorphic pupal phase, all neuroblasts terminate division via apoptosis or cell cycle exit (Bello *et al.*, 2003; Maurange and Gould, 2005; Maurange *et al.*, 2008; Siegrist *et al.*, 2010; Truman and Bate, 1988). Consequently, no neural proliferation is observed in the adult animal. While initially poorly investigated, postembryonic neuroblasts and their associated lineages have attracted much interest in the past decade, thanks to the appearance of new genetic tools (Lee and Luo, 2001). Because neuroblasts and their associated lineages can now be both individually followed and genetically manipulated throughout development, they represent an outstanding model to investigate the processes governing temporal specification of NSCs.

## 3. The Sequence of "Temporal Transcription Factors" During Embryonic Neurogenesis

### 3.1. The Hb → Kr → Pdm → Cas → Grh sequence

A key observation that emerged from early lineage-tracing studies is that the different types of neuroblast progeny, identified by the expression of specific combinations of neural genes, are generated in an invariant order (Higashijima *et al.*, 1996). In addition, it was uncovered that most embryonic neuroblasts sequentially express a series of three cross-regulating transcription factors: the zinc finger protein Hunchback (Hb), the Pou-domain proteins Pdm1 and 2 that are coexpressed from the same locus, and the zinc finger protein Castor (Cas; Kambadur *et al.*, 1998). When one of these factors is expressed in the neuroblast, it is inherited and maintained in the maturing daughter cell. The sequential expression of Hb → Pdm → Cas in neuroblasts therefore allows each newly born GMC to inherit a different factor and could contribute to the generation of diversity in a lineage (Kambadur *et al.*, 1998). This temporal specification hypothesis was validated by a later study, from Chris Doe's lab, demonstrating that the sequential expression of the three previously described transcription factors in neuroblasts, plus a new zinc finger protein Kruppel (Kr), determines the order of appearance of different types of progeny (Grosskortenhaus *et al.*, 2006; Isshiki *et al.*, 2001). These factors are expressed in an invariable order (Hb → Kr → Pdm → Cas) in most neuroblasts of the nerve cord during embryogenesis (Fig. 8.1A). Because each transcription factor in the series is expressed at a different time during development and specifies the identity

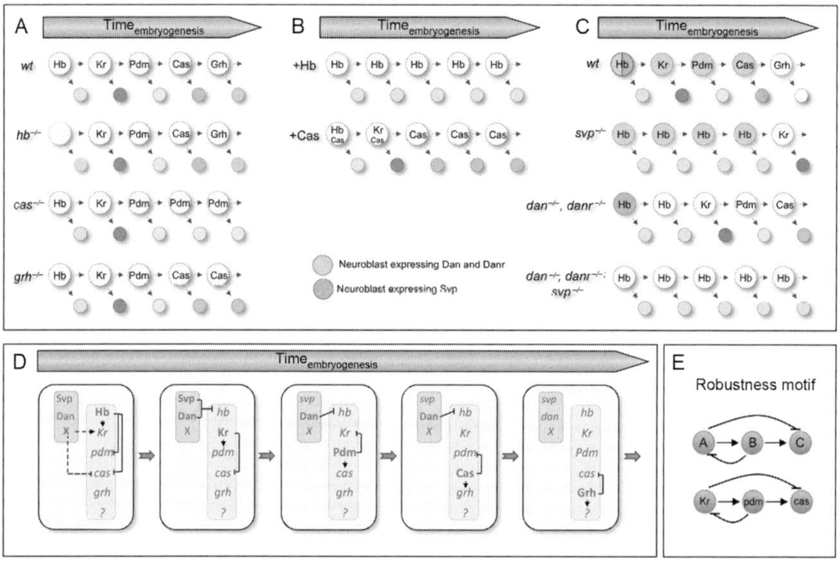


**Figure 8.1** The temporal transcription factor series in embryonic *Drosophila* neuroblasts. Wild-type (*wt*) *Drosophila* neuroblasts (large circles) express five temporal transcription factors (TTFs), at different times during embryogenesis, in the following sequence: Hunchback (Hb) → Kruppel (Kr) → Pdm → Castor (Cas) → Grainyhead (Grh). (A) Each TTF is associated with a postmitotic progeny (small circles) of a defined identity symbolized by the color code (blue, red, yellow, green, and pink). Loss-of-function of a TTF either leads to the skipping of one temporal identity (shown in grey for $hb^{-/-}$) or to stalled temporal series progression (in $cas^{-/-}$ and $grh^{-/-}$). The later phenotype is associated with supernumerary progenies of the previous temporal identity that are generated at the expense of later identities. (B) Continuous misexpression of a TTF also leads to a stalled progression of the series leading to supernumerary progenies with the corresponding temporal identity. (C) The orphan nuclear receptor Seven-up (Svp) and the Pipsqueak domain proteins Dan and Danr function synergistically to repress *hb* transcription and favor the switch to the following temporal identities. In mutants, an excess of early identity neurons are produced. (D) Deployment over time of the temporal series in a neuroblast. In green are shown the factors expressed at a given time. TTFs and factors facilitating transitions are separated in two distinct boxes. Included is the hypothetic factor *x* whose expression and regulatory activity on *Kr* and *cas* could explain why progression of the series is not affected by the loss of Hb, Kr, or Pdm. Positive and negative cross-regulatory interactions between factors are shown. (E) A robustness motif of positive and negative regulations ensures effective sequential expression of TTFs. (For interpretation of the references to color in this figure legend, the reader is referred to the Web version of this chapter.)

of a specific set of progeny in the lineage, this sequence of transcription factor is often referred as the "temporal series." A fundamental property of the temporal transcription factors (TTFs) is that they do not drive specification toward a specific cell identity. For example, a TTF can equally specify an interneuron, a motoneuron, or a glia, depending on the spatial identity of the neuroblast in which it is expressed. This was beautifully demonstrated in the case of Hb that specifies the first-born motoneuron in lineages 4–2 and 7–1, the first-born motoneuron and interneuron in lineage 1–1, while specifying the first-born glia in lineage 6–4 (Isshiki *et al.*, 2001). Consequently, loss of Hb leads to a loss of the first neuronal identity in lineage 4–2, 7–1, and 1–1, and a loss of the first glia in lineage 6–4, while continuous misexpression of Hb leads to ectopic progeny with the first identity (Fig. 8.1A and B). Another property of TTFs is that they cross-regulate such that their loss or misexpression interferes with the sequential expression of the other factors (Fig. 8.1B and D). Based on these criteria, recent experiments suggest that a fifth transcription factor, the bHLH protein Grainyhead (Grh), that appears after Cas in neuroblasts, is likely to be an additional TTF (Fig. 8.1A and D). Indeed, Grh (i) is activated by Cas in most, if not all, neuroblasts, (ii) can repress *cas*, and (iii) specifies different types of neural fates (Baumgardt *et al.*, 2009). However, in contrast to the other TTFs, Grh is permanently expressed in neuroblasts once activated. This has lead to the assumption that the temporal series may terminate by the expression of Grh, but as we will see later, this is unlikely to be the case.

## 3.2. Cross-regulatory interactions between TTFs promote sequential expression

Remarkable cross-regulatory interactions exist between the different members of the temporal series and appear essential for the robustness of the sequential expression. Loss- and gain-of-function experiments have revealed the ability of a TTF to initiate activation of the next factor in the series. In addition, cross-repression motifs have been observed such that Hb represses *pdm* and *cas* transcription, Kr represses *cas*, Pdm represses *kr*, Cas represses *pdm*, and Grh represses *cas* (Fig. 8.1; Baumgardt *et al.*, 2009; Grosskortenhaus *et al.*, 2006; Isshiki *et al.*, 2001; Kambadur *et al.*, 1998; Tran and Doe, 2008). In fact, a recent study (Nakajima *et al.*, 2010) has used a Boolean-type computational model to demonstrate that a particular set of positive and negative regulations in the Hb → Kr → Pdm → Cas → Grh sequence provides an extremely robust system ensuring controlled progression of the series and discrete transitions. In this "robustness" motif (Fig. 8.1E), a TTF activates the next factor but represses the previous and the second next factor. Here, negative interactions ensure that, at a given time, earlier TTFs are rapidly shut off and TTFs after the next one are repressed to prevent precocious expression. A direct consequence of these regulatory interactions is that artificially maintaining continuous expression of a TTF prevents progression through the sequence by blocking expression of late TTFs (Fig. 8.1B). On the other hand, the loss of Cas results in the maintenance of Pdm expression in the neuroblast during the following divisions (Fig. 8.1A), while Grh fails to be activated (Maurange *et al.*, 2008; Tran and Doe, 2008). Such manipulations of the temporal series lead to a block of the sequential dynamic and "freeze" the neuroblast in a constant temporal identity, leading to continuous generation of the same type of progeny (Fig. 8.1A and B).

## 3.3. Facilitators of TTF transitions

Although the positive and negative cross-regulatory interactions that have been described between Hb, Kr, Pdm, Cas, and Grh ensure robust sequential expression (Fig. 8.1D), they are not sufficient to recapitulate all sequential events observed *in vivo*.

First, no TTFs are known to feedback negatively on Hb expression raising the question of how its timely downregulation is controlled. The answer was provided by the identification of the orphan nuclear receptor Seven-Up (Svp) as an *hb* transcriptional repressor (Kanai *et al.*, 2005). Indeed, Svp accumulates toward the end of the Hb expression window in neuroblasts, and its loss induces a failure to repress *hb*, leading to an excess of progeny with an early identity (Fig. 8.1C and D). In fact, *svp* is actively transcribed from very early on in the neuroblast, but its efficient translation requires mitosis which seems to promote the nuclear export of *svp* mRNA (Mettler *et al.*, 2006). Consequently, after the first division, a large amount of svp mRNA diffuses into the cytoplasm through an unknown mechanism, allowing a high level of protein to be rapidly produced leading to efficient repression of *hb*, and progression to the Kr-mediated temporal identity. This mitosis-dependent expression of Svp fits well with previous observations showing that Hb downregulation requires cell division (Grosskortenhaus *et al.*, 2005). Because of its ability to promote the Hb → Kr transition, Svp has been classified as a switching factor. Importantly, Svp does not fit the criteria to be classified as a TTF as its misexpression does not induce the specification of neurons with a particular temporal identity. Recently, two additional factors have been discovered that act redundantly to promote the Hb → Kr switch. Distal antenna (Dan) and Distal antenna-related (Danr) are two DNA-binding proteins of the Pipsqueak domain family, located in the same genomic region (Kohwi *et al.*, 2011). Their deletion delays the Hb → Kr switch, leading to an excess of progeny with an early identity (Fig. 8.1C and D). The Dan proteins seem to function in concert with Svp as their effect is synergistic (Fig. 8.1C). Given the longer window of expression of Dan, and the ability of some Pipsqueak domain proteins to recruit Polycomb group complexes to the DNA (Huang *et al.*, 2002), Dan and Danr might operate at the chromatin level to promote and/or maintain the Svp-mediated repression of *hb*.

A second aspect concerns the fact that known cross-regulatory interactions cannot account for all phenotypes obtained in the various loss-of-function studies. Indeed, loss of Hb, Kr, or Pdm has only a minor effect on the progression through the series (Fig. 8.1A; Grosskortenhaus *et al.*, 2006; Isshiki *et al.*, 2001). This has led to the assumption that an upstream timer may be superimposed to activate Kr, Pdm, and Cas at the correct time when the previous TTF is lacking. Based on computational reconstruction, a recent study (Nakajima *et al.*, 2010) predicts that a single additional factor *x* in the network would be sufficient to reproduce the sequential expression pattern observed in both *wt* and TTF mutant embryos. According to this model, factor *x* would be expressed in a short temporal window at the beginning of the series (overlapping the Hb window) and would ensure low Kr activation (Kr is usually coexpressed at low levels with Hb in early neuroblasts) and Cas repression during this time, with the default state of *pdm* and *cas* being transcription in neuroblasts (Fig. 8.1D). *In vivo* identification of this hypothetic factor would provide a great example of the potential contribution of system-level approaches in the dissection of the temporal gene regulatory network.

All available data suggests that Hb, Kr, Pdm, and Cas are regulated at the transcriptional levels (Grosskortenhaus *et al.*, 2005; Kambadur *et al.*, 1998; Mettler *et al.*, 2006; Tran *et al.*, 2011); however, translational control of TTF mRNA and postmodifications of the proteins may still be implicated in fine-tuning TTF levels or activity.

# 4. Temporal Specification During Postembryonic Neurogenesis

## 4.1. Birth-order-dependent identity of postembryonic neurons

Until recently, the existence and necessity of an elaborated mechanism such as the temporal series to promote the diversity of postembryonic neurons remained questionable, as many neuroblasts appear to only generate interneurons after embryogenesis. However, several examples in defined lineages attest that postembryonic neuronal diversity is significant and is determined by birth order. First, in the central brain, a structure involved in learning and memory, called the mushroom body (MB), is composed of four classes of neurons with distinct morphologies. These neurons are generated in a stereotypical order by common neuroblasts, the last three classes being exclusively produced during larval and pupal stages (Lee *et al.*, 1999). Second, in the olfactory system, about 150 projection neurons are generated from three different neuroblasts, located in the central brain. These neurons project their dendrites to the antennal lobes constituted of about 50 different glomeruli each, and their axons to different structures in the central brain. Interestingly, many of these neurons are produced during postembryonic stages and the identity of the glomeruli they project depends on the moment they are born (Jefferis *et al.*, 2001). In a recent *tour-de-force*, using a newly established technique for clonal analysis, Tzumin Lee and his colleagues have reconstituted the full lineage composition of the anterodorsal neuroblast, one of the three neuroblasts that generate these projection neurons (Yu *et al.*, 2010). They found that the anterodorsal neuroblast first generates 18 different types of neurons during embryonic stages, one every cell cycle. During larval stages, 22 morphologically distinct classes of neurons are added in controlled numbers and in an invariant chronological order, suggesting that there is an exquisite postembryonic temporal specification mechanism. Third, in the nerve cord, a set of neuroblasts generate postembryonically motoneurons that innervate different leg muscles depending on their birth order (Baek and Mann, 2009; Brierley *et al.*, 2009). These three examples illustrate that complex temporal specification mechanisms must operate in the different regions of the postembryonic CNS.

## 4.2. TTFs in postembryonic neuroblasts

In the quest to dissect the postembryonic temporal system, we have recently tested the possibility that postembryonic neuroblasts also sequentially express transcription factors as a way to regulate temporal specification. However, Hb, Kr, and Pdm are not redeployed in larval neuroblasts. Nevertheless, a subset of neuroblasts transiently express Cas after leaving quiescence in early larvae, and this pulse of Cas is rapidly followed by a short pulse of Svp (Maurange *et al.*, 2008; Fig. 8.2A). Concomitantly, we identified two BTB–zinc finger transcription factors, Chinmo and Broad, as being pan-lineage markers for early and late neuronal identity, respectively. Accordingly, each neuroblast first generates Chinmo$^+$ neurons from embryonic to mid-larval stages; indeed Chinmo$^+$ neurons include all subpopulations of embryonic neurons, the individual fate of which is specified by the sequential expression of TTFs (as in Fig. 8.1). From mid-larval stages, neuroblasts switch to generate Broad$^+$ neurons (from the Z4 isoform; Maurange *et al.*, 2008; Zhou *et al.*, 2009; Fig. 8.2A). As the pulse of Svp in postembryonic neuroblasts happens a few hours before the Chinmo-to-Broad switch occurs, we investigated its role in this temporal transition. We found that neuroblast clones lacking the postembryonic pulse of Svp continuously generate Chinmo$^+$ neurons and fail to generate Broad$^+$ neurons (Fig. 8.2B). Similar results are found when expression of Cas is maintained or eliminated in the neuroblast (Fig. 8.2B). Thus, these manipulated postembryonic neuroblasts fail to progress to later temporal competences and continuously generate an early type of neuron (Chinmo$^+$). Interestingly, Chinmo was formerly identified as a neuronal fate determinant in MB neurons (Doe, 2006; Zhu *et al.*, 2006). The regulation of Chinmo by TTFs thus directly impacts on the identity of neurons. Several conclusions can be deduced from this set of experiments. First, the embryonic temporal series appears to be prolonged during postembryonic stages to promote

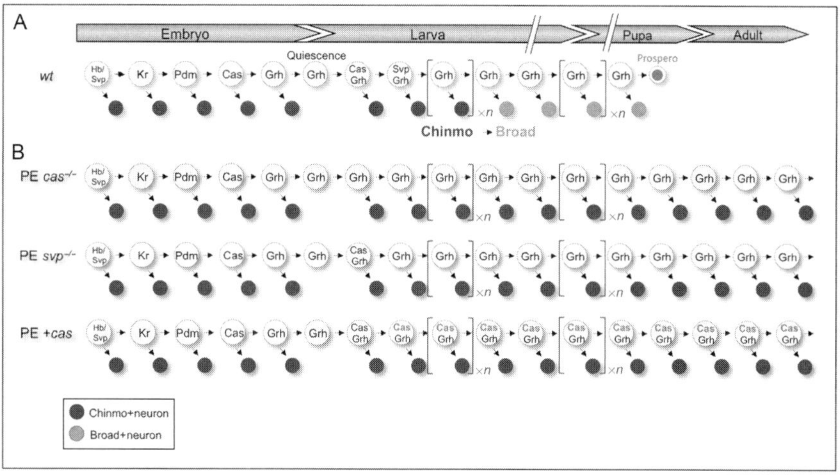


**Figure 8.2** The temporal series continues postembryonically and regulates the end of neurogenesis. (A) Most, if not all, neuroblasts first generate Chinmo$^+$ neurons during embryonic and early larval stages, they then switch to producing Broad$^+$ (Br$^+$) neurons during late larval and pupal stages. Some postembryonic neuroblasts in the thorax and central brain transiently express Cas and then Svp, shortly after their reentry into the cell cycle. The Chinmo → Broad switch happens a few cell cycles after Svp expression. Therefore, Chinmo labels early-born neurons while Broad labels late-born neurons. All thoracic and central brain neuroblasts (except MB neuroblasts) stop dividing 24h after pupa formation. Termination of neuroblast division involves a pulse of nuclear Prospero. Note that the number of neuroblast divisions represented on the scheme is arbitrary. TTF pulses may last longer than a single cell cycle. In addition, the total number of divisions undergone by neuroblasts varies according to their identity. (B) The Chinmo → Broad switch depends on the correct progression of the temporal series. Elimination of the postembryonic (PE) pulses of Cas (PE *cas*$^{-/-}$) or Svp (PE *svp*$^{-/-}$) prevents the switch, leading to neuroblasts continuously generating the early types of neurons (Chinmo$^+$). In addition, these neuroblasts do not stop dividing and continue generating an excess of progeny in adults. Similar results are obtained by blocking progression toward a late temporal identity through the sustained expression of Cas (+PE Cas). (For color version of this figure, the reader is referred to the Web version of this chapter.)

birth-order-dependent neuronal diversity. Given the large number of different neuronal identities uncovered in some postembryonic lineages—22 in the anterodorsal lineage (Yu *et al.*, 2010)—it is likely that many novel TTFs remain to be identified. Second, whether postembryonic Svp acts as a TTF or as a switching factor similar to its function during the embryonic Hb → Kr transition is unclear (Kanai *et al.*, 2005). As neither Hb nor Kr are expressed around this time, these results suggest new, yet unknown, targets for postembryonic Svp. Third, as the Chinmo → Broad switch happens a few cell cycles after the Svp pulse, it is unlikely that Svp directly regulates this transition suggesting that the TTF that mediates the Chinmo → Broad switch remains to be identified.

## 5. Extrinsic Versus Intrinsic Regulation of Temporal Series Progression

Switching factors and transcriptional cross-regulatory interactions play major roles in ensuring sequential progression of TTF expression. It remained to be investigated whether this process is influenced by extrinsic cues. For this purpose, isolated embryonic neuroblasts were cultivated *in vitro*. In these conditions, they exhibit the same sequential expression as neuroblasts in their natural environment (Brody and Odenwald, 2000; Grosskortenhaus *et al.*, 2005). This suggests that transitions through the first five factors (Hb, Kr, Pdm, Cas, Grh) are lineage intrinsically regulated and do not depend on inductive signals from surrounding embryonic tissues. In addition, unlike the Hb → Kr switch, the Kr → Pdm → Cas → Grh transitions do not require cell cycle progression, and so far, all results support the hypothesis of transitions mainly governed by transcriptional regulatory interplays (Grosskortenhaus *et al.*, 2005). It is unclear whether extrinsic inputs are involved in the control of temporal progression during later developmental stages (after Grh). Intriguingly, a recent study has found that TTF transitions are suspended in neuroblasts entering quiescence at the end of embryonic stages (Tsuji *et al.*, 2008). Moreover, this temporal freezing happens independently of the TTF expressed at the time of quiescence. This block is relieved and transitions proceed again when neuroblasts reactivate the cell cycle at the end of the first instar larva. Interestingly, two studies have recently demonstrated that neuroblast reentry into the cell cycle was conditioned by the availability of sufficient nutrients in the food, as sensed by the fat body and relayed to the neuroblast by glial insulin (Chell and Brand, 2010; Sousa-Nunes *et al.*, 2011). In this context, it will be interesting to investigate whether the block and reactivation of TTF switching is directly linked to the cell cycle or is independently regulated by an upstream input from the insulin/TOR pathway. On the same note, it is still unknown whether signaling from the steroid hormone ecdysone, that is known to trigger global developmental transitions (Thummel, 2001), impacts this process to coordinate neuronal production with specific stages of development.

## 6. The Temporal Series Schedules the End of Neurogenesis

Depending on their identity, neuroblasts adopt different strategies to terminate neurogenesis toward the end of development, either through apoptosis or through cell cycle exit. However, both these events appear to be scheduled by the same underlying timing mechanism.

## 6.1. Cell cycle exit in thoracic and central brain neuroblasts

Until recently, the mechanisms regulating the end of neuroblast divisions and their disappearance before adulthood were unknown. Given that a neuroblast should continue to divide until it has generated its complete repertoire of progeny, we were intrigued by the possibility that the sequential expression of TTFs could be used as a timer to coordinate the end of neurogenesis with the generation of the latest type of progeny produced in the lineage. To test this hypothesis, we investigated whether the termination of neurogenesis was affected in neuroblasts blocked in an early temporal identity. Wild-type neuroblasts in thoracic and brain segments normally stop dividing 1 day after the beginning of pupal stages (with the exception of MB neuroblasts that divide longer; Fig. 8.2A). Consequently, no neural proliferation is observed in the adult CNS (Truman and Bate, 1988). Strikingly, we found that neuroblasts lacking postembryonic Cas or Svp, or continuously misexpressing Cas, never stop dividing, even during adulthood, and generate an excess of neurons, albeit of early identity (Chinmo$^{+}$, see above; Fig. 8.2B). Thus, termination of division is linked to the expression of a late temporal factor, and consequently, manipulated neuroblasts blocked in a "young" temporal identity never stop dividing (Maurange *et al.*, 2008). In essence, connecting (i) promotion of neural diversity and (ii) termination of neurogenesis to the same timing mechanism ensures that neuroblasts only stop dividing when they have generated their complete set of progeny. It is still unclear how a presumptive late TTF triggers the last neuroblast division. Wild-type neuroblasts during their last division, in pupal stages, exhibit nuclear Prospero (Fig. 8.2A; Maurange *et al.*, 2008), a transcription factor that promotes differentiation in the GMC but that is normally tethered to the cytoplasmic membrane of the neuroblast by the asymmetric division machinery (Choksi *et al.*, 2006; Knoblich, 2008). This delocalization of Prospero from the membrane to the nucleus does not occur in temporally manipulated neuroblasts that persist in adults. In this context, it is possible that the last TTF in wild-type neuroblasts contributes to destabilizing the machinery normally maintaining Prospero at the membrane in neuroblasts, allowing its translocation to the neuroblast nucleus and cell cycle exit. Future studies will be needed to identify both the last TTF and its mode of action.

## 6.2. Apoptosis in abdominal neuroblasts

In contrast to brain and thoracic neuroblasts, abdominal postembryonic neuroblasts stop dividing at mid-larval stages, following an apoptotic event mediated by expression of the Hox/homeotic gene *abdominal-A* (*abd-A*; Bello *et al.*, 2003; Fig. 8.3A). In this context, we could show that the timing of neuroblast death also depends on the proper progression to a

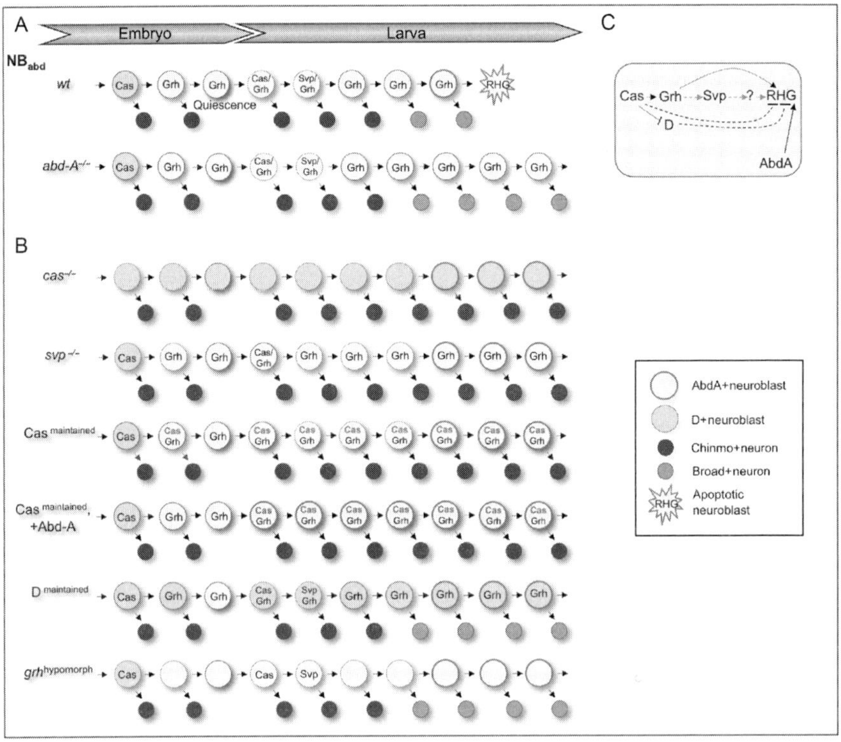


**Figure 8.3** Regulation of abdominal neuroblast apoptosis by the temporal series and Abd-A. (A) In the abdominal segments of the larval CNS, a neuroblast resumes quiescence by expressing Cas followed by Svp. In wild-type (*wt*) conditions, progression of the temporal series allows the Chinmo → Broad switch around mid-larval stages. A few cell cycles later, the neuroblast undergoes apoptosis upon Abd-A expression and subsequent induction of proapoptotic genes (RHG genes). (B) When the temporal series is blocked by either removing the postembryonic pulse of Cas or Svp or by maintaining Cas, both the Chinmo → Broad switch and apoptosis are prevented. However, in that situation, the endogenous Abd-A pulse is retained and even an additional dose of Abd-A cannot trigger apoptosis. Therefore, the temporal series installs the competence to undergo Abd-A-mediated apoptosis. This process requires two temporal inputs. The first temporal input is provided by transient Cas through the activation of Grh and the repression of D. The second input is provided by the later pulse of Svp. (C) Summary of cross-regulatory interactions controlling abdominal neuroblast apoptosis. Feed-forward loops initiated by Cas or Grh are likely to be involved in the regulation of RHG genes. (For color version of this figure, the reader is referred to the Web version of this chapter.)

late, yet unknown, temporal identity that follows the second pulse of Svp (Fig. 8.3B; Maurange *et al.*, 2008). Thus, in such neuroblasts, the temporal series appears to install the competence to respond to Abd-A, allowing precocious termination of neuroblast division and a smaller lineage. It is still unclear how the last TTF then triggers neuroblast death. However, the competence to stop dividing is also determined by the action of previous TTFs in the series. This aspect will be discussed in the next part of the review where I address the progress made regarding the mechanistic principles orchestrated by TTFs.

## 7. The Temporal Series and the Regulation of Neuroblast Competence

### 7.1. Neuroblasts transit throughout various competence windows

Several studies have demonstrated that in the course of development, mammalian NSCs transit through various competence windows during which they respond to given cues by generating specific types of progeny. For example, heterochronic transplantation experiments have shown that mouse NSCs will respond differently to the same extrinsic cue if applied at different developmental times (Desai and McConnell, 2000; Livesey and Cepko, 2001). The mechanisms that permit neural progenitors to change their competence over time are still poorly understood and are the subject of intense research in all types of animal models. *Drosophila* neuroblasts too transit through various competence windows, as demonstrated by the fact that a temporal factor expressed at different times in the same neuroblast does not have the same impact on the neuroblast and its subsequent progeny. For example, pioneering studies have shown that experimental pulses of each of the two first TTFs, Hb and Kr, are only able to induce the U neuronal fate (defined by expression of the terminal factor Even-Skipped (Eve) in the NB7-1 lineage) when they are expressed before the fifth neuroblast division (Fig. 8.4A). If the pulses are performed later, NB7-1 fails to generate ectopic U neurons (Cleary and Doe, 2006; Pearson and Doe, 2003). Thus, the neuroblast response to a TTF will differ depending on the competence state at the moment it is expressed. In principle, this system allows the same TTF to be used several times during different competence windows to specify different neuronal fates. Moreover, as we shall see in the next paragraphs, TTFs themselves orchestrate neuroblast transitions throughout various competence windows.

### 7.2. Maintenance of a TTF blocks transition to the next competence window

Not only can Hb specify a particular neural identity when expressed in a neuroblast during a defined competence window, but it can also regulate the competence state. Indeed, as previously mentioned, the competence of neuroblast 7–1 (NB7-1) to respond to Hb by generating U neurons is lost after five divisions. However, artificially maintaining Hb from its initial

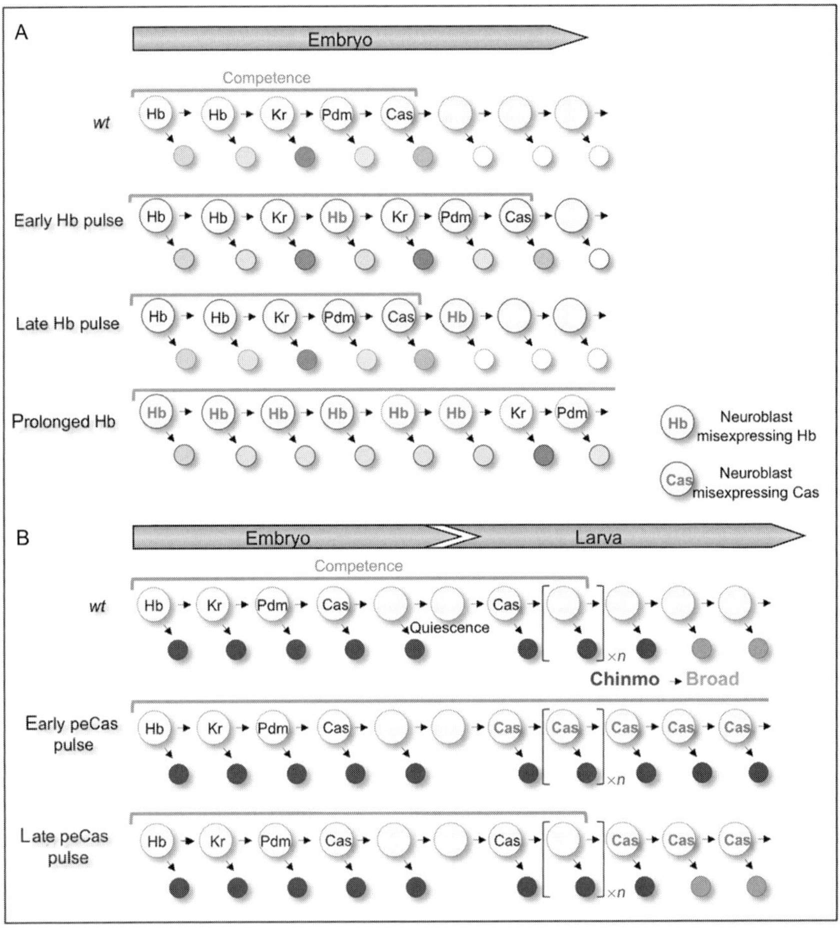


**Figure 8.4** Regulation of neuroblast competence by TTFs. (A) During the first five divisions, NB7-1 has the competence to respond to the sequential expression of Hb $\rightarrow$ Kr $\rightarrow$ Pdm $\rightarrow$ Cas by generating five so-called U neurons. The five U neurons have distinct identities (small circles in different colors) but express in common the transcription factor Even-Skipped (Eve). This competence is limited to the first five divisions. Consequently, an early ectopic pulse of Hb, induced during this window, promotes the generation of an additional early U neuron, whereas a late ectopic pulse, after the fifth neuroblast division, does not lead to the generation of a U neuron. In addition, sustained expression of Hb, but not Kr, can extend the window of competence, leading to an excess of U neurons. (B) Larval neuroblasts first generate Chinmo$^+$ neurons and then switch to generate Broad$^+$ neurons. If Cas expression is artificially sustained from early on, Cas extends the neuroblast competence to generate Chinmo+neurons. If Cas is ectopically expressed from a later stage, it is then unable to affect the transition to the next competence state. (For interpretation of the references to color in this figure legend, the reader is referred to the Web version of this chapter.)

window of expression prolongs the window of competence and leads to NB7-1 generating an excess of U neurons (Fig. 8.4A; Cleary and Doe, 2006; Pearson and Doe, 2003). In contrast, sustained expression of Kr cannot efficiently prolong this competence window (Cleary and Doe, 2006). It now turns out that the repressing activity of Hb is essential to sustain the first competence window. Indeed, misexpression of a form of Hb that can only act as a transcriptional activator cannot prolong the competence state (Tran and Doe, 2008; Tran *et al.*, 2011). In fact, Hb maintains the first competence window by the constant repression of *pdm*, and other genes, and only downregulation of Hb permits neuroblasts to transit to the next competence window through the activation of these genes (Tran and Doe, 2008; Tran *et al.*, 2011). In another example, we have shown that Cas can also maintain its initial competence window (Maurange *et al.*, 2008). Indeed, during a first neurogenic period that lasts from embryogenesis to mid-larval stages, neuroblasts generate progeny that express the transcription factor Chinmo. In a second neurogenic period, neuroblasts generate progeny that express the transcription factor Broad. In response to the artificial maintenance of Cas expression after its initial pulse of expression, neuroblasts will continuously generate $Chinmo^+$ progeny and will not acquire the competence to generate $Broad^+$ neurons (Fig. 8.4B). In contrast, we could show that this ability of neuroblasts to respond to Cas is progressively lost as neuroblasts age. Consequently, older neuroblasts in which a long pulse of Cas is reintroduced fail to maintain the competence to produce $Chinmo^+$ neurons and can transit to the following competence window. This shows that, as with Hb, there is only a limited window during which a neuroblast can respond to Cas and that downregulation of Cas allows the neuroblast to exit this competence state. Cas target genes and the combinatorial transcription factor code that defines this Cas competence state are not known.

## 7.3. TTFs initiate feed-forward loops that install the next competence state

The windows of expression of Hb and Cas need to be transient to permit transition to later competence states. Further studies have shown that TTFs can directly regulate the nature of the next competence state. In particular, feed-forward loops (FFLs) triggered by TTFs are used to generate diversity in neural fates.

### 7.3.1. FFLs to promote neural diversity

In an elegant study, Stefan Thor and his group (Baumgardt *et al.*, 2009) have concentrated on the lineage generated by thoracic NB5-6 and resolved its complete temporal sequence. They found that in this neuroblast, Cas is expressed in two embryonic pulses lasting for several cell cycles. The first pulse occurs from late stage 11 to late stage 12 during which several neurons of unknown identity are generated. At the end of this pulse, Cas activates in the neuroblast three transcription factors, Collier (Col), Squeeze (Sqz), and Grh, the latter one being expressed at low levels at this stage (Fig. 8.5A). Cas is then transiently switched off during stage 13 when NB5-6 generates neuron Ap1 (Fig. 8.5C), the first progeny of a group of four distinct neurons called the "Ap cluster." Cas is afterward reactivated in a second pulse during stage 14. During this second window of expression, Cas acts in a FFL (Shoval and Alon, 2010) with its previous target Sqz to activate Nab in the neuroblast (Figs. 8.5A and 8.6). The presence of Cas, Sqz, and Nab in NB5-6 induces newborn progeny to adopt the Ap2/3 fate. At the end of the second Cas pulse, accumulating levels of Grh contribute to the establishment of the last neuronal fate of the lineage, Ap4, before NB5-6 stop dividing.

Thus, in NB5-6, Cas is expressed in two pulses during which it specifies neurons with distinct identity. During the first pulse, Cas establishes the neuroblast competence to respond to the second pulse. This is done by installing a new combinatorial code through a FFL involving Sqz and Nab (Figs. 8.5 and 8.6). Therefore, FFL motifs triggered by TTFs appear essential to promote neural diversity.

Although Sqz and Nab can discriminate the various Cas temporal windows, and therefore provide some temporal information, they have been defined as subtemporal TTFs rather than TTFs (Baumgardt *et al.*, 2009). The reasons for this are that they do not seem to be involved in cross-regulatory interactions with TTFs, they may only be expressed by a subset of neuroblasts—although Nab seems to be quite widely expressed in postembryonic neuroblasts (C. Maurange, unpublished data)—and therefore may only be responsible for specifying only certain types of cell fates. However, investigation on additional lineages will be necessary to unambiguously define their hierarchical positioning in the temporal gene regulatory network.

### 7.3.2. FFLs to schedule neuroblast apoptosis

Another essential role of Cas is to establish the later competence of mid-larval neuroblast to undergo Abd-A-mediated cell death (Bello *et al.*, 2003; Cenci and Gould, 2005; Maurange *et al.*, 2008). For this purpose, two temporal inputs are necessary. The first input is provided by the embryonic/early larval pulse of Cas that permanently activates *grh* and represses the Sox gene *Dichaete* (*D*) in neuroblasts. Indeed, abdominal neuroblasts missing Cas or Grh, or misexpressing D fail to undergo apoptosis later on (Fig. 8.3A and B). In addition, while early Cas expression is required, it needs to be transient, as sustained Cas prevents apoptosis even in the presence of an extra dose of Abd-A. Therefore, a $\text{Cas}^{-}\text{D}^{-}\text{Grh}^{+}$ combinatorial code is necessary for timely neuroblast apoptosis. In addition, a second temporal input is provided through the postembryonic pulse of Svp. Indeed, loss of

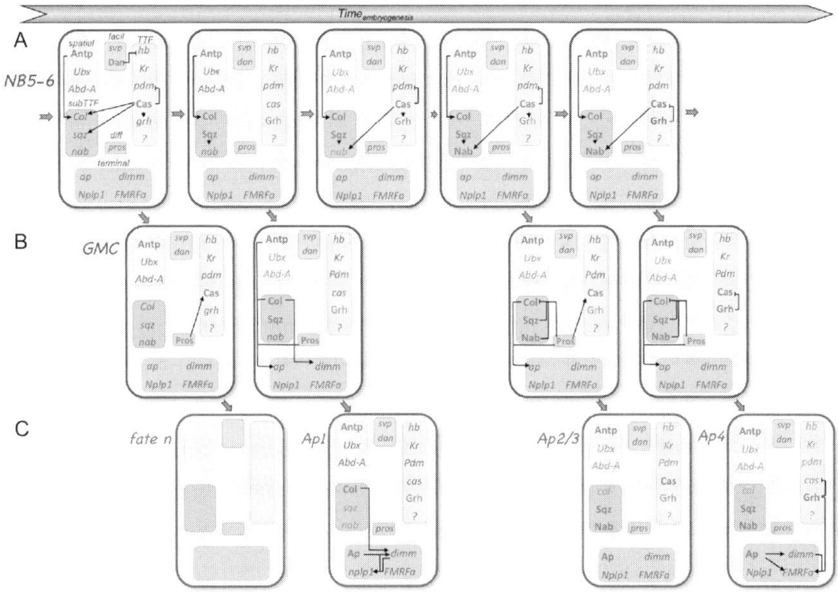


**Figure 8.5** Deployment of the temporal gene regulatory network over time in the embryonic NB5-6 neuroblast and its progeny. (A) The temporal gene regulatory network is composed of different "modules" of genes that are regulated by, or cooperate with, the temporal series to specify the identity of the various progeny generated over time. Proteins expressed at a given time are shown in green. First, temporal transcription factors (*TTF* module) are sequentially expressed, thanks to positive and negative cross-regulatory interactions and facilitators (*facilit* module). The scheme shows NB5-6 from its first pulse of Castor (Cas) expression that promotes a progeny of an unknown identity (*fate n* in C). Cas then activates the subtemporal TTFs *collier* (*col*) and *squeeze* (*sqz*; *sub-TTF* module) and the TTF *grainyhead* (*grh*). Col expression also necessitates input from the Hox factor Antp (*spatial* module). Activation of low levels of Grh and repression of Cas permit transition to the next temporal window ($Grh^{low}$ $Col^{+}$ $Sqz^{+}$) that will specify the new neuronal fate *Ap1* (*Ap1* is the first neuron from the Ap cluster to be generated). Then, a second pulse of Cas in the neuroblast feeds forward with Sqz to activate *nab*, thus defining a novel combinatorial code ($Grh^{low}$, $Col^{+}$, $Cas^{+}$, $Sqz^{+}$, $Nab^{+}$) to specify the *Ap2/3* neurons. Later on, high levels of Grh ensure the appearance of a new code ($Grh^{high}$, $Col^{+}$, $Sqz^{+}$, $Nab^{+}$) in NB5-6 and production of the *Ap4* neuron. (B) Expression of temporal and subtemporal TTFs persists in the GMC. The presence of the differentiation transcription factor Prospero (*diff* module) is believed to help maintaining TTF expression in daughter cells, thus "freezing" progression of the series. In addition, cooperation of Prospero with TTFs and sub-TTFs allows expression of terminal differentiation genes (*terminal* module) in differentiated neurons. (C) Activation of various terminal genes results from the cooperative action of TTFs, sub-TTFs, spatial factors, and differentiation factors. The combination of expressed terminal genes defines neuronal identity (fate n; Ap1; Ap2/3; Ap4). (For color version of this figure, the reader is referred to the Web version of this chapter.)

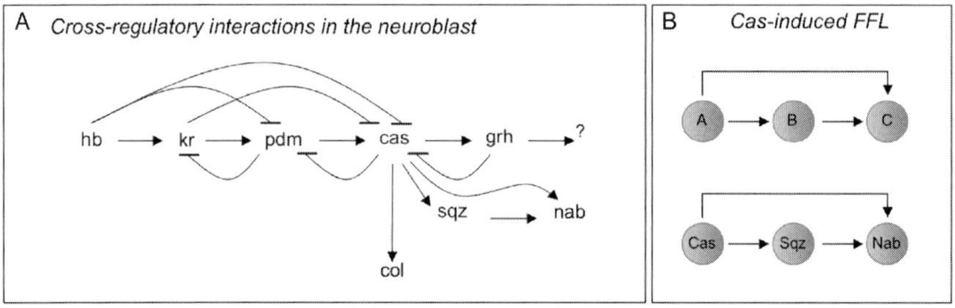


**Figure 8.6** In the temporal gene regulatory network, Cas feeds forward to promote neural diversity. (A) Recapitulation of the temporal gene network in NB5-6. (B) Feed-forward loop model (FFL). Cas feeds forward with Sqz to activate Nab. This loop motif is used to create new combinatorial codes in the neuroblast, installing novel competence windows that permit the same TTF to specify different neuronal fates when expressed at different times in a given neuroblast. (For color version of this figure, the reader is referred to the Web version of this chapter.)

postembryonic Svp, although not affecting the $Cas^{-}D^{-}Grh^{+}$ state, prevents later apoptosis (Fig. 8.3B). Therefore, Cas through activation of *grh* and repression of *D* establishes competence for later temporal factors to promote the end of neuroblast division upon Abd-A expression (Maurange *et al.*, 2008). Here again, FFLs are likely to be involved as sustained expression of Grh and later, yet unknown, TTFs converge with Abd-A to trigger apoptosis (Fig. 8.3C).

All together, these studies support the emerging paradigm that the transient expression of early TTFs installs the molecular context/combinatorial code that guides the action of subsequent TTFs.

# 8. Integration of Temporal and Spatial Cues Ensures Unique Lineage Size and Composition

## 8.1. Control of lineage size through Hox and temporal factor integration

Most embryonic neuroblasts of the nerve cord appear to progress through an identical sequence of TTFs (with some exceptions as certain TTFs may be skipped in some neuroblasts; Benito-Sipos *et al.*, 2010). Yet, they generate very diverse lineages because their spatial identity is different, determined by distinct combinations of transcription factors. Several studies have started to unveil how temporal and spatial determinants intersect to promote diversity in the developing CNS. Hox/homeotic genes are transcription factors that determine spatial identity along the anteroposterior axis. In this context, they appear to cooperate with temporal factors to modulate the timing at which homologous neuroblasts from different segments enter quiescence or apoptosis during development. We have already seen how the temporal series installs the competence to undergo apoptosis upon Abd-A expression in mid-larval neuroblasts located in the abdominal CNS (Fig. 8.3; Maurange *et al.*, 2008). Accordingly, the final number of progeny generated by abdominal neuroblasts is smaller than that generated by their thoracic homologs that do not express Hox genes postembryonically. The resulting anterior-prominent CNS also reflects the locations of the major sensory and motor organs, such as antennae, eyes, wings, and legs. Recently, the proapoptotic gene *grim* has been identified as responsible for postembryonic abdominal neuroblast death (Tan *et al.*, 2011). Genetic dissection of the upstream 22kb enhancer region may reveal how TTFs, Hox, and probably other spatial factors cooperate to finely regulate *grim* expression.

Besides the role of Abd-A in postembryonic neuroblasts, additional studies have now shown that Hox factors cooperate with different TTFs in embryonic neuroblasts to schedule their entry into quiescence before larval stages. In thoracic NB3-3, the Hox factor Antennapedia (Antp) is responsible for inducing quiescence earlier than in the homologous abdominal neuroblast expressing Abdominal-A (Abd-A; Tsuji *et al.*, 2008). Precisely, Antp makes thoracic NB3-3 enter quiescence during the second pulse of Cas expression. In contrast, Abd-A misexpression can force thoracic NB3-3 to enter quiescence later, after the second pulse of Svp that follows the Cas expression window. Thus, spatial information provided by Hox genes integrates with temporal cues to determine the number of divisions undergone by each embryonic neuroblast. However, it is not a general property of Antp to promote cell cycle exit earlier than Abd-A in embryonic neuroblasts. Indeed, in NB5-6, Antp promotes cell cycle exit in thoracic segments during the late Cas window, while ultrabithorax (Ubx) and Abd-A promote cell cycle exit during the earlier Pdm window in more posterior segments (Karlsson *et al.*, 2010; Fig. 8.7). Therefore, in addition to Hox genes, other spatial factors, such as those encoded by segmentation and dorsoventral genes, must act in concert to schedule entry into quiescence, allowing each neuroblast to stop dividing at a defined time. The fact that all neuroblasts do not enter quiescence at the same temporal identity during embryogenesis engenders a slight asynchronicity in the progression of the series that is passed on in postembryonic neuroblasts once they resume dividing (Maurange *et al.*, 2008; Tsuji *et al.*, 2008).

## 8.2. Control of lineage composition through Hox and temporal factor integration

Hox genes not only participate in the regulation of lineage size but also influence lineage composition. For example, NB5-6 in the anterior subesophageal and brain segments generates a large lineage that does

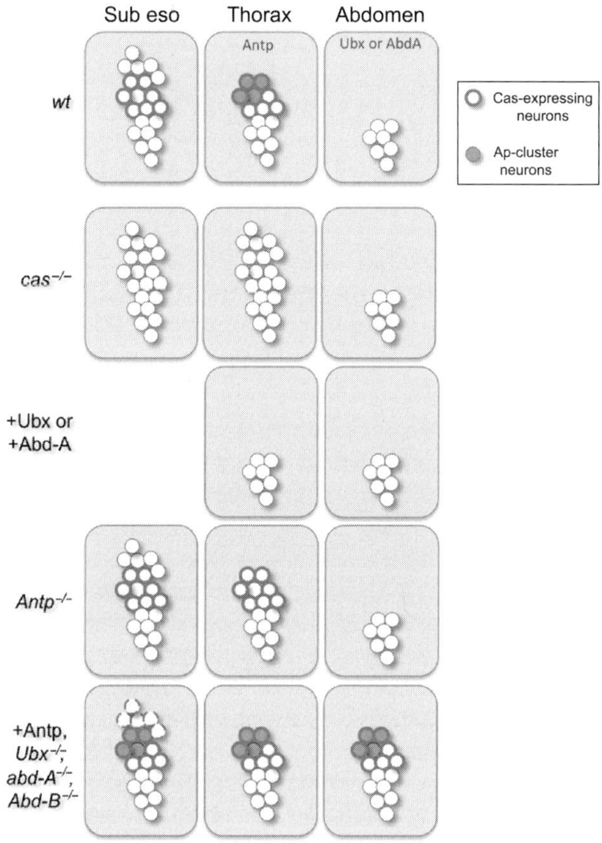


**Figure 8.7** Integration of Hox and temporal identity factors regulates NB5-6 lineage size and composition. Different Hox factors (Antp, Ubx, Abd-A) are expressed in the segments of the nerve cord (subesophageal, thoracic, and abdominal) along the anterior–posterior axis. The scheme depicts the size and composition of the lineage generated by NB5-6 in the different segments at the end of embryogenesis. Each circle represents a differentiated neural cell. The size and composition of the lineage is regulated by the integration of the spatial and temporal information conveyed by Hox and temporal factors, respectively. Cas is expressed during restricted time windows in the neuroblast and is inherited in daughter cells (circles highlighted in green). Generation of the "Ap cluster" of neurons (orange circles) in thoracic lineages depends on both the presence of Cas and Antp. In abdominal lineages, the presence of posterior Hox genes such as Abd-A or Ubx induces neuroblast cell cycle exit before $Cas^+$ neurons are being generated. Loss of Abd-A and Ubx associated with the misexpression of Antp can convert the subesophageal lineage into a thoracic-like lineage, forcing the generation of the Ap cluster. (For interpretation of the references to color in this figure legend, the reader is referred to the Web version of this chapter.)

not contain the so-called Ap cluster of neurons generated by the homologous, more posterior, thoracic neuroblasts (Fig. 8.7). Although the TTF Cas, known to be required for the specification of these neurons (Baumgardt *et al.*, 2009; Fig. 8.5), is also expressed in the late anterior NB5-6, Antp is not. However, misexpression of Antp in these anterior segments triggers the appearance of the Ap cluster (Fig. 8.7; Karlsson *et al.*, 2010). This demonstrates that the spatial factor Antp cooperates with the temporal factor Cas to determine neuronal identity in the thoracic lineages. Further results indicate that one of the genes of convergence is the transcription factor Collier, required to determine "Ap cluster" identity (Fig. 8.5; Baumgardt *et al.*, 2009; Karlsson *et al.*, 2010). Activation of *collier* expression also requires the Hox cofactors, Homothorax and Extradenticles, whose expression is also spatially regulated. However, the nature of the molecular interactions between the temporal and spatial factors and the DNA surrounding *collier* remain unknown, and in this context, it represents an excellent model to further dissect how spatiotemporal inputs regulate gene expression pattern. In conclusion, integration of spatial and temporal cues in neuroblasts defines both the size and cell composition of the descendant lineage.

Although integration of spatial and temporal cues permits each neuroblast to generate a unique set of progeny with different identities, a recent study has also reported that a defined type of progeny can arise from different combinations of spatiotemporal cues. This counterintuitive observation was made when the authors demonstrated that different combinatorial codes converge to activate the same terminal differentiation gene in different lineages (Losada-Perez *et al.*, 2010). As seen above, the generation of the "Ap cluster" neurons is specific to NB5-6 located in the thoracic segments. Among this cluster, one neuron, Ap4, expresses the neuropeptide FMRFa (Fig. 8.5). Specification of the FRMFa-expressing neuron from NB5-6 depends on a regulatory cascade initiated by Cas and spatial determinants such as the Hox factor Antp. Strikingly, the authors find that FMRFamidergic cells are also generated by another type of neuroblast (possibly NB3-1), during another temporal window defined by Hb (Losada-Perez *et al.*, 2010). In addition, these cells share none of the factors defining the combinatorial code for production of Ap4 cells (with the exception of the neuroendocrine regulator Dimm). Therefore, the *cis*-regulatory DNA region that controls expression of the FMRFa neuropeptide appears to be able to integrate multiple combinations of spatiotemporal inputs to activate expression in neurons of distinct origin. Characterization of the different enhancers of this gene may shed light on how this is achieved.

## 9. Integration of Differentiation and Temporal Cues in Neuroblast Progeny

Most TTFs are very transiently expressed in neuroblasts allowing the sequential expression of all members of the series. However, their expression persists longer in the daughter cells where transition to the next factor is blocked (Isshiki *et al.*, 2001; Kambadur *et al.*, 1998). In the differentiating progeny, it is believed that the persisting TTF contributes to the activation of specific cell fate genes. How is the temporal series differentially regulated in the progeny and in the neuroblast? How are TTFs able to activate different set of genes in neuroblasts and in daughter cells? A study by the lab of Andrea Brand suggests that the homeodomain transcription factor Prospero may play an essential role in these processes (Choksi *et al.*, 2006). Prospero is one of the determinants that asymmetrically segregate to the basal pole of the neuroblast during mitosis to be specifically inherited by the GMC. Once in the GMC, this transcription factor translocates to the nucleus to trigger transcription programs promoting cell cycle exit and differentiation into neurons and glia. Genome-wide investigation of Prospero targets has shown that Prospero has the ability to both activate and repress transcription (Choksi *et al.*, 2006). Interestingly, Prospero binds in the vicinity of *hb*, *Kr*, and *pdm* and seems to be required to maintain their expression in the progeny. We can speculate that, this way, Prospero contributes to prevent transition to the next TTF and "freezes" the temporal identity of the daughter cell (Fig. 8.5B). On the other hand, in the neuroblast, "self-renewal transcription factors" such as Snail, Asense, and Deadpan, that are repressed by Prospero in the GMC (Choksi *et al.*, 2006), may cooperate with TTFs to promote temporal series progression. In addition, other genes that are activated by Prospero include neuronal identity genes such as *apterous* (*ap*), and *even-skipped* (*eve*). *Ap* is a target of Collier, itself activated by Cas in thoracic NB5-6 (Baumgardt *et al.*, 2009). However, while both *cas* and *collier* are initially activated in the neuroblast, Ap expression is restricted to the progeny. It can therefore be assumed that Collier present in the progeny after the asymmetric division cooperates with nuclear Prospero to trigger *ap* transcription (Fig. 8.5B and C). The presence of Prospero in the progeny of NB7-1 may be similarly required to activate *eve* in the five U neurons generated in response to the sequential expression of embryonic TTFs (Isshiki *et al.*, 2001). The presence of Prospero may also explain why, as seen in lineage 5–6, Nab and Squeeze repress Collier when coexpressed in a neuron, but not when coexpressed in the neuroblast (Fig. 8.5A and B; Baumgardt *et al.*, 2009). Therefore, it appears that transient Prospero expression in the GMC promotes a binary switch that changes the molecular context in which TTFs evolve, allowing different downstream targets to be activated in neuroblasts or in differentiating progeny. The nature of the molecular interactions allowing (i) TTFs and self-renewal factors in neuroblasts or (ii) TTFs and Prospero or other differentiation factors in differentiating progeny to trigger the appropriate gene expression program awaits exploration.

## 10. Mammalian Orthologs

So far, no temporal series similar to the Hb → Kr → Pdm → Cas → Grh has been identified in mammals. However, some studies suggest that there is a great deal of evolutionary conservation in the mechanisms regulating NSC temporal specification (Jacob *et al.*, 2008). We will briefly mention here two remarkable examples. First, cortical NSCs appear to transiently express the nuclear receptors COUP-TFI and II around embryonic day 12.5 during the early steps of corticogenesis (Naka *et al.*, 2008). Interestingly, NSCs lacking COUP-TFI and II produce an excess of early types of neurons (deep layer neurons) and fail to generate later types of neurons (destined for the superficial upper layers). Further, the neurogenic to gliogenic transition that appears toward the end of embryonic development is also blocked. Therefore, cortical NSCs mutant for COUP-TFI and II appear to be blocked in an early temporal identity. It is striking that *COUP-TF* genes are orthologs to *svp* in *Drosophila* which loss leads to a similar phenotype—failure of the NSC to generate late types of progeny.

Another illustration of this conservation was brought to light by the study of the role of Ikaros in retinal neural progenitors (Elliott *et al.*, 2008). Retinal progenitors are multipotent and can give rise to different types of retinal cell types. Ikaros is expressed in early retinal progenitors and its misexpression is sufficient to induce an excess of early types of progeny while its loss reduces the number of these cells. Intriguingly, *ikaros* is the mouse ortholog of *hb* and its function in conferring early temporal identity to retinal progenitors suggests a conserved role between mouse and *Drosophila*. A more systematic investigation of mammalian orthologs of the *Drosophila* TTFs may thus reveal further conserved players in the temporal specification of NSCs.

## 11. Conclusion

We have seen here how the sequential expression of TTFs in *Drosophila* NSCs provides a powerful strategy not only to promote neural diversity but also to regulate the total number of progeny to be produced in the adult CNS. We begin to understand how regulatory network motifs initiated by the TTFs series (such as FFLs) install various competence states in neuroblasts over time. We are also starting to comprehend how multiple gene networks (temporal, spatial, differentiation) cooperate to generate different outputs in neuroblasts and in their progeny. In which directions could we now concentrate our efforts?

Ultimately, and for potential therapeutic purposes, it will be important to understand how the temporal specification of old, or adult, NSCs could be reprogrammed to force regeneration of desired types of neurons and glia. Because neuroblasts transit through various competence states, reexpressing an early TTF in an old neuroblast is not sufficient to induce regeneration of the initial early type of progeny. Therefore, resetting the temporal identity of neuroblasts will require deciphering the combinatorial transcription factor codes that define the various competence states, and figuring out how these can be reinstalled in old neuroblasts to provide the adequate competency. This task will involve identification of new TTFs, particularly the presumptive postembryonic ones, their target genes, and their mode of action. It is also likely that chromatin factors are involved, and their role in constraining or perpetuating the action of TTFs should be investigated. The development of technologies allowing genome-wide investigation of transcription factor binding sites from single or few cells would help resolving the interplay between chromatin, temporal, and spatial factors, and their integration in neuroblasts or their progeny. Computational modeling of the temporal gene regulatory network should parallel these approaches and be used to predict solutions for controlled temporal reprogramming of old NSCs.

In the past years, new types of neuroblasts have been characterized in *Drosophila*, that either divide for longer than the typical ones (e.g., MB neuroblasts divide for almost 72h longer; Sousa-Nunes *et al.*, 2010) or do not originate from the same neuroepithelium (e.g., medulla neuroblasts in the optic lobe; Egger *et al.*, 2007), or generate intermediate progenitors with a higher proliferation potential than typical GMCs (Bello *et al.*, 2008; Boone and Doe, 2008; Bowman *et al.*, 2008). How different is the temporal specification system in these atypical neuroblasts? Does it involve additional TTFs that are not used in the other types of neuroblasts?

Neuroblasts that are mutants for some genes affecting asymmetric divisions generate progeny that fail to differentiate and continue dividing, rapidly producing tumors in the developing CNS (Knoblich, 2010). These tumors are composed of cells expressing the usual neuroblast markers. However, in contrast to their *wt* counterparts, neuroblast-like tumor cells are unable to stop dividing at the end of development and continue to proliferate in the adult. As the temporal series regulates the number of neuroblast divisions during development, could its deregulation be involved in the overproliferation abilities of neuroblast-like cells?

Finally, we need to rapidly learn more about the extent of conservation between insect and mammalian temporal specification systems. Ultimately, the future will tell how much impact cracking the temporal specification code in *Drosophila* neuroblasts, will have on our ability to safely manipulate human NSCs for medical applications, but for now, more mechanistic breakthroughs are certainly to be expected from this powerful model system.

## ACKNOWLEDGMENTS

I am grateful to Vincent Bertrand, Robert Kelly, Elodie Lanet, Arièle Legendre, Samir Merabet, and Benjamin Prud'homme for critical comments and useful discussions on the chapter. C. M. research is supported by the CNRS, Université de la Méditerranée, Fondation pour la Recherche Médicale and Association pour la Recherche contre le Cancer. C. M. is an associate PI of Eurosystem.

## REFERENCES

Baek, M., and Mann, R. S. (2009). Lineage and birth date specify motor neuron targeting and dendritic architecture in adult Drosophila. *J. Neurosci.* **29,** 6904–6916.

Bate, C. M. (1976). Embryogenesis of an insect nervous system. I. A map of the thoracic and abdominal neuroblasts in Locusta migratoria. *J. Embryol. Exp. Morphol.* **35,** 107–123.

Baumgardt, M., Karlsson, D., Terriente, J., Diaz-Benjumea, F. J., and Thor, S. (2009). Neuronal subtype specification within a lineage by opposing temporal feed-forward loops. *Cell* **139,** 969–982.

Bello, B. C., Hirth, F., and Gould, A. P. (2003). A pulse of the Drosophila Hox protein Abdominal-A schedules the end of neural proliferation via neuroblast apoptosis. *Neuron* **37,** 209–219.

Bello, B. C., Izergina, N., Caussinus, E., and Reichert, H. (2008). Amplification of neural stem cell proliferation by intermediate progenitor cells in Drosophila brain development. *Neural Dev.* **3,** 5.

Benito-Sipos, J., Estacio-Gomez, A., Moris-Sanz, M., Baumgardt, M., Thor, S., and Diaz-Benjumea, F. J. (2010). A genetic cascade involving klumpfuss, nab and castor specifies the abdominal leucokinergic neurons in the Drosophila CNS. *Development* **137,** 3327–3336.

Boone, J. Q., and Doe, C. Q. (2008). Identification of Drosophila type II neuroblast lineages containing transit amplifying ganglion mother cells. *Dev. Neurobiol.* **68,** 1185–1195.

Bossing, T., Udolph, G., Doe, C. Q., and Technau, G. M. (1996). The embryonic central nervous system lineages of Drosophila melanogaster. I. Neuroblast lineages derived from the ventral half of the neuroectoderm. *Dev. Biol.* **179,** 41–64.

Bowman, S. K., Rolland, V., Betschinger, J., Kinsey, K. A., Emery, G., and Knoblich, J. A. (2008). The tumor suppressors Brat and Numb regulate transit-amplifying neuroblast lineages in Drosophila. *Dev. Cell* **14,** 535–546.

Brand, A. H., and Livesey, F. J. (2011). Neural stem cell biology in vertebrates and invertebrates: More alike than different? *Neuron* **70**(4), 719–729. DOI: 10.1016/j.neuron.2011.05.016. Review.

Brierley, D. J., Blanc, E., Reddy, O. V., Vijayraghavan, K., and Williams, D. W. (2009). Dendritic targeting in the leg neuropil of Drosophila: The role of midline signalling molecules in generating a myotopic map. *PLoS Biol.* **7,** e1000199.

Broadus, J., Skeath, J. B., Spana, E. P., Bossing, T., Technau, G., and Doe, C. Q. (1995). New neuroblast markers and the origin of the aCC/pCC neurons in the Drosophila central nervous system. *Mech. Dev.* **53,** 393–402.

Brody, T., and Odenwald, W. F. (2000). Programmed transformations in neuroblast gene expression during Drosophila CNS lineage development. *Dev. Biol.* **226,** 34–44.

Cenci, C., and Gould, A. P. (2005). Drosophila Grainyhead specifies late programmes of neural proliferation by regulating the mitotic activity and Hox-dependent apoptosis of neuroblasts. *Development* **132,** 3835–3845.

Chell, J. M., and Brand, A. H. (2010). Nutrition-responsive glia control exit of neural stem cells from quiescence. *Cell* **143,** 1161–1173.

Choksi, S. P., Southall, T. D., Bossing, T., Edoff, K., de Wit, E., Fischer, B. E., van Steensel, B., Micklem, G., and Brand, A. H. (2006). Prospero acts as a binary switch between self-renewal and differentiation in Drosophila neural stem cells. *Dev. Cell* **11,** 775–789.

Cleary, M. D., and Doe, C. Q. (2006). Regulation of neuroblast competence: Multiple temporal identity factors specify distinct neuronal fates within a single early competence window. *Genes Dev.* **20,** 429–434.

Desai, A. R., and McConnell, S. K. (2000). Progressive restriction in fate potential by neural progenitors during cerebral cortical development. *Development* **127,** 2863–2872.

Dessaud, E., McMahon, A. P., and Briscoe, J. (2008). Pattern formation in the vertebrate neural tube: A sonic hedgehog morphogen-regulated transcriptional network. *Development* **135,** 2489–2503.

Doe, C. Q. (1992). Molecular markers for identified neuroblasts and ganglion mother cells in the Drosophila central nervous system. *Development* **116,** 855–863.

Doe, C. Q. (2006). Chinmo and neuroblast temporal identity. *Cell* **127,** 254–256.

Doe, C. Q. (2008). Neural stem cells: Balancing self-renewal with differentiation. *Development* **135,** 1575–1587.

Doe, C. Q., and Goodman, C. S. (1985). Early events in insect neurogenesis. I. Development and segmental differences in the pattern of neuronal precursor cells. *Dev. Biol.* **111,** 193–205.

Egger, B., Boone, J. Q., Stevens, N. R., Brand, A. H., and Doe, C. Q. (2007). Regulation of spindle orientation and neural stem cell fate in the Drosophila optic lobe. *Neural Dev.* **2,** 1.

Elliott, J., Jolicoeur, C., Ramamurthy, V., and Cayouette, M. (2008). Ikaros confers early temporal competence to mouse retinal progenitor cells. *Neuron* **60,** 26–39.

Gaspard, N., Bouschet, T., Hourez, R., Dimidschstein, J., Naeije, G., van den Ameele, J., Espuny-Camacho, I., Herpoel, A., Passante, L., Schiffmann, S. N., Gaillard, A., and Vanderhaeghen, P. (2008). An intrinsic mechanism of corticogenesis from embryonic stem cells. *Nature* **455,** 351–357.

Grosskortenhaus, R., Pearson, B. J., Marusich, A., and Doe, C. Q. (2005). Regulation of temporal identity transitions in Drosophila neuroblasts. *Dev. Cell* **8,** 193–202.

Grosskortenhaus, R., Robinson, K. J., and Doe, C. Q. (2006). Pdm and Castor specify late-born motor neuron identity in the NB7-1 lineage. *Genes Dev.* **20,** 2618–2627.

Higashijima, S., Shishido, E., Matsuzaki, M., and Saigo, K. (1996). Eagle, a member of the steroid receptor gene superfamily, is expressed in a subset of neuroblasts and regulates the fate of their putative progeny in the Drosophila CNS. *Development* **122,** 527–536.

Huang, D. H., Chang, Y. L., Yang, C. C., Pan, I. C., and King, B. (2002). Pipsqueak encodes a factor essential for sequence-specific targeting of a polycomb group protein complex. *Mol. Cell. Biol.* **22,** 6261–6271.

Isshiki, T., Pearson, B., Holbrook, S., and Doe, C. Q. (2001). Drosophila neuroblasts sequentially express transcription factors which specify the temporal identity of their neuronal progeny. *Cell* **106,** 511–521.

Jacob, J., Maurange, C., and Gould, A. P. (2008). Temporal control of neuronal diversity: Common regulatory principles in insects and vertebrates? *Development* **135,** 3481–3489.

Jefferis, G. S., Marin, E. C., Stocker, R. F., and Luo, L. (2001). Target neuron prespecification in the olfactory map of Drosophila. *Nature* **414,** 204–208.

Kambadur, R., Koizumi, K., Stivers, C., Nagle, J., Poole, S. J., and Odenwald, W. F. (1998). Regulation of POU genes by castor and hunchback establishes layered compartments in the Drosophila CNS. *Genes Dev.* **12,** 246–260.

Kanai, M. I., Okabe, M., and Hiromi, Y. (2005). Seven-up controls switching of transcription factors that specify temporal identities of Drosophila neuroblasts. *Dev. Cell* **8,** 203–213.

Karlsson, D., Baumgardt, M., and Thor, S. (2010). Segment-specific neuronal subtype specification by the integration of anteroposterior and temporal cues. *PLoS Biol.* **8,** e1000368.

Knoblich, J. A. (2008). Mechanisms of asymmetric stem cell division. *Cell* **132,** 583–597.

Knoblich, J. A. (2010). Asymmetric cell division: Recent developments and their implications for tumour biology. *Nat. Rev. Mol. Cell Biol.* **11,** 849–860.

Kohwi, M., Hiebert, L. S., and Doe, C. Q. (2011). The pipsqueak-domain proteins Distal antenna and Distal antenna-related restrict Hunchback neuroblast expression and early-born neuronal identity. *Development* **138,** 1727–1735.

Lee, T., and Luo, L. (2001). Mosaic analysis with a repressible cell marker (MARCM) for Drosophila neural development. *Trends Neurosci.* **24,** 251–254.

Lee, T., Lee, A., and Luo, L. (1999). Development of the Drosophila mushroom bodies: Sequential generation of three distinct types of neurons from a neuroblast. *Development* **126,** 4065–4076.

Livesey, F. J., and Cepko, C. L. (2001). Vertebrate neural cell-fate determination: Lessons from the retina. *Nat. Rev. Neurosci.* **2,** 109–118.

Losada-Perez, M., Gabilondo, H., del Saz, D., Baumgardt, M., Molina, I., Leon, Y., Monedero, I., Diaz-Benjumea, F., Torroja, L., and Benito-Sipos, J. (2010). Lineage-unrelated neurons generated in different temporal windows and expressing different combinatorial codes can converge in the activation of the same terminal differentiation gene. *Mech. Dev.* **127,** 458–471.

Maurange, C., and Gould, A. P. (2005). Brainy but not too brainy: Starting and stopping neuroblast divisions in Drosophila. *Trends Neurosci.* **28,** 30–36.

Maurange, C., Cheng, L., and Gould, A. P. (2008). Temporal transcription factors and their targets schedule the end of neural proliferation in Drosophila. *Cell* **133,** 891–902.

Mettler, U., Vogler, G., and Urban, J. (2006). Timing of identity: Spatiotemporal regulation of hunchback in neuroblast lineages of Drosophila by Seven-up and Prospero. *Development* **133,** 429–437.

Molyneaux, B. J., Arlotta, P., Menezes, J. R., and Macklis, J. D. (2007). Neuronal subtype specification in the cerebral cortex. *Nat. Rev. Neurosci.* **8,** 427–437.

Naka, H., Nakamura, S., Shimazaki, T., and Okano, H. (2008). Requirement for COUP-TFI and II in the temporal specification of neural stem cells in CNS development. *Nat. Neurosci.* **11,** 1014–1023.

Nakajima, A., Isshiki, T., Kaneko, K., and Ishihara, S. (2010). Robustness under functional constraint: The genetic network for temporal expression in Drosophila neurogenesis. *PLoS Comput. Biol.* **6,** e1000760.

Pearson, B. J., and Doe, C. Q. (2003). Regulation of neuroblast competence in Drosophila. *Nature* **425,** 624–628.

Pearson, B. J., and Doe, C. Q. (2004). Specification of temporal identity in the developing nervous system. *Annu. Rev. Cell Dev. Biol.* **20,** 619–647.

Schmid, A., Chiba, A., and Doe, C. Q. (1999). Clonal analysis of Drosophila embryonic neuroblasts: Neural cell types, axon projections and muscle targets. *Development* **126,** 4653–4689.

Schmidt, H., Rickert, C., Bossing, T., Vef, O., Urban, J., and Technau, G. M. (1997). The embryonic central nervous system lineages of Drosophila melanogaster. II. Neuroblast lineages derived from the dorsal part of the neuroectoderm. *Dev. Biol.* **189,** 186–204.

Shen, Q., Wang, Y., Dimos, J. T., Fasano, C. A., Phoenix, T. N., Lemischka, I. R., Ivanova, N. B., Stifani, S., Morrisey, E. E., and Temple, S. (2006). The timing of cortical neurogenesis is encoded within lineages of individual progenitor cells. *Nat. Neurosci.* **9,** 743–751.

Shoval, O., and Alon, U. (2010). SnapShot: Network motifs. *Cell* **143,** 326-e1.

Siegrist, S. E., Haque, N. S., Chen, C. H., Hay, B. A., and Hariharan, I. K. (2010). Inactivation of both Foxo and reaper promotes long-term adult neurogenesis in Drosophila. *Curr. Biol.* **20,** 643–648.

Skeath, J. B., and Thor, S. (2003). Genetic control of Drosophila nerve cord development. *Curr. Opin. Neurobiol.* **13,** 8–15.

Sousa-Nunes, R., Cheng, L. Y., and Gould, A. P. (2010). Regulating neural proliferation in the Drosophila CNS. *Curr. Opin. Neurobiol.* **20,** 50–57.

Sousa-Nunes, R., Yee, L. L., and Gould, A. P. (2011). Fat cells reactivate quiescent neuroblasts via TOR and glial insulin relays in Drosophila. *Nature* **471,** 508–512.

Tan, Y., Yamada-Mabuchi, M., Arya, R., St Pierre, S., Tang, W., Tosa, M., Brachmann, C., and White, K. (2011). Coordinated expression of cell death genes regulates neuroblast apoptosis. *Development* **138,** 2197–2206.

Technau, G. M., Berger, C., and Urbach, R. (2006). Generation of cell diversity and segmental pattern in the embryonic central nervous system of Drosophila. *Dev. Dyn.* **235,** 861–869.

Thummel, C. S. (2001). Molecular mechanisms of developmental timing in C. elegans and Drosophila. *Dev. Cell* **1,** 453–465.

Tran, K. D., and Doe, C. Q. (2008). Pdm and Castor close successive temporal identity windows in the NB3-1 lineage. *Development* **135,** 3491–3499.

Tran, K. D., Miller, M. R., and Doe, C. Q. (2011). Recombineering hunchback identifies two conserved domains required to maintain neuroblast competence and specify early-born neuronal identity. *Development* **137,** 1421–1430.

Truman, J. W., and Bate, M. (1988). Spatial and temporal patterns of neurogenesis in the central nervous system of Drosophila melanogaster. *Dev. Biol.* **125,** 145–157.

Tsuji, T., Hasegawa, E., and Isshiki, T. (2008). Neuroblast entry into quiescence is regulated intrinsically by the combined action of spatial Hox proteins and temporal identity factors. *Development* **135,** 3859–3869.

Yu, H. H., Kao, C. F., He, Y., Ding, P., Kao, J. C., and Lee, T. (2010). A complete developmental sequence of a Drosophila neuronal lineage as revealed by twin-spot MARCM. *PLoS Biol.* **8,** e1000461.

Zhou, B., Williams, D. W., Altman, J., Riddiford, L. M., and Truman, J. W. (2009). Temporal patterns of broad isoform expression during the development of neuronal lineages in Drosophila. *Neural Dev.* **4,** 39.

Zhu, S., Lin, S., Kao, C. F., Awasaki, T., Chiang, A. S., and Lee, T. (2006). Gradients of the Drosophila Chinmo BTB–zinc finger protein govern neuronal temporal identity. *Cell* **127,** 409–422.

CHAPTER NINE

# Transcriptional Switches Direct Plant Organ Formation and Patterning

Miguel A. Moreno-Risueno,[1] Jaimie M. Van Norman,[1] *and* Philip N. Benfey

## Contents



Department of Biology and Duke Center for Systems Biology, Duke University, Durham, North Carolina, USA

[1] These authors contributed equally to this work.

*Current Topics in Developmental Biology*, Volume 98
ISSN 0070-2153, DOI: 10.1016/B978-0-12-386499-4.00009-4

## Abstract

Development of multicellular organisms requires specification of diverse cell types. In plants, development is continuous and because plant cells are surrounded by rigid cell walls, cell division and specification of daughter cell fate must be carefully orchestrated. During embryonic and postembryonic plant development, the specification of cell types is determined both by positional cues and cell lineage. The establishment of distinct transcriptional domains is a fundamental mechanism for determining different cell fates. In this review, we focus on four examples from recent literature of switches operating in cell fate decisions that are regulated by transcriptional mechanisms. First, we highlight a transcriptional mechanism involving a mobile transcription factor in formation of the two ground tissue cell types in roots. Specification of vascular cell types is then discussed, including new details about xylem cell-type specification via a mobile microRNA. Next, transcriptional regulation of two key embryonic developmental events is considered: establishment of apical–basal polarity in the single-celled zygote and specification of distinct root and shoot stem cell populations in the plant embryo. Finally, a dynamic transcriptional mechanism for lateral organ positioning that integrates spatial and temporal information into a repeating pattern is summarized.

## 1. Introduction

Plant growth and development constitute a continuous process. The plant embryo does not contain most of the organs found in adult plants; instead, they have a simple structure composed of an embryonic root or radicle, one or two embryonic leaves or cotyledons, and a connecting stem or hypocotyl (Esau, 1977). Importantly, the two primary stem cell populations (meristems) are formed during embryogenesis, which will give rise to all adult organs. Thus, growth and development are largely postembryonic with new organs being forming throughout the plant's entire life. In addition, plants do not have a fixed body plan so individuals of the same species can have a variable number of organs. In contrast, development in most animals is more finite; the number of organs is strictly defined and organ formation is generally limited to embryogenesis. Plants are also exposed to a vast range of environmental conditions during development. As immobile organisms, plants must integrate endogenous and exogenous cues and respond in an accurate and timely manner to form and pattern organs.

Organ patterning relies on specification of different cell types and tissues with each cell type having specialized features. Plant cells are constrained by interconnected cell walls that prevent cellular movement. Therefore, a plant cell must integrate information about its relative position from neighboring cells and the lineage from which it is derived to make cell fate decisions. Thus, cues required for cell fate specification can be positional, inherited, or rely on both the ancestry and the position of the cell. For instance, positional cues in plants include hormones, short peptides, mobile transcriptional regulators, and, as recently reported, microRNAs. In addition, some transcription factors (TFs) are differentially inherited and/or expressed after cell division, which can also establish distinct transcriptional domains that determine new cell fates. Here, we discuss several recent examples of transcriptional regulators that act as switches for cell fate specification during development.

## 2. Cell Fate Specification in the Cortex–Endodermal Cell Lineage

The outer tissues of the *Arabidopsis thaliana* root are organized in concentric cell layers around the stele. From the stele outward there are two ground tissue layers, with the endodermis immediately adjacent to the stele followed by the cortex and the exterior epidermal layer (Fig. 9.1A). The two ground tissue cell types are generated through asymmetric division of a single stem cell lineage, the cortex/endodermal initial (CEI). The CEI undergoes a transverse asymmetric division to renew itself and generate a CEI daughter (CEID). The CEID then undergoes another asymmetric cell division, this time in a longitudinal orientation, to produce one cell each in the endodermal and cortical cell layers (Fig. 9.1B; Benfey *et al.*, 1993; Di Laurenzio *et al.*, 1996; Dolan *et al.*, 1993; Scheres *et al.*, 1994). The asymmetric division of the CEID and the switch to endodermal or cortical fate in the daughter cells are regulated by a transcriptional mechanism that links patterning, development, and the cell cycle.

The asymmetric division of the CEID is regulated by the activity of two TFs, SHORT ROOT (SHR) and SCARECROW (SCR). These proteins are both members of the GRAS family of transcriptional regulators (Benfey *et al.*, 1993; Di Laurenzio *et al.*, 1996; Helariutta *et al.*, 2000; Pysh *et al.*, 1999). *shr* and *scr* mutant plants each have only a single layer of ground tissue because the CEID fails to undergo the longitudinal asymmetric cell division. In *shr* mutants, the single ground tissue layer has some cortical cell features but no endodermal features. Whereas in *scr,* the mutant layer has both endodermal and cortical cell features (Fig. 9.1B; Benfey *et al.*, 1993; Di Laurenzio *et al.*, 1996; Helariutta *et al.*, 2000). Because these two genes are both required for CEID division but only SHR appears to be necessary for specification of endodermal fate, SHR was predicted to be upstream of SCR in the ground tissue developmental pathway. This hypothesis was

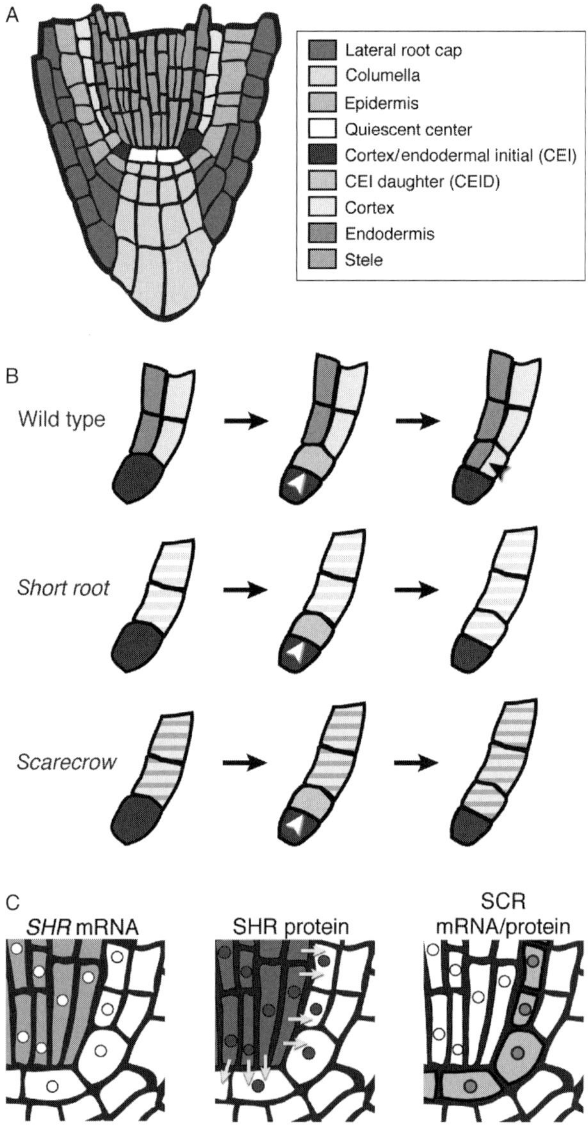


**Figure 9.1** Ground tissue formation in the root. (A) Schematic of a longitudinal section of the Arabidopsis root tip. Individual or groups of cell types are depicted in different colors. (B) Cut away of the ground tissue cell types from (A) with an emphasis on the asymmetric divisions and cellular defects in *short root (shr)* and *scarecrow (scr)* mutants. (B, upper panel) Ground tissue formation in wild type. (B upper panel, left to right) The CEI (dark green) divides transversely (white arrowheads) to regenerate itself and produce the CEID (light green). The CEID then divides longitudinally (black arrowhead) to generate the cells of the endodermis (blue) and cortex (yellow). (B, center panel) Ground tissue formation in *short root*. In the absence of the longitudinal

confirmed by epistasis of *shr* to *scr* and the decrease in *SCR* expression in *shr* plants (Helariutta *et al.*, 2000). These data indicate that SHR and SCR act together to regulate the asymmetric CEID division and specification of endodermal and cortical fates in the daughter cells.

## 2.1. SHR reveals a twist on transcriptional regulation of cell fate specification

Differences in *SHR* mRNA and SHR protein localization suggested that a nonautonomous transcriptional mechanism functioned to specify the endodermis. *SHR* transcripts are restricted to the stele, whereas SHR protein is found both in the stele and the immediately adjacent cell layer, which includes the CEI, CEID, and endodermis (Fig. 9.1C). The subcellular localization of SHR changes between different root tissues: in the stele SHR is nuclear and cytoplasmic, whereas in the adjacent layer SHR is nuclear. Ectopic expression of *SHR* in other root cell types results in formation of endodermal features (Helariutta *et al.*, 2000; Nakajima *et al.*, 2001). These results revealed that TFs were not strictly cell autonomous but could function as intercellular signaling molecules with the ability to directly activate a new transcriptional program in neighboring cells.

The capacity for SHR to function as a positional cue as well as a transcriptional switch in root patterning is evident when *SHR* is ectopically expressed (Helariutta *et al.*, 2000; Nakajima *et al.*, 2001; Sena *et al.*, 2004). Ectopic expression of *SHR* in the adjacent cell layer increased the number of cell layers between the epidermis and stele; these layers exhibited cellular markers of endodermal identity, including expression of *SCR* (Nakajima *et al.*, 2001). Additionally, ectopic SHR expression in cell types outside the stele, such as the epidermis, can induce these cells to exhibit endodermal features (Sena *et al.*, 2004). This suggests that SHR movement from the stele is not a prerequisite for its activity in specifying endodermal cell fate. However, SHR movement across only one cell layer is highly regulated appearing to require both cytoplasmic and nuclear localization prior to trafficking out of the stele, suggesting that regulation of SHR movement

asymmetric cell division, a single layer of ground tissue with cortical cell features (yellow with white stripes) forms. (B, lower panel) Ground tissue formation in *scarecrow*. The longitudinal asymmetric cell division also does not occur; however, the single layer of ground tissue exhibits both endodermal and cortical cell features (yellow with blue stripes). (C) Schematic of a portion of the Arabidopsis root tip, focusing on the localization of *SHR* mRNA, SHR protein and *SCR* mRNA and protein. Yellow arrows depict SHR protein movement from the stele into the adjacent cell layer. Note that SHR and SCR proteins are colocalized in the nuclei (small circles within the cells) of the adjacent cell layer. This figure was adapted from Petricka *et al.* (2009). (See Color Insert.)

is important for its function (Gallagher and Benfey, 2009; Gallagher *et al.*, 2004). SCR was implicated in limiting SHR movement because ectopic SHR movement and decreased nuclear localization were observed in the *scr* mutant background (Cui *et al.*, 2007; Heidstra *et al.*, 2004). Additionally, SHR expressed in the epidermis was more cytoplasmic and showed movement into the adjacent (inner) mutant cell layer in *scr* mutants (Sena *et al.*, 2004). These observations reveal the interdependent nature of SHR and SCR in root radial patterning, including an additional role for SCR in endodermal specification via modulation of SHR intercellular movement and subcellular localization.

The observation that cell-specific factors could limit SHR movement provided a straightforward mechanism for forming a single endodermal tissue layer. *In vivo* molecular evidence supporting this hypothesis, with SCR as a key component in this process, has been obtained. SHR and SCR proteins physically interact and have many common transcriptional targets suggesting that they form a transcriptional regulatory complex (Cui *et al.*, 2007; Levesque *et al.*, 2006; Sozzani *et al.*, 2010). Additionally, binding of SHR and SCR to the *SCR* promoter was detected by chromatin immunoprecipitation experiments. *SCR* expression is reduced in both *shr* and *scr* mutants suggesting that *SCR* expression is controlled by a SHR–SCR-dependent positive feedback loop (Cui *et al.*, 2007; Levesque *et al.*, 2006). This predicts that relatively high levels of SCR would be necessary to interact with SHR to sequester it into the nucleus. This prediction was examined using RNA interference lines with variable *SCR* expression levels. Plants with reduced *SCR* mRNA levels revealed SHR movement into adjacent cell layers, which led to ectopic endodermal cell fate specification (Cui *et al.*, 2007). These data indicate that SCR functions to limit SHR movement via nuclear sequestration to one cell layer outside the stele, therefore preventing excess endodermis formation.

Together, these observations have resulted in the following model for SHR–SCR function in asymmetric cell division and cell fate specification in the ground tissue. First, both nuclear and cytoplasmic localization of SHR in the stele promotes SHR movement to the adjacent cell layer (Gallagher and Benfey, 2009). In the adjacent cell layer, SHR interacts with SCR and is nuclear-localized. The SHR–SCR complex activates *SCR* transcription forming a positive feedback loop that sequesters all the SHR protein in the nuclei, thereby restricting endodermal cell fate specification to a single layer (Cui *et al.*, 2007). SHR and SCR then activate transcription of downstream targets, which leads to the asymmetric division in the ground tissue initials and specification of the endodermal cell layer (Cui *et al.*, 2007; Levesque *et al.*, 2006). In addition, other players, such as the zinc finger proteins JACKDAW, MAGPIE, and NUTCRACKER, have been tied to the SHR/SCR model for regulation of asymmetric division of the CEID cells (Levesque *et al.*, 2006; Welch *et al.*, 2007).

Recently, this spatial model for asymmetric cell division and endodermal specification has been expanded to include temporal components. Tissue-specific inducible versions of SHR and SCR were utilized to examine the temporal progression of asymmetric cell divisions and the changes in gene expression induced by these proteins. The timing of cell division after SHR or SCR induction coincided with induced expression of direct target genes. Remarkably, a component of the cell cycle machinery, a D-type cyclin, is directly regulated by SHR and SCR and involved in asymmetric CEID division providing an unexpected direct link between patterning and the cell cycle (Sozzani *et al.*, 2010). Thus, the model for regulating CEI/CEID cell division likely involves more intricate transcriptional regulatory complexes and is possibly more tunable than previously understood. Another open question involves the shutdown of the SHR/SCR transcriptional switch following the asymmetric division of the CEID. Why does not SHR movement into the endodermal cells cause them to divide like the CEID, forming additional cell layers? This suggests that the transcriptional network mediated by SHR and SCR is a dynamic switch that functions to integrate ancestry and positional cues in ground tissue development.

## 3. Cell Fate Switches During Vascular Tissue Development

Plant vasculature comprises many different tissues with the predominant ones being the xylem and phloem (Brady *et al.*, 2007; Esau, 1977). Xylem is the water-conducting tissue, while phloem specializes in nutrient transport. In above and below ground organs, the spatial organization of the xylem and phloem is different (Fig. 9.2). In roots, these tissues are typically arranged in a central cylinder while in the above ground organs they are arranged in bundles that are stereotypic in number and disposition. In the Arabidopsis root, primary xylem develops from vascular stem cells (procambial cells; Ohashi-Ito and Fukuda, 2010) and is made up of a single row of cells that extends across the central vascular cylinder (Figs. 9.2D and 9.3A). In addition, the two most peripheral xylem cells, which are in contact with the pericycle, are subsequently specified into protoxylem with spiral thickenings of the secondary cell walls. The remaining cells, in the central part of the row, form metaxylem with pitted secondary cell walls. Recent findings in Arabidopsis have identified novel TFs (Cano-Delgado *et al.*, 2010; Zhang *et al.*, 2011) that confer different xylem cell identities and appear to act as a multistep transcriptional switch that integrates signals and positional information from surrounding tissues (Fig. 9.2).

Xylem can also be formed during secondary development by the activity of the fascicular and interfascicular cambium (Agusti *et al.*, 2011). Secondary

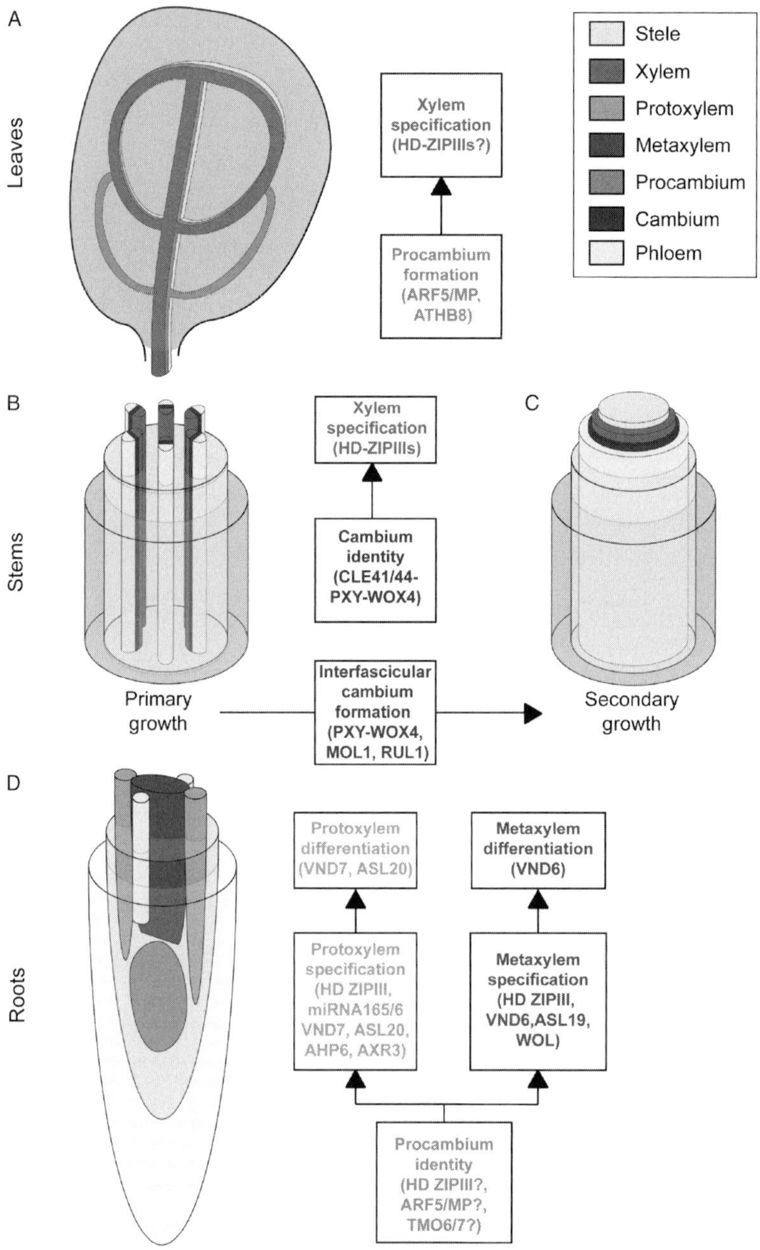


**Figure 9.2** Schematic of vascular patterning in Arabidopsis. (A) Patterning of leaf vasculature occurs through establishment of procambial cells and subsequent specification of vascular tissues from these cells. Regulators of procambium formation during venation are indicated. (B) Vascular tissues in stems are organized into vascular bundles. These bundles are comprised of xylem toward the inside and phloem toward the outside separated by cambial cells. Xylem is specified from cambium by specific

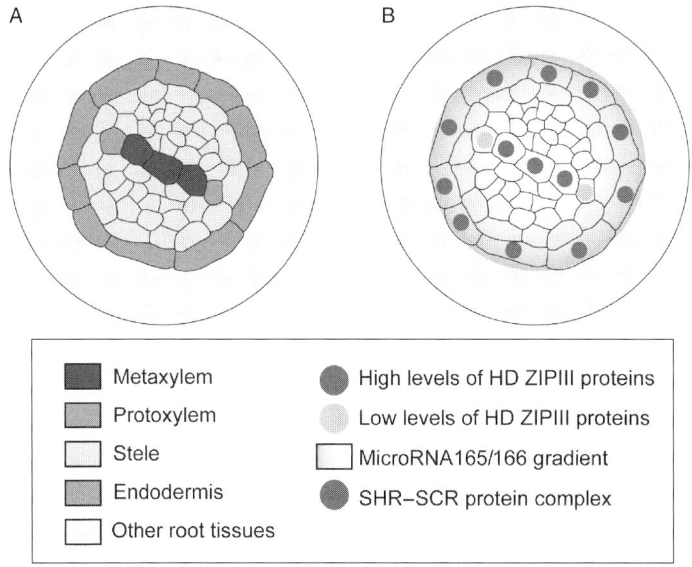


**Figure 9.3** Non-cell autonomous specification of xylem cell types by SHORT ROOT and SCARECROW. (A) Schematic of metaxylem and protoxylem tissues in a transverse section of the Arabidopsis root. Note that the endodermis forms a concentric layer of cells, whereas protoxylem and metaxylem constitute a single row of internal cells surrounded by other cells of the stele. (B) Specification of protoxylem versus metaxylem involves movement of SHORT ROOT to the endodermis where, together with SCARECROW, activates the microRNA165 and 166. The microRNA165/166 then travels back to the stele forming a gradient that targets the transcripts of CLASS III HOMEODOMAIN LEUCINE ZIPPER (HD-ZIPIII) transcription factors. This generates different levels of HD-ZIPIII proteins that determine cell fate in a dose-dependent manner. Low levels of HD-ZIPIII specify protoxylem and high levels specify metaxylem. (See Color Insert.)

xylem is normally more complex than primary xylem, however, in both xylem types, water conduction is carried out by the tracheary elements: vessels and tracheids (Esau, 1977). In Arabidopsis, secondary vascular growth is observed in the root, hypocotyl, and inflorescence stems (Zhang *et al.*, 2011), although it is more typically associated with perennial plants, like Poplar. Remarkably, recent studies indicate that there are conserved

regulators that switch cell fate. (C) During secondary growth, a ring of vascular tissue is generated through formation of cambium that closes the spaces between bundles. Common regulators during primary and secondary growth are indicated. (D) Vascular tissues in the root are organized in a central cylinder. Xylem constitutes a symmetry arch with metaxylem being specified toward the center of the arch and two protoxylem poles toward the outside. Regulators of xylem fate act as multistep switch specifying metaxylem and protoxylem. (See Color Insert.)

regulatory mechanisms for vascular development in Arabidopsis and Poplar, despite their evolutionary distance (Zhang *et al.*, 2011).

## 3.1. Specification of procambium and xylem identity

During Arabidopsis postembryonic root development, xylem specification requires a set of five homeodomain leucine zipper III (HD-ZIPIII) TFs: PHABULOSA (PHB), PHAVOLUTA (PHV), REVOLUTA (REV), CORONA (CNA), and ATHB8. Seedlings with loss-of-function (LOF) mutations for all the HD-ZIPIIIs TFs fail to form xylem, indicating that these regulators determine *de novo* xylem formation (Carlsbecker *et al.*, 2010). In addition, the quintuple *HD-ZIPIII* mutant appears to have broader morphological defects including reduced vascular cell number, which suggests that HD-ZIPIIIs might also regulate maintenance of procambium identity. In agreement with these functions, *HD-ZIP III* transcripts are accumulated in the root procambium, and their expression patterns overlap in those cells that will give rise to xylem (Carlsbecker *et al.*, 2010; Miyashima *et al.*, 2011). *PHB* and *REV* have broad expression in the stele and *PHV*, *CNA*, and *ATHB8* are more specifically expressed in the xylem precursor cells; however, *phv cna athb8* triple mutants still make xylem (Carlsbecker *et al.*, 2010). This suggests that additional regulators participate in xylem cell fate specification or are sufficient for determination of procambium identity.

In the shoot, the HD-ZIPIIIs have been also shown to be involved in the specification of xylem (Ilegems *et al.*, 2010), and ATHB8 has been proposed to have a role in procambium formation during leaf venation and in interfascicular cambium formation (Agusti *et al.*, 2011; Donner *et al.*, 2009; Ohashi-Ito and Fukuda, 2010). ATHB8 and the TF AUXIN RESPONSIVE FACTOR 5/MONOPTEROS (ARF5/MP) are necessary for procambial cell identity during leaf venation and ATHB8 is a direct target of ARF5/MP. These genes are part of a feedback loop that requires localized transport and signaling of the plant hormone auxin (Donner *et al.*, 2009; Ohashi-Ito and Fukuda, 2010). As expected for regulators of procambium specification, mutations in ARF5/MP cause a strong reduction in the number of leaf vascular bundles and *athb8* LOF mutants appear to have defects in selection of cells that will acquire preprocambial state. However, the double mutant *athb8 arf5/mp* exhibits only a slight reduction in vein pattern complexity compared to *mp/arf5* single mutants, while single *athb8* mutants do not show detectable changes in leaf vein patterns (Donner *et al.*, 2009). Thus, additional regulators redundant with ATHB8 and downstream of MP are likely involved in procambial cell determination. As ATHB8 participates in cambium and/or xylem identity in both shoot and roots, it is possible that other regulators, such as ARF5/MP might be also shared between shoots and roots to specify vascular tissues. Future studies might reveal a role for MP and its downstream targets in establishment of procambial cell identity in root tissues. In support of this, several MP direct targets, such as TARGET OF MP 7 (TMO7) and TMO6, which encode a bHLH and a Dof TF, respectively, are preferentially expressed in procambial cells and their precursors in the embryonic root (Schlereth *et al.*, 2010). It is thus, tempting to speculate that these TFs have a role in regulating root vascular fates.

In another pathway, the KANADI genes, which are expressed in phloem, repress procambium identity. This has been proposed to occur by regulation of auxin transport, likely mediated by repression of the transporter PIN-FORMED 1 (PIN1). Ectopic expression of KANADI 1 in provascular cells represses the activity and polar localization of PIN1 (Ilegems *et al.*, 2010). Supporting a role for plant hormones in regulation of vascular cell fates, xylem transported auxin and its interaction through a negative feedback loop with cytokinin, which is in turn transported through the phloem, is required to correctly specify vascular tissues (Bishopp *et al.*, 2011a,b). Therefore, specification of procambial cells appears to require the interplay of different types of TFs that not only regulate cell identity but also procambium cell number by promoting differentiation into xylem or phloem.

## 3.2. Signaling pathways involved in maintenance of cambium identity during primary and secondary growth

During primary growth, the receptor-like kinase PHLOEM INTERCALATED WITH XYLEM (PXY) and the TF WUSCHEL-RELATED HOMEOBOX 4 (WOX4) are expressed in procambium and cambium and respond to CLE41/44 peptides secreted from adjacent phloem tissues. In *pxy* mutants, procambial cells differentiate into xylem, while in *wox4* mutants, procambial cells are reduced to a single layer. These results indicate that procambial cells fail to proliferate and/or self-renew (Hirakawa *et al.*, 2010, 2011). In addition, *cle41* mutants have fewer procambial cells and exogenous application of CLE41 inhibits xylem specification from procambial cells. Therefore, CLE41/44-PXY-WOX4 signaling system maintains procambial/cambial identity by using positional information from adjacent (phloem) tissues to activate cell proliferation and repress xylem cell fate.

During secondary growth, differentiated cells are respecified and change their identity to become interfascicular cambium. In the shoot, the interfascicular cambium is formed between the primary vascular bundles generating a ring of cambium that will specify xylem and phloem toward the inside and outside, respectively (Fig. 9.2). This secondary growth coordinately enlarges shoot girth. Genes involved in primary growth also have roles in secondary growth (Agusti *et al.*, 2011; Hirakawa *et al.*, 2011). Likewise, *PXY, WOX4,* and *ATHB8* expression is detected in the fascicular or interfascicular cambium, and *pxy* mutants fail to establish a closed cambium ring in the stem. In addition, two novel receptor-like kinases have been identified to be involved in secondary growth (Agusti *et al.*, 2011). MORE LATERAL

GROWTH 1 (MOL1) has been proposed to repress cambium formation, while REDUCED IN LATERAL GROWTH 1 (RUL1) has been predicted to function as an activator in the process. Genetic and expression analyses showed that MOL1 functions in the same pathway as PXY and WOX4 while RUL1 likely works independently of PXY. Therefore, it appears that a common set of players are required for specification of cambial cells and/or maintenance of their identity during primary and secondary growth.

## 3.3. Xylem specification requires autonomous and nonautonomous transcriptional regulators

A subsequent step in xylem development involves differentiation into either protoxylem or metaxylem (Fig. 9.3A). Surprisingly, the endodermal regulators SHR and SCR function non-cell autonomously in controlling protoxylem specification in the stele (Carlsbecker *et al.*, 2010; Miyashima *et al.*, 2011). In *shr* and *scr* mutants, all xylem cells incorrectly differentiate into metaxylem. *SCR* is specifically expressed in the endodermis, while SHR is present both in vascular tissues and in the endodermis. However, only endodermis-specific activity of SHR is required for xylem patterning. This was demonstrated by ectopic expression of *SHR* in the *shr* mutant followed by examination of protoxylem formation. When *SHR* was expressed in the ground tissue of *shr*, protoxylem formation was restored in the stele, whereas *SHR* expression in the stele did not rescue xylem patterning. In addition, it was shown that this developmental mechanism requires movement of the microRNA165/166 from the endodermis to the stele where it targets the aforementioned HD-ZIP III TFs to specify protoxylem (Fig. 9.3B).

This regulatory loop was unraveled through the identification of a new allele of *PHB* that has a point mutation in the microRNA165/166 binding site. This gain-of-function mutant fails to form protoxylem. A connection with SHR was made through the observation that the PHB protein encoded by the allele resistant to microRNA165/166 degradation was broadly expressed both in *shr* and wild-type roots; in contrast to the normal PHB protein, whose expression was reduced in wild-type plants but not in *shr* mutants. It was also shown that microRNA165a and 166b are direct targets of SHR and SCR and that their endodermal expression largely depended on these two TFs. In agreement with the hypothesized movement and non-cell autonomous function of microRNA165/166, their activity was found to be high in the stele although they are generated in the endodermis. Further, ectopic expression of microRNA165 in the mutant ground tissue layer of *shr* rescued protoxylem formation in the stele. Thus, SHR activates microRNA165/166 in the endodermis, the microRNA then moves into the stele to restrict PHB protein accumulation to the central part of the vascular cylinder, where metaxylem is specified. In contrast, lower levels of PHB protein in the outer xylem cells specify protoxylem.

However, protoxylem cell fate cannot be explained solely by PHB protein accumulation, as *phb* LOF mutants do not show defects in protoxylem fate specification. This is because the other four *HD-ZIPIII*s (*PHV, REV, CNA,* and *ATHB8*) are also targets of the microRNA165/166 and act redundantly with PHB in the specification of metaxylem and protoxylem. All HD-ZIPIIIs have been shown to be directly involved in protoxylem/metaxylem specification through different combinations of LOF mutants that show ectopic formation of protoxylem instead of metaxylem. For instance, *athb8 phb* as well as any combination of quadruple mutations show ectopic protoxylem formation. Further support for their redundant role in protoxylem/metaxylem specification is provided by the quadruple mutant *phb phv rev shr*, which rescues the xylem patterning defects of *shr* (Carlsbecker *et al.*, 2010; Miyashima *et al.*, 2011).

HD-ZIPIIIs are not the only regulators of xylem fate that have been identified. Other regulators of protoxylem/metaxylem specification include WOODENLEG (Mahonen *et al.*, 2000) and ARABIDOPSIS HISTIDINE PHOSPHOTRANSFER PROTEIN 6 (Mahonen *et al.*, 2006) in the cytokinin signaling pathway; AUXIN RESISTANT 3 (Bishopp *et al.*, 2011a), in the auxin signaling pathway, as well as a group of seven TFs designated as VASCULAR-RELATED NAC-DOMAIN (VND; Kubo *et al.*, 2005; Zhang *et al.*, 2011). The VND TFs were identified through microarray analyses of cells induced to differentiate into xylem tracheary elements under specific *in vitro* conditions. Expression analyses of these TFs showed that they were specifically expressed in vascular tissues. Intriguingly, VND6 expression was restricted to root metaxylem precursor cells, while VND7 was expressed in immature protoxylem cells. The ectopic expression of VND6 and VND7 switched the fate of various cell types into metaxylem and protoxylem, respectively. Based on morphology, VND6 specified xylem elements with reticulate and/or pitted wall thickening similar to metaxylem, whereas VND7 produced xylem cells with annular and/or spiral wall thickening similar to protoxylem. However, *vnd6* and *vnd7* mutants did not show obvious morphological defects, which may be attributed to genetic redundancy with other VNDs or other regulators (Kubo *et al.*, 2005).

Transcriptional analyses revealed that VND6 and VND7 regulate genes involved in tracheary element specification, such as secondary cell wall formation and programmed cell death genes. Additionally, VND7 also regulates proteolytic enzyme encoding genes (Ohashi-Ito *et al.*, 2010; Yamaguchi *et al.*, 2011). Further examination of these data might reveal the molecular basis of protoxylem and metaxylem identity and their subsequent differentiation. In addition, a number of secondary cell wall formation genes identified as regulated by VND6 and VND7 were different from those downstream of the xylem fiber regulators, SECONDARY WALL-ASSOCIATED NAC-DOMAIN PROTEIN 1 (SND1) and NAC SECONDARY WALL THICKENING PROMOTING FACTOR 3 (NST3; Ohashi-Ito *et al.*, 2010; Yamaguchi *et al.*, 2011). Fibers and tracheary elements constitute distinct xylem types; however, both require secondary cell wall formation during their differentiation process. The VND6/7 transcriptomic analyses suggest that unique genes or pathways may be functioning in secondary cell wall formation in these two xylem types. Finally, ASYMMETRIC LEAVES 2 LIKE (ASL) 19/LATERAL ORGAN BOUNDARIES DOMAIN (LBD) 30 and ASL20/LBD18 are TFs downstream of VND6 and VND7 (Soyano *et al.*, 2008). ASL19 and ASL20 are expressed in immature tracheary elements and their expression depends on VND6 and VND7. Ectopic expression of *ASL19* and *ASL20* specifies tracheary elements similar to *VND6* and 7. For *ASL20*, this appears to occur not only by activation of a number of VND downstream targets but also, unexpectedly, by activation of VND7 expression itself. Thus, ASL20 and VND7 appear to function in a regulatory feedback loop that is able to specify xylem fate.

In conclusion, specification of different xylem cell fates is regulated by many transcriptional regulators functioning in different pathways and linked, in some cases, to non-cell autonomous regulators signaling from adjacent tissues. This tight regulation likely ensures proper patterning of the water-conducting tissues, which are vital for plant survival.

## 4. Transcriptional Regulation of Apical–Basal Cell Fate Determination after Zygotic Division in Arabidopsis

Embryogenesis is the process by which a unicellular zygote undergoes elaborate changes in cell number, fate, and morphology to ultimately form the mature embryo of a multicellular organism. Initial establishment of polarity is a critical step in embryogenesis in many organisms and typically involves activation of distinct transcriptional programs to drive differential development of the two embryonic axes. In many organisms, the single-celled zygote undergoes an asymmetric division that is critical in establishing embryo polarity. Initial embryo polarity is typically anterior–posterior (head–tail) in animals and apical–basal (shoot–root) in plants.

In Arabidopsis, the egg cell is polarized with the nucleus and the majority of cytoplasm localized apically and vacuoles localized basally (Fig. 9.4A). After fertilization, the zygote dramatically elongates and, like the egg cell, the nucleus is localized apically and a large vacuole forms basally (Faure *et al.*, 2002). The zygote then undergoes an asymmetric division to produce two daughter cells with distinctly different sizes and developmental fates. The smaller apical cell will give rise to a majority of the embryo (Fig. 9.4A). The basal cell will generate an extraembryonic support structure, the suspensor, which connects the embryo to the maternal tissue

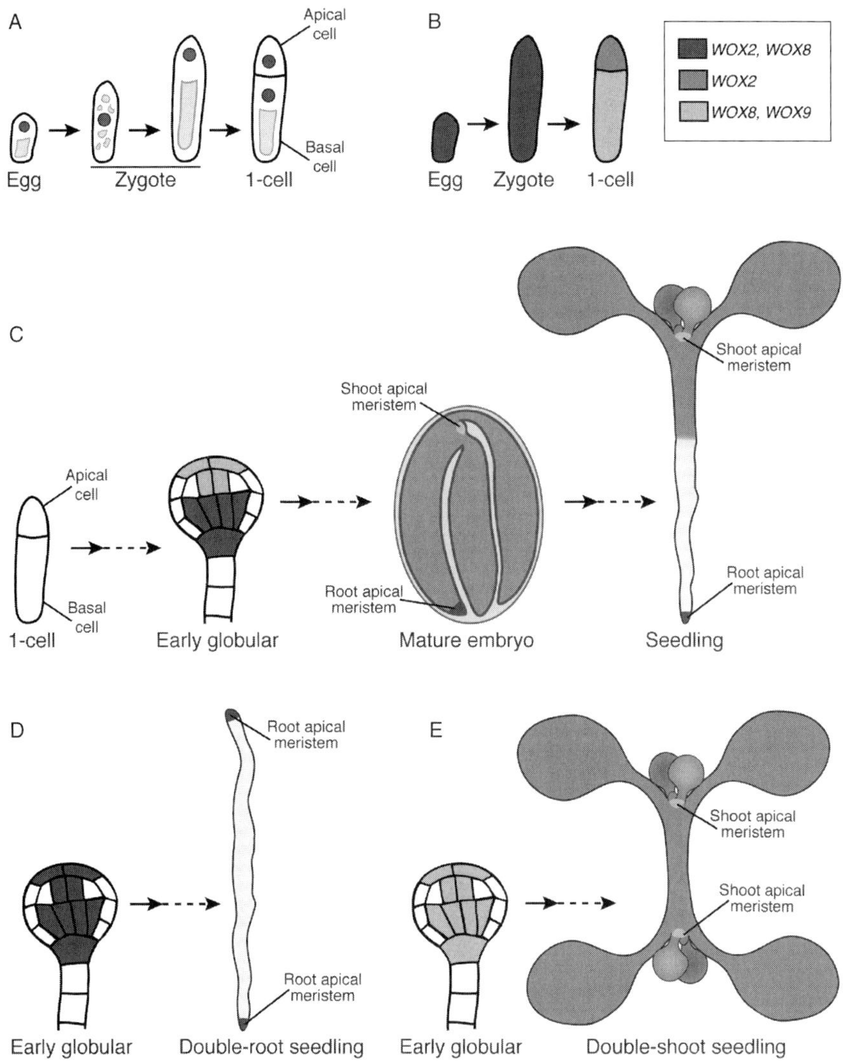


**Figure 9.4** Apical–basal polarity and specification of meristem fate in the Arabidopsis embryo through establishment of distinct transcriptional domains. (A) Schematic of early embryo development from the unfertilized egg cell through the first zygotic division. (A, left to right) The egg cell is polarized with the nucleus (dark gray) at the apical end and a vacuole at the basal end (light gray). After fertilization, the zygote is transiently symmetrical, then it elongates and is repolarized. The first zygotic division is asymmetric and producing an apical and basal cell with distinct fates. Note that embryo stages are based on the cell number in the apical domain, thus the first zygotic division results in one-cell stage embryo. (B) Schematic of the *WOX2/8/9* gene expression patterns in the egg cell through the first zygotic division. (C) Schematic of developmental snapshots from the one-celled embryo to the seedling focused on specification

(Jurgens, 2001; Scheres *et al.*, 1994). Given the similar polarity of the egg cell and zygote, it was unclear whether egg cell polarity was required for zygotic polarity. Additionally, the transcriptional mechanisms regulating specification of cell fate after the first zygotic cell division were unknown. Recent findings have begun to clarify some of these questions.

### 4.1. Homeodomain TFs establish apical–basal polarity after the asymmetric zygotic division

*WUSCHEL-RELATED HOMEOBOX* (*WOX*) genes are a plant-specific family of TFs that demarcate distinct transcriptional domains along the apical–basal embryo axis starting with the first zygotic division. Expression of *WOX2* and *WOX8* is found in both the egg cell and single-celled zygote. Following the first asymmetric division of the zygote, expression of *WOX2* and *WOX8* becomes restricted to the apical and basal cells, respectively (Fig. 9.4B; Haecker *et al.*, 2004). Expression of another *WOX* gene, *WOX9,* is then activated in the basal cell (Haecker *et al.*, 2004; Wu *et al.*, 2007). Consistent with their expression patterns, these WOX genes have critical roles in development of the apical and basal lineages. Embryos of *wox2* mutants show cell division defects in the apical domain, while *wox8 wox9* double mutants, have abnormal cell divisions in both the apical and basal cell lineages. Additionally, several markers of apical cell fate, including *WOX2,* are undetectable in *wox8 wox9* embryos (Breuninger *et al.*, 2008; Haecker *et al.*, 2004; Wu *et al.*, 2007). These results indicate WOX8/WOX9 function in the basal lineage is required for apical lineage development via WOX2 expression and also suggests signaling between the lineages is important for early embryo patterning (Breuninger *et al.*, 2008). Ectopic *WOX2* expression in *wox8 wox9* partially rescued later developmental defects in both the apical and basal lineages, indicating the apical defects in *wox8 wox9* embryos can be attributed to loss of *WOX2* expression. Unexpectedly, these zygotes showed defects earlier in development, they

---

of the root and shoot apical meristems. (C, left to right) Apical–basal polarity is established after the first zygotic division. By the early globular embryo stage, *PLETHORA*s (*PLT*s; blue) and the *HD-ZIPIIIs* (orange) expression is restricted to the domains that will give rise to the root (blue) and shoot (orange) apical meristems, respectively. In the mature embryo, these meristems are quiescent; however, become active after germination. The root meristem (blue) will give rise to all the cells of the primary root and the shoot meristem (orange) will give rise to all the cells of the aerial organs. (D) In the *topless-1* (*tpl-1*) mutant, the *PLT* expression domain has expanded into more apical regions and *HD-ZIPIII* expression is absent leading to a double-root phenotype. (E) Expansion of *HD-ZIPIII* expression into more basal embryo regions represses *PLT* expression leading to a double-shoot phenotype. (See Color Insert.)

failed to elongate and the first zygotic division was more symmetrical. Together, these results reveal a critical role for these WOX genes in the establishment of distinct transcriptional domains that determine differential fate decisions in the apical and basal cells.

## 4.2. Transcriptional activation of WOX8/9 is required to break zygotic symmetry and specify the basal cell lineage

In animals, transcription is often inhibited during early embryogenesis with important developmental transcripts inherited from the gametes or maternally supplied after fertilization. In addition, transcript localization within the egg or zygote can be important for embryo polarity. To address whether plant embryos utilize similar mechanisms to regulate zygotic polarity, the mechanisms directly regulating *WOX8* expression were investigated. This revealed a candidate transcriptional regulator of *WOX8* expression called WRKY2. WRKY2 is a member of the plant-specific family of WRKY proteins, which are zinc finger-containing proteins. *WOX8* and *WRKY2* have overlapping expression patterns in the early embryo and *wrky2* mutants show reduced expression of a *WOX8* transcriptional reporter. The *WOX8* promoter contains a known WRKY binding site, the W-box, and mutations in the W-box reduced expression of reporter constructs containing this *cis*-element. The *WOX9* promoter also contains a W-box and, like *WOX8,* its expression is significantly reduced in *wrky2* mutant embryos (Ueda *et al.*, 2011). Together these results indicate that WRKY2 is required to activate or maintain expression of *WOX8* and *WOX9* in the basal cell lineage, as well as in the single-celled zygote.

Mutation of *WRKY2* resulted in a more symmetrical first zygotic division although the *wrky2* mutant egg cell maintained its polarity (Ueda *et al.*, 2011). This phenotype suggested that characterizing WRKY2 function would provide insight into the relationship between egg cell and zygotic polarity. Careful examination revealed that after fertilization, the wild-type zygote is transiently symmetrical with the nucleus at the center and small vacuoles distributed throughout the cell. The single-celled zygote then expands and repolarizes with an apically localized nucleus and one large vacuole localized basally (Ueda *et al.*, 2011). Similar to wild type, *wrky2* mutants maintain egg cell polarity and the zygote shows transient symmetry. However, *wrky2* zygotes fail to repolarize and the majority does not undergo an asymmetric division. These defects were attributed to *WOX8* misregulation after fertilization as *WOX8* expression independent of WRKY2 partially restores asymmetric division of *wrky2* zygotes (Ueda *et al.*, 2011). This indicates that polarity of the egg cell and zygote can be uncoupled and that zygotic polarity depends on the transcriptional activity of *WRKY2* and *WOX8*.

*WRKY2* is expressed in both gametes but appears to be required only zygotically. This raised questions about whether zygotic activity of WRKY2 is sufficient for normal embryogenesis. Reciprocal crosses between *wrky2* and wild type revealed that normal embryogenesis occurred when functional *WRKY2* was inherited from either gamete, indicating that its activity post-fertilization is sufficient for embryo patterning (Ueda *et al.*, 2011). However, *WRKY2* expression in both gametes suggests that WRKY2 protein or transcript may be inherited by the zygote. Because only the female gamete expresses *WOX8*, this expression can be used to determine if WRKY2 activates *WOX8* transcription immediately after fertilization or whether *WOX8* transcripts are maternally inherited. In zygotes from wild type or *wrky2* egg cells fertilized with *wrky2* pollen, carrying a *WOX8* reporter, *WOX8* expression was higher when a functional copy of *WRKY2* was inherited from the egg cell (Ueda *et al.*, 2011). This result indicates that *WOX8* transcript inheritance is not necessary; instead WRKY2-driven transcriptional activation of *WOX8* in the zygote is sufficient for normal embryogenesis. Thus, WRKY2 directly activates a transcription switch required for zygotic polarization and asymmetric division leading to differential cell fate decisions in the resulting daughter cells. How the apical and basal lineages shutdown expression of *WOX8/9* and *WOX2*, respectively, remain open questions in the mechanics of this switch.

## 4.3. The impact of cell-to-cell signaling networks on early embryo patterning and its relationship to WOX transcriptional regulation

Despite the importance of the WOX transcriptional domains in determination of distinct apical and basal lineage development in the Arabidopsis embryo, plant cell fate decisions often rely more on positional cues. There are two main signaling pathways known to function in early embryogenesis in plants, a mitogen-activated protein (MAP) kinase cascade and a plant hormone signaling pathway. Mutation of the MAPKK kinase, YODA (YDA), results in embryos that do not elongate and the first zygotic division is more symmetrical, closely resembling embryos ectopically expressing *WOX2* in a *wox8 wox9* mutant background (Lukowitz *et al.*, 2004). This similarity suggests that *YDA* and *WOX8/9* function in a common pathway. However, *yda wox8 wox9* triple mutant embryos arrest development after the first nearly symmetrical zygotic division (Breuninger *et al.*, 2008). These data indicate that there are at least two independent pathways regulating asymmetry in the first zygotic division: a kinase cascade including YDA and a transcriptional mechanism via WOX8/9.

The plant hormone auxin has roles in a broad range of developmental processes (see Jenik *et al.*, 2007; Leyser, 2006 for recent reviews) including embryo patterning. Directional movement of auxin from cell-to-cell is a key feature of this signaling pathway; a family of transmembrane efflux carriers, called PINs, mediates auxin movement through specific membrane localization (Galweiler *et al.*, 1998; Petrasek *et al.*, 2006). PIN7 is localized on the apical membrane of cells in the basal lineage indicating auxin movement toward the apical cells. Additionally, expression of a synthetic auxin-dependent transcriptional reporter, *DIRECT REPEAT5* (*DR5*), occurs only in the apical lineage during early embryo stages (Friml *et al.*, 2003). However, at early embryo stages the role for auxin is not entirely clear, as mutants with defects in auxin transport generally have phenotypes that are very weak and not fully penetrant (Friml *et al.*, 2003), while *wox8 wox9* mutants exhibit more severe embryo defects. Additionally, neither *WOX2* nor *WOX8* expression in early embryos is affected by exogenous auxin or by mutants affecting auxin transport suggesting that *WOX2/8* expression is independent of auxin distribution (Breuninger *et al.*, 2008; Ueda *et al.*, 2011). These data suggest that regulation of apical/basal cell fate specification via the *WOX* genes occurs earlier and upstream of auxin in the early embryo.

## 5. Antagonism Between Transcriptional Regulators Specifies Two Distinct Stem Cell Populations in the Embryo

In plant embryos, apical–basal polarity is particularly important as the two main stem cell populations (meristems) are formed at opposite ends of this axis (Fig. 9.4C). Because plant growth and development is indeterminate, formation of the root and shoot apical meristems is essential for growth of all plant organs below and above ground, respectively. Two classes of transcriptional regulators have been identified as key regulators of root and shoot meristem fate specification in the embryo: the PLETHORA (PLT) family of AP2-domain proteins and the previously mentioned HD-ZIPIII TFs. There are four related *PLT* genes expressed in the root meristem of the embryo and mature root. Mutation of multiple *PLT*s results in rootless seedlings and embryo lethality; importantly, *PLT* overexpression induces formation of ectopic root meristems, indicating that PLTs have a master regulatory role in embryonic root meristem formation (Aida *et al.*, 2004; Galinha *et al.*, 2007). HD-ZIPIII proteins participate in various shoot developmental processes such as specification of the central domain of the shoot apical meristem and lateral organ polarity (reviewed in Engstrom *et al.*, 2004). Similar to the *PLT* genes, only mutation of multiple *HD-ZIPIII* genes results in defective embryogenesis (Prigge *et al.*, 2005). The crucial role for HD-ZIPIIIs in shoot fate specification and their antagonistic relationship with PLTs in the embryo was recently revealed through functional characterization of the TOPLESS (TPL) protein.

The key role for the transcriptional corepressor TPL in apical–basal embryo patterning is evident as *tpl-1* mutants conditionally produce embryos in which the shoot meristem is replaced by a second root meristem

(Fig. 9.4D; Long *et al.*, 2006, 2002). In these *tpl-1* embryos, *PLT* transcripts ectopically accumulated in the apical embryo domain, suggesting a role for TPL in repressing *PLT* expression apically. The double-root phenotype was not observed in *tpl-1 plt1 plt2* triple mutants consistent with the master regulatory role of the *PLT*s in root meristem formation (Fig. 9.4D). Additionally, *PLT1* and *PLT2* were identified as direct targets of TPL indicating that *PLT* repression in the apical domain is required for normal embryogenesis (Smith and Long, 2010). It seemed plausible from *tpl-1* phenotype that TPL or TPL-related proteins may also regulate genes specifically involved in shoot meristem fate in the embryo, which could lead to a double-shoot phenotype. In a screen for suppressors of the *tpl-1* double-root phenotype, a mutation in an *HD-ZIPII* TF gene was identified. This mutation disrupts a microRNA-binding site leading to excess transcript accumulation. Genetic analyses of multiple mutant embryos revealed that *PLT*s act as negative regulators of *HD-ZIPIII* expression in the basal domain and that *HD-ZIPIII*s block ectopic *PLT* expression in the apical domain. These data indicate that the *PLT* and *HD-ZIPIII* pathways act antagonistically in embryonic meristem formation. Finally, a second shoot meristem was formed in place of a root meristem as a consequence of *HD-ZIPIII* expression in the basal embryo domain (Fig. 9.4E; Smith and Long, 2010). These results indicate that transcriptional activity of these factors are necessary and sufficient for meristem formation and their mutual antagonism is also central to the formation of two distinct stem cell populations.

## 6. Specification and Positioning of Organs Forming Postembryonically

Plant postembryonic development initially relies on the activity of primary meristems, which are already present in the embryo. Primary meristems generate the main stem, leaves, flowers, and the primary root. Positioning of leaves and flowers takes place in the shoot apical meristem and ultimately requires that subsets of meristematic cells are specified to become new organs emerging from the flanks of the primary meristem (Hamant *et al.*, 2010). In contrast, subsequent growth of branches and lateral roots (LRs) is coordinated by the activity of lateral meristems. Because lateral meristems generating LRs are specified *de novo*, proper patterning requires not only cell fate specification but also correct positioning of the newly specified populations of cells along the primary root. Because primary root growth is continuous, positioning of LRs has to integrate spatial and temporal information. Recent findings in Arabidopsis show that positioning of LRs is mediated by a time-keeping mechanism that appears to involve oscillating gene expression (Moreno-Risueno *et al.*, 2010).

## 6.1. Positioning of leaves and flowers

In Arabidopsis, new aerial organ positioning and determination requires the hormone auxin. Leaves and flowers are positioned around the growing stem with a certain angle relative to the previous one (phyllotaxis), which likely maximizes light harvesting or pollination. Phyllotaxis takes place in the shoot apical meristem and involves formation of local gradients of auxin (Reinhardt *et al.*, 2003). These gradients form through the activity of intercellular transporters, such as PIN1, whose expression is, in turn, activated by auxin. Models of this feedback mechanism indicate that different auxin maxima may be formed in specific subsets of cells at precise angles, normally 137.5°, following a helical curve around the main axis. Live confocal imaging revealed that these subsets of cells reporting an auxin maximum initiate new lateral organs and showed temporal correlations between expression of PIN1 and known regulators of meristem identity and function (Hamant *et al.*, 2010; Heisler *et al.*, 2005). Additionally, differential growth of cells during phyllotaxis appears to generate biomechanical signals. These signals, in turn, feed back into this morphogenetic process (Hamant *et al.*, 2008) and may mediate the link between auxin and its transport (Heisler *et al.*, 2010).

## 6.2. Oscillating gene expression is involved in positioning LRs

In roots, evidence for oscillating gene expression playing a role in positioning lateral organs came from the observation of the dynamic expression of the DR5 marker gene (De Smet *et al.*, 2007; Moreno-Risueno *et al.*, 2010). Live imaging of DR5 fused to a Luciferase reporter allowed real-time expression analyses and showed a temporal and spatial relationship between periodic pulses of DR5 expression and the subsequent generation of LRs (Moreno-Risueno *et al.*, 2010). These periodic pulses of expression take place over a region of the Arabidopsis root tip termed the oscillation zone (OZ). During an oscillation cycle, DR5 expression is first observed at the more proximal or rootward region of the OZ, and over time, its expression increases and moves shootward within the OZ. Then, DR5 expression shuts down and a new oscillation begins. Growth of the root continuously displaces the OZ further from the shoot but coincident with the physical location where a DR5 oscillation occurred, a prebranch site is observed outside of the OZ (Fig. 9.5). Prebranch sites are marked by static points of DR5 expression, and subsequently, LR primordia are formed at prebranch sites as shown by lineage analyses and microscopy. Selection of subsets of cells in the primary root that become competent to generate a new organ (prebranch sites) is therefore reported by the DR5 oscillation. A remarkable feature of prebranch site formation is its capacity to compensate for variation in temperature and other environmental conditions (Moreno-Risueno *et al.*, 2010). This indicates that formation of prebranch sites acts as a

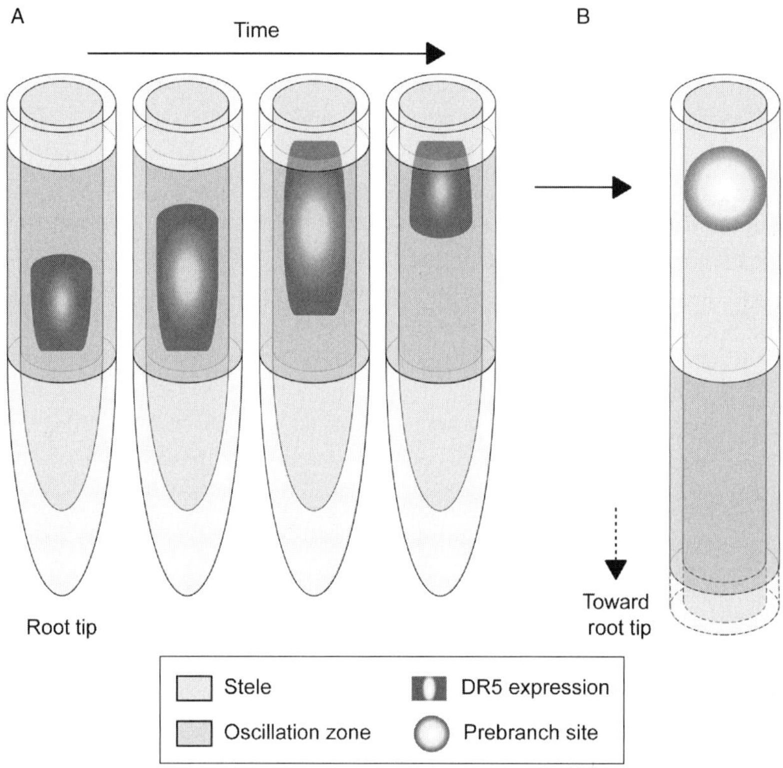


**Figure 9.5** The DR5 oscillation marks position of future lateral roots through establishment of prebranch sites. (A) DR5 expression in the oscillation zone (OZ) increases and moves shootward over time. At the beginning of one oscillation cycle, DR5 is expressed at the proximal part of the OZ (rootward). Over time DR5 expression increases and moves toward the distal or shootward part of the OZ. At the end of the cycle, DR5 expression is turned off. A new oscillation then begins in the OZ. Similar expression patterns have been observed for genes oscillating in phase with DR5. (B) Following a DR5 oscillation in the OZ, a prebranch site is established. Prebranch sites are observed outside the OZ but their locations coincide with the region of the root where an oscillation was observed. Prebranch sites mark the position of future lateral roots. (See Color Insert.)

biological clock and/or time-keeping mechanism, and it was consequently named the lateral root clock.

DR5 is also used as a marker for the transcriptional readout of auxin signaling. Thus, it was initially proposed that the changing expression of DR5 in the OZ was due to formation of a local auxin maximum or to increased auxin sensitivity in this region (De Smet *et al.*, 2007). This would lead to the prediction that priming of cell populations to become LRs is determined by local accumulation of auxin acting as a switch of cell fate, in a fashion somewhat analogous to phyllotaxis (Heisler *et al.*, 2005). However, other auxin-responsive promoters did not oscillate in the OZ and some genes that showed oscillating expression did not respond to auxin. This indicates that oscillations in DR5 and in other genes cannot be entirely explained by changing auxin levels in the OZ. Furthermore, exogenous auxin treatments localized to the OZ did not induce prebranch site formation through the initiation of a DR5 oscillation (Moreno-Risueno *et al.*, 2010). Intriguingly, a gain-of-function mutation of INDOLE ACETIC ACID FACTOR 28 (IAA28) that confers resistance to auxin, likely because the mutant protein is not degraded by auxin, appeared to affect the DR5 oscillation and is required for normal LR formation (De Rybel *et al.*, 2010; Dreher *et al.*, 2006; Rogg *et al.*, 2001). However, IAA28 expression has not been shown to oscillate, and appears to be complementary to the graded distribution of auxin at the root tip (Petersson *et al.*, 2009). Thus, auxin appears to be required for positioning of LRs but appears insufficient to trigger this developmental mechanism independently. Future studies might reveal a connection between auxin and oscillating gene expression in recruiting specific cell populations in time and space.

Further insight into the morphogenetic mechanism positioning lateral organs in the Arabidopsis root was obtained by microarray analyses of the OZ at various discrete points during the DR5 oscillation (Moreno-Risueno *et al.*, 2010). Approximately 2000 genes showed a similar oscillatory pattern as DR5, and about 1400 were shifted one phase and therefore showed an antiphase oscillatory pattern. Gene expression in two different oscillatory phases was confirmed by real-time imaging of predicted oscillating TF genes fused to a Luciferase reporter. In addition, the expression of these oscillating genes also propagated along the OZ and in some cases passed outside this developmental region. LOF mutants for TFs oscillating in both phases showed reduced numbers and irregular positioning of prebranch sites and LRs. This indicates that both phases are required for lateral organ positioning, suggesting that this mechanism operates as a complex and interconnected network.

## 6.3. A developmental switch might operate in combination with oscillating gene expression to position LRs

Oscillating gene expression is involved in another developmental mechanism in which repeating units are specified along an elongating axis: the segmentation clock of vertebrates (Krol *et al.*, 2011). In both the segmentation and the LR clocks, there are two sets of genes oscillating in opposite phases and their expression propagates along an elongating axis (the presomitic mesoderm and the primary root). In addition, based on their periodic and compensatory nature, both mechanisms can be described as biological clocks that convert time into precise spatial developmental patterns (Moreno-Risueno and Benfey, 2011). During vertebrate segmentation, a model for how cells can be recruited to form somites (vertebrae precursors) in a succession of discrete groups in time and space is given by the clock and wavefront model (Cooke and Zeeman, 1976). In this model, the clock is defined as an oscillator shared by all presomitic cells to which they are entrained and synchronized on a developmental time scale. The wavefront is defined as a point of irreversible, rapid cell change that moves down the longitudinal axis of the embryo. The interaction between the oscillator and the wavefront creates a pattern by selecting (in time) cells oscillating between a permissive and nonpermissive phase to undergo rapid alteration.

Further development of the clock and wavefront model has shown that mutually inhibitory gradients of retinoic acid (RA) and fibroblast growth factor (FGF) along the presomitic mesoderm can generate and position a sharp morphogen threshold (Goldbeter *et al.*, 2007). This threshold separates two stable steady states based on abrupt changes in levels of FGF and RA. The segmentation clock (the oscillating genes), in combination with two different developmental states, has been proposed to synchronously activate segmentation genes in successive discrete populations of cells. This mechanism, thus, combines a developmental switch with oscillating gene expression (segmentation clock) to precisely pattern somites during embryogenesis. In the root clock, oscillating gene expression is required for proper LR positioning along the primary root (Moreno-Risueno *et al.*, 2010). However, it is unknown how oscillating gene expression in the root clock successively selects populations of cells to form prebranch sites. Given the similarity with somitogenesis, it is tempting to speculate that a developmental switch is working in combination with the oscillating genes of the root clock. Future studies may demonstrate how similar or disparate these clock mechanisms are, despite the obvious evolutionary distance.

## 7. Concluding Remarks

Plants specify and pattern new organs both embryonically and postembryonically as part of normal development. This is largely achieved by transcriptional regulators functioning as switches for cell fate specification. Over the past few years, a number of transcriptional regulators have been identified in Arabidopsis that integrate positional information and cues relative to cell ancestry or lineage to coordinate patterning and development. In addition, different regulators act in parallel pathways to specify cell fates in response to various endogenous and, in some cases, environmental stimuli. This tight regulation ensures proper patterning under a wide range of conditions, which can help explain the developmental plasticity observed in plants.

Despite recent progress, there are still numerous unresolved developmental and mechanistic questions: How do the transcriptional switches work at the molecular level? What *cis*-motifs are responsible for specific protein–DNA interactions? And what is the nature of the transcriptional changes produced by these cell fate regulators? Some studies have begun to address these questions by molecular, genetic, and genome-wide approaches. For instance, WRKY2 is known to bind to the W-box in the regulatory sequences of its downstream target WOX8 to establish embryo polarity, and SHR binds to the promoter of the cell cycle regulator CYCD6;1 to activate an asymmetric division required for ground tissue patterning and specification. Future work might address how different pathways involved in cell fate determination function at the molecular level in response to variable inputs both from endogenous and exogenous sources.

## ACKNOWLEDGMENTS

We apologize to those researchers whose work we could not cover due to space limitations. We thank H. Cederholm, A. Iyer-Pascuzzi, R. Sozzani, C. Topp, and C. Winter for critical reading of this chapter. J. M. V. N. is supported by a NIH NRSA postdoctoral fellowship. Work in the Benfey lab is supported by grants from the NIH, NSF, and DARPA.

## REFERENCES

Agusti, J., Lichtenberger, R., Schwarz, M., Nehlin, L., and Greb, T. (2011). Characterization of transcriptome remodeling during cambium formation identifies MOL1 and RUL1 as opposing regulators of secondary growth. *PLoS Genet.* **7,** e1001312.

Aida, M., Beis, D., Heidstra, R., Willemsen, V., Blilou, I., Galinha, C., Nussaume, L., Noh, Y. S., Amasino, R., and Scheres, B. (2004). The PLETHORA genes mediate patterning of the Arabidopsis root stem cell niche. *Cell* **119,** 109–120.

Benfey, P. N., Linstead, P. J., Roberts, K., Schiefelbein, J. W., Hauser, M. T., and Aeschbacher, R. A. (1993). Root development in Arabidopsis: Four mutants with dramatically altered root morphogenesis. *Development* **119,** 57–70.

Bishopp, A., Help, H., El-Showk, S., Weijers, D., Scheres, B., Friml, J., Benkova, E., Mahonen, A. P., and Helariutta, Y. (2011a). A mutually inhibitory interaction between auxin and cytokinin specifies vascular pattern in roots. *Curr. Biol.* **21,** 917–926.

Bishopp, A., Lehesranta, S., Vaten, A., Help, H., El-Showk, S., Scheres, B., Helariutta, K., Mahonen, A. P., Sakakibara, H., and Helariutta, Y. (2011b). Phloem-transported cytokinin regulates polar auxin transport and maintains vascular pattern in the root meristem. *Curr. Biol.* **21,** 927–932.

Brady, S. M., Orlando, D. A., Lee, J. Y., Wang, J. Y., Koch, J., Dinneny, J. R., Mace, D., Ohler, U., and Benfey, P. N. (2007). A high-resolution root spatiotemporal map reveals dominant expression patterns. *Science* **318,** 801–806.

Breuninger, H., Rikirsch, E., Hermann, M., Ueda, M., and Laux, T. (2008). Differential expression of WOX genes mediates apical–basal axis formation in the Arabidopsis embryo. *Dev. Cell* **14,** 867–876.

Cano-Delgado, A., Lee, J. Y., and Demura, T. (2010). Regulatory mechanisms for specification and patterning of plant vascular tissues. *Annu. Rev. Cell Dev. Biol.* **26,** 605–637.

Carlsbecker, A., Lee, J. Y., Roberts, C. J., Dettmer, J., Lehesranta, S., Zhou, J., Lindgren, O., Moreno-Risueno, M. A., Vaten, A., Thitamadee, S., *et al.* (2010). Cell signalling by microRNA165/6 directs gene dose-dependent root cell fate. *Nature* **465,** 316–321.

Cooke, J., and Zeeman, E. C. (1976). A clock and wavefront model for control of the number of repeated structures during animal morphogenesis. *J. Theor. Biol.* **58,** 455–476.

Cui, H., Levesque, M. P., Vernoux, T., Jung, J. W., Paquette, A. J., Gallagher, K. L., Wang, J. Y., Bilou, I., Scheres, B., and Benfey, P. N. (2007). An evolutionarily conserved mechanism delimiting SHR movement defines a single layer of endodermis in plants. *Science* **316,** 421–425.

De Rybel, B., Vassileva, V., Parizot, B., Demeulenaere, M., Grunewald, W., Audenaert, D., Van Campenhout, J., Overvoorde, P., Jansen, L., Vanneste, S., *et al.* (2010). A novel aux/IAA28 signaling cascade activates GATA23-dependent specification of lateral root founder cell identity. *Curr. Biol.* **20,** 1697–1706.

De Smet, I., Tetsumura, T., De Rybel, B., Frey, N. F., Laplaze, L., Casimiro, I., Swarup, R., Naudts, M., Vanneste, S., Audenaert, D., *et al.* (2007). Auxin-dependent regulation of lateral root positioning in the basal meristem of Arabidopsis. *Development* **134,** 681–690.

Di Laurenzio, L., Wysocka-Diller, J., Malamy, J. E., Pysh, L., Helariutta, Y., Freshour, G., Hahn, M. G., Feldmann, K. A., and Benfey, P. N. (1996). The SCARECROW gene regulates an asymmetric cell division that is essential for generating the radial organization of the Arabidopsis root. *Cell* **86,** 423–433.

Dolan, L., Janmaat, K., Willemsen, V., Linstead, P., Poethig, S., Roberts, K., and Scheres, B. (1993). Cellular organisation of the Arabidopsis thaliana root. *Development* **119,** 71–84.

Donner, T. J., Sherr, I., and Scarpella, E. (2009). Regulation of preprocambial cell state acquisition by auxin signaling in Arabidopsis leaves. *Development* **136,** 3235–3246.

Dreher, K. A., Brown, J., Saw, R. E., and Callis, J. (2006). The Arabidopsis Aux/IAA protein family has diversified in degradation and auxin responsiveness. *Plant Cell* **18,** 699–714.

Engstrom, E. M., Izhaki, A., and Bowman, J. L. (2004). Promoter bashing, microRNAs, and Knox genes. New insights, regulators, and targets-of-regulation in the establishment of lateral organ polarity in Arabidopsis. *Plant Physiol.* **135,** 685–694.

Esau, K. (1977). Anatomy of Seed Plants. Wiley, New York.

Faure, J. E., Rotman, N., Fortune, P., and Dumas, C. (2002). Fertilization in Arabidopsis thaliana wild type: Developmental stages and time course. *Plant J.* **30,** 481–488.

Friml, J., Vieten, A., Sauer, M., Weijers, D., Schwarz, H., Hamann, T., Offringa, R., and Jurgens, G. (2003). Efflux-dependent auxin gradients establish the apical–basal axis of Arabidopsis. *Nature* **426,** 147–153.

Galinha, C., Hofhuis, H., Luijten, M., Willemsen, V., Blilou, I., Heidstra, R., and Scheres, B. (2007). PLETHORA proteins as dose-dependent master regulators of Arabidopsis root development. *Nature* **449,** 1053–1057.

Gallagher, K. L., and Benfey, P. N. (2009). Both the conserved GRAS domain and nuclear localization are required for SHORT–ROOT movement. *Plant J.* **57,** 785–797.

Gallagher, K. L., Paquette, A. J., Nakajima, K., and Benfey, P. N. (2004). Mechanisms regulating SHORT–ROOT intercellular movement. *Curr. Biol.* **14,** 1847–1851.

Galweiler, L., Guan, C., Muller, A., Wisman, E., Mendgen, K., Yephremov, A., and Palme, K. (1998). Regulation of polar auxin transport by AtPIN1 in Arabidopsis vascular tissue. *Science* **282,** 2226–2230.

Goldbeter, A., Gonze, D., and Pourquie, O. (2007). Sharp developmental thresholds defined through bistability by antagonistic gradients of retinoic acid and FGF signaling. *Dev. Dyn.* **236,** 1495–1508.

Haecker, A., Gross-Hardt, R., Geiges, B., Sarkar, A., Breuninger, H., Herrmann, M., and Laux, T. (2004). Expression dynamics of WOX genes mark cell fate decisions during early embryonic patterning in Arabidopsis thaliana. *Development* **131,** 657–668.

Hamant, O., Heisler, M. G., Jonsson, H., Krupinski, P., Uyttewaal, M., Bokov, P., Corson, F., Sahlin, P., Boudaoud, A., Meyerowitz, E. M., *et al.* (2008). Developmental patterning by mechanical signals in Arabidopsis. *Science* **322,** 1650–1655.

Hamant, O., Traas, J., and Boudaoud, A. (2010). Regulation of shape and patterning in plant development. *Curr. Opin. Genet. Dev.* **20,** 454–459.

Heidstra, R., Welch, D., and Scheres, B. (2004). Mosaic analyses using marked activation and deletion clones dissect Arabidopsis SCARECROW action in asymmetric cell division. *Genes Dev.* **18,** 1964–1969.

Heisler, M. G., Ohno, C., Das, P., Sieber, P., Reddy, G. V., Long, J. A., and Meyerowitz, E. M. (2005). Patterns of auxin transport and gene expression during primordium development revealed by live imaging of the Arabidopsis inflorescence meristem. *Curr. Biol.* **15,** 1899–1911.

Heisler, M. G., Hamant, O., Krupinski, P., Uyttewaal, M., Ohno, C., Jonsson, H., Traas, J., and Meyerowitz, E. M. (2010). Alignment between PIN1 polarity and microtubule orientation in the shoot apical meristem reveals a tight coupling between morphogenesis and auxin transport. *PLoS Biol.* **8,** e1000516.

Helariutta, Y., Fukaki, H., Wysocka-Diller, J., Nakajima, K., Jung, J. W., Sena, G., Hauser, M.-T., and Benfey, P. N. (2000). The SHORT-ROOT gene controls radial patterning of the Arabidopsis root through radial signaling. *Cell* **101,** 555–567.

Hirakawa, Y., Kondo, Y., and Fukuda, H. (2010). TDIF peptide signaling regulates vascular stem cell proliferation via the WOX4 homeobox gene in Arabidopsis. *Plant Cell* **22,** 2618–2629.

Hirakawa, Y., Kondo, Y., and Fukuda, H. (2011). Establishment and maintenance of vascular cell communities through local signaling. *Curr. Opin. Plant Biol.* **14,** 17–23.

Ilegems, M., Douet, V., Meylan-Bettex, M., Uyttewaal, M., Brand, L., Bowman, J. L., and Stieger, P. A. (2010). Interplay of auxin, KANADI and Class III HD-ZIP transcription factors in vascular tissue formation. *Development* **137,** 975–984.

Jenik, P. D., Gillmor, C. S., and Lukowitz, W. (2007). Embryonic patterning in Arabidopsis thaliana. *Annu. Rev. Cell Dev. Biol.* **23,** 207–236.

Jurgens, G. (2001). Apical–basal pattern formation in Arabidopsis embryogenesis. *EMBO J.* **20,** 3609–3616.

Krol, A. J., Roellig, D., Dequeant, M. L., Tassy, O., Glynn, E., Hattem, G., Mushegian, A., Oates, A. C., and Pourquie, O. (2011). Evolutionary plasticity of segmentation clock networks. *Development* **138,** 2783–2792.

Kubo, M., Udagawa, M., Nishikubo, N., Horiguchi, G., Yamaguchi, M., Ito, J., Mimura, T., Fukuda, H., and Demura, T. (2005). Transcription switches for protoxylem and metaxylem vessel formation. *Genes Dev.* **19,** 1855–1860.

Levesque, M. P., Vernoux, T., Busch, W., Cui, H., Wang, J. Y., Bilou, I., Hassan, H., Nakajima, K., Matsumoto, N., Lohmann, J. U., *et al.* (2006). Whole-genome analysis of the SHORT-ROOT developmental pathway in Arabidopsis. *PLoS Biol.* **4,** e143.

Leyser, O. (2006). Dynamic integration of auxin transport and signalling. *Curr. Biol.* **16,** R424–R433.

Long, J. A., Woody, S., Poethig, S., Meyerowitz, E. M., and Barton, M. K. (2002). Transformation of shoots into roots in Arabidopsis embryos mutant at the TOPLESS locus. *Development* **129,** 2797–2806.

Long, J. A., Ohno, C., Smith, Z. R., and Meyerowitz, E. M. (2006). TOPLESS regulates apical embryonic fate in Arabidopsis. *Science* **312,** 1520–1523.

Lukowitz, W., Roeder, A., Parmenter, D., and Somerville, C. (2004). A MAPKK kinase gene regulates extra-embryonic cell fate in Arabidopsis. *Cell* **116,** 109–119.

Mahonen, A. P., Bonke, M., Kauppinen, L., Riikonen, M., Benfey, P. N., and Helariutta, Y. (2000). A novel two-component hybrid molecule regulates vascular morphogenesis of the Arabidopsis root. *Genes Dev.* **14,** 2938–2943.

Mahonen, A. P., Bishopp, A., Higuchi, M., Nieminen, K. M., Kinoshita, K., Tormakangas, K., Ikeda, Y., Oka, A., Kakimoto, T., and Helariutta, Y. (2006). Cytokinin signaling and its inhibitor AHP6 regulate cell fate during vascular development. *Science* **311,** 94–98.

Miyashima, S., Koi, S., Hashimoto, T., and Nakajima, K. (2011). Non-cell-autonomous microRNA165 acts in a dose-dependent manner to regulate multiple differentiation status in the Arabidopsis root. *Development* **138,** 2303–2313.

Moreno-Risueno, M. A., and Benfey, P. N. (2011). Time-based patterning in development: The role of oscillating gene expression. *Transcription* **2,** 124–129.

Moreno-Risueno, M. A., Van Norman, J. M., Moreno, A., Zhang, J., Ahnert, S. E., and Benfey, P. N. (2010). Oscillating gene expression determines competence for periodic Arabidopsis root branching. *Science* **329,** 1306–1311.

Nakajima, K., Sena, G., Nawy, T., and Benfey, P. N. (2001). Intercellular movement of the putative transcription factor SHR in root patterning. *Nature* **413,** 307–311.

Ohashi-Ito, K., and Fukuda, H. (2010). Transcriptional regulation of vascular cell fates. *Curr. Opin. Plant Biol.* **13,** 670–676.

Ohashi-Ito, K., Oda, Y., and Fukuda, H. (2010). Arabidopsis VASCULAR-RELATED NAC-DOMAIN6 directly regulates the genes that govern programmed cell death and secondary wall formation during xylem differentiation. *Plant Cell* **22,** 3461–3473.

Petersson, S. V., Johansson, A. I., Kowalczyk, M., Makoveychuk, A., Wang, J. Y., Moritz, T., Grebe, M., Benfey, P. N., Sandberg, G., and Ljung, K. (2009). An auxin gradient and maximum in the Arabidopsis root apex shown by high-resolution cell-specific analysis of IAA distribution and synthesis. *Plant Cell* **21,** 1659–1668.

Petrasek, J., Mravec, J., Bouchard, R., Blakeslee, J. J., Abas, M., Seifertova, D., Wisniewska, J., Tadele, Z., Kubes, M., Covanova, M., *et al.* (2006). PIN proteins perform a rate-limiting function in cellular auxin efflux. *Science* **312,** 914–918.

Petricka, J. J., Van Norman, J. M., and Benfey, P. N. (2009). Symmetry breaking in plants: Molecular mechanisms regulating asymmetric cell divisions in Arabidopsis. *Cold Spring Harb. Perspect. Biol.* **1,** a000497.

Prigge, M. J., Otsuga, D., Alonso, J. M., Ecker, J. R., Drews, G. N., and Clark, S. E. (2005). Class III homeodomain-leucine zipper gene family members have overlapping, antagonistic, and distinct roles in Arabidopsis development. *Plant Cell* **17,** 61–76.

Pysh, L. D., Wysocka-Diller, J. W., Camilleri, C., Bouchez, D., and Benfey, P. N. (1999). The GRAS gene family in Arabidopsis: Sequence characterization and basic expression analysis of the SCARECROW-LIKE genes. *Plant J.* **18,** 111–119.

Reinhardt, D., Pesce, E. R., Stieger, P., Mandel, T., Baltensperger, K., Bennett, M., Traas, J., Friml, J., and Kuhlemeier, C. (2003). Regulation of phyllotaxis by polar auxin transport. *Nature* **426,** 255–260.

Rogg, L. E., Lasswell, J., and Bartel, B. (2001). A gain-of-function mutation in IAA28 suppresses lateral root development. *Plant Cell* **13,** 465–480.

Scheres, B., Wolkenfelt, H., Willemsen, V., Terlouw, M., Lawson, E., Dean, C., and Weisbeek, P. (1994). Embryonic origin of the *Arabidopsis* primary root and root meristem initials. *Development* **120,** 2475–2487.

Schlereth, A., Moller, B., Liu, W., Kientz, M., Flipse, J., Rademacher, E. H., Schmid, M., Jurgens, G., and Weijers, D. (2010). MONOPTEROS controls embryonic root initiation by regulating a mobile transcription factor. *Nature* **464,** 913–916.

Sena, G., Jung, J. W., and Benfey, P. N. (2004). A broad competence to respond to SHORT ROOT revealed by tissue-specific ectopic expression. *Development* **131,** 2817–2826.

Smith, Z. R., and Long, J. A. (2010). Control of Arabidopsis apical–basal embryo polarity by antagonistic transcription factors. *Nature* **464,** 423–426.

Soyano, T., Thitamadee, S., Machida, Y., and Chua, N. H. (2008). ASYMMETRIC LEAVES2-LIKE19/LATERAL ORGAN BOUNDARIES DOMAIN30 and ASL20/LBD18 regulate tracheary element differentiation in Arabidopsis. *Plant Cell* **20,** 3359–3373.

Sozzani, R., Cui, H., Moreno-Risueno, M. A., Busch, W., Van Norma, J. M., Vernoux, T., Brady, S. M., Dewitte, W., Murray, J. A. H., and Benfey, P. N. (2010). Spatiotemporal regulation of cell-cycle genes by SHORTROOT links patterning and growth. *Nature* **466,** 128–132.

Ueda, M., Zhang, Z., and Laux, T. (2011). Transcriptional activation of Arabidopsis axis patterning genes WOX8/9 links zygote polarity to embryo development. *Dev. Cell* **20,** 264–270.

Welch, D., Hassan, H., Blilou, I., Immink, R., Heidstra, R., and Scheres, B. (2007). Arabidopsis JACKDAW and MAGPIE zinc finger proteins delimit asymmetric cell division and stabilize tissue boundaries by restricting SHORT-ROOT action. *Genes Dev.* **21,** 2196–2204.

Wu, X., Chory, J., and Weigel, D. (2007). Combinations of WOX activities regulate tissue proliferation during Arabidopsis embryonic development. *Dev. Biol.* **309,** 306–316.

Yamaguchi, M., Mitsuda, N., Ohtani, M., Ohme-Takagi, M., Kato, K., and Demura, T. (2011). VASCULAR-RELATED NAC-DOMAIN7 directly regulates the expression of a broad range of genes for xylem vessel formation. *Plant J.* **66,** 579–590.

Zhang, J., Elo, A., and Helariutta, Y. (2011). Arabidopsis as a model for wood formation. *Curr. Opin. Biotechnol.* **22,** 293–299.

CHAPTER TEN

# Transcriptional Control of Cell Fate Specification: Lessons from the Fly Retina

Xiao-jiang Quan,[*,†] Ariane Ramaekers,[*,†] *and* Bassem A. Hassan[*,†,‡]

## Contents




## Abstract

It is now widely recognized that as cells of developing tissues transition through successive states of decreasing pluripotency into a state of terminal differentiation, they undergo significant changes in their gene expression profiles. Interestingly, these successive states of increasing differentiation are marked by the spatially and temporally restricted expression of sets of transcription factors. Each wave of transcription factors not only signals the arrival of a given stage in cellular differentiation, but it is also necessary for the activation of the next set of transcription factors, creating the appearance of a smooth, directed, and deterministic genetic program of cellular differentiation. Until recently, however, it was largely unknown which genes, besides each other, these transcription factors were activating. Thus, the molecular definition of any given step of differentiation, and how it gave rise to the following step remained unclear. Recent advances in transcriptomics, bioinformatics, and molecular



[*] Laboratory of Neurogenetics, VIB Center for the Biology of Disease, VIB, KU Leuven School of Medicine, Leuven, Belgium

[†] Center for Human Genetics, KU Leuven School of Medicine, Leuven, Belgium

[‡] Janelia Farm Research Campus, Howard Hughes Medical Institute, Ashburn, Virginia, USA


*Current Topics in Developmental Biology*, Volume 98

ISSN 0070-2153, DOI: 10.1016/B978-0-12-386499-4.00010-0

genetics resulted in the identification of numerous transcription factor target genes (TGs). These advances have opened the door to using similar approaches in developmental biology to understand what the transcriptional cascades of cellular differentiation might be. Using the development of the *Drosophila* eye as a model system, we discuss the role of transcription factors and their TGs in cell fate specification and terminal differentiation.

## 1. Introduction

The *Drosophila melanogaster* retina has been used for many years as a valuable experimental system for studying complex cell fate choices, cell–cell communication, and tissue patterning. The *D. melanogaster* compound eye consists of ~800 highly ordered eyelets called ommatidia. Each ommatidium contains 20 cells arranged in the same configuration but in opposite orientations either side of the dorsal–ventral equator. These 20 cells include 8 photoreceptor neurons (R1–R8) forming the core of the ommatidium, 6 pigment cells, 4 cone cells, and 2 mechanosensory bristles (Wolff and Ready, 1993). The adult compound eye arises from the patterning, determination, and differentiation during the larval stage of most of the posterior portion of a monolayered epithelium called the eye-antennal disc. The disc is derived from a single primordium of ~20 cells that become segregated at late embryonic stages (Garcia-Bellido and Merriam, 1969). After the initial expansion of these cells and the formation of the disc during early larval life, a wave of differentiation accompanied by a visible trough called the morphogenetic furrow (MF) sweeps across the disc from posterior to anterior leaving differentiated cells behind it (i.e., posterior to it) (Ready *et al.*, 1976; Wolff and Ready, 1991).

Cascades of transcription factor activities are necessary for the determination of the eye field, the specification of different cell types and the differentiation and specialization of these cells. Many of these will be discussed in further detail in this review (Fig. 10.1). In addition, several signaling pathways are required for these differentiation events. The initiation of furrow movement requires Hedgehog (Hh) pathway activity and the furrow's movement is positively regulated by Decapentaplegic (Dpp) signaling, the major TGFb pathway in *Drosophila*, and negatively regulated by the Wnt molecule Wingless (Wg) (Dominguez and Hafen, 1997; Heberlein *et al.*, 1993; Ma and Moses, 1995; Ma *et al.*, 1993; Treisman and Rubin, 1995). Notch and epidermal growth factor receptor (EGFR) signaling play key roles in the patterning the developing ommatidia and in cell fate specification (Baker and Yu, 1997; Cagan and Ready, 1989; Kumar and Moses, 2001; Li and Baker, 2001). Although the signaling pathways themselves will not be described in any detail, how these pathways relate to the activities of transcription factors will be dealt with further.

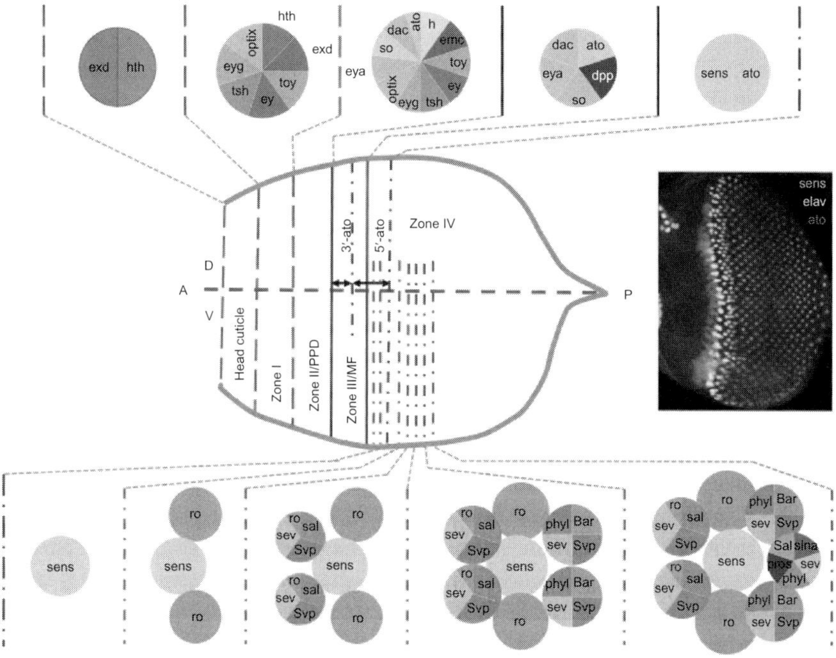


**Figure 10.1** A generalized schematic representing the expression code of factors discussed in this review that have been found to be necessary for photoreceptor specification of *Drosophila*. The different zones corresponding to the head cuticle and successive developmental stages of the eye field are represented. Pie charts indicate the specific combinations of factors expressed at each stage. Subdivision of Zone IV corresponds to successive steps of photoreceptor R1–R8 formation (PPD, preproneural domain; MF, morphogenetic furrow). (See Color Insert.)

The specification of the different cell types proceeds in a highly stereotyped temporal sequence (Brennan and Moses, 2000). The R8 photoreceptor is the first cell to be specified. Subsequently, the R2/R5 and R3/R4 pairs are recruited. Next, the remaining undifferentiated cells undergo a round of cell division known as the second mitotic wave. This is followed by the specification of the R1/R6 pair, completing the outer set of R cells (R1–R6). Finally, the second of the two inner R cells—the other being R8—the UV-sensitive R7 photoreceptor is specified, followed by four nonneuronal cone cells. During pupal development, the pigment cells and mechanosensory bristles are added at the periphery to complete ommatidial development. Remaining undifferentiated cells are removed by apoptosis. The nonneuronal accessory cells provide structural support and isolate each ommatidium from its neighbors. Also during pupal development, the R cells acquire their characteristic apical membrane specializations called rhabdomeres, which house the photosensitive Rhodopsin molecules (Wernet and Desplan, 2004).

This well described, elegant, series of events has made the fruit fly retina a powerful system in which to understand the molecular and genetic mechanisms underlying tissue patterning and cell fate specification. Because a large number of transcription factors and their modes of action have been identified as part of these efforts, we argue that the fly retina is an excellent *in vivo* system in which to study the role of transcription factors in cell fate specification and differentiation.

## 2. Eye(less) to Eye

The specification of the eye primordium is setup through the antagonistic activities of the Wnt (Wg) and TGFb (Dpp) signaling pathways (Dominguez and Casares, 2005; Kenyon *et al.*, 2003). This results in the activation of a set of mostly nuclear factors known as the retinal determination (RD) genes. RD genes cross regulate one another and this acts as what is known as the RD network (Kumar, 2009). Several of these transcription factors, such as the Pax6 homologues Eyeless (Ey) and Twin of Eyeless (Toy), are necessary and sufficient for eye specification. Loss of RD gene function results in flies that completely lack the eye field, and instead simply show head cuticle where the eye should be. Conversely, ectopic expression of many RD genes produces ectopic eyes on various appendages such as antennae, legs, and wings (Bonini *et al.*, 1997; Jang *et al.*, 2003; Shen and Mardon, 1997; Weasner *et al.*, 2007). Interestingly, Pax6 genes appear to specify eyes in most, if not all, animal species suggesting that all eyes have a common evolutionary origin. In support of this idea, expression of human Pax6 in *Drosophila* results in ectopic fly eyes (Halder *et al.*, 1995).

In addition to activating one another, RD genes must also regulate the expression of many other genes. The majority of the effort to identify RD target genes (TGs) has centered on Ey, mostly for historical reasons, as this was the first Pax6 homologue to be discovered and extensively studied. Ey has several putative direct TGs (Aerts *et al.*, 2010; Ostrin *et al.*, 2006), although only very few have been validated *in vivo*. Further, we currently know little about the contribution of the other members of the RD network to the regulation of these genes. Finally, because Ey TG studies chiefly utilized gain of function approaches it is difficult to ascertain whether the putative TGs are specifically those of Ey or are targets of other TFs with related binding sites, such as Toy. Briefly then, while the links between the various RD genes are reasonably well understood, the output of the network is generally less clear.

One crucial output of the RD network and the Wg/Dpp antagonism is the activation of a highly conserved, evolutionarily ancient basic-helix–loop–helix

TF called Atonal (Ato) (Dominguez, 1999; Tanaka-Matakatsu and Du, 2008; Zhang *et al.*, 2006). Ato activity is essential for the subsequent differentiation of all retinal cells. Ato mutants display a complete failure of any of the eye disc cells to express any specification markers and this is followed by apoptosis, resulting in flies that have no eye (Jarman *et al.*, 1995). This phenotype is distinct from the RD network loss function phenotype as Ato mutant flies have a distinct "slit eye" in the correct position which, however, is essentially completely devoid of specialized structures. Ectopic expression of Ato does not result in ectopic eyes or eye tissue. Rather it promotes the formation of context dependent sensory organs, which means that Ato acts to promote the neural character of eye tissue, as opposed to retinal fate *per se*. Thus, the initiation of Ato expression marks the transition from determination of the eye field to the sequential differentiation of all the various subtypes of cells that will make up the functional eye. How Ato activity orchestrates this transition is the subject of the following section.

## 3. And Then There Was Atonal!

Figure 10.2 summarizes some of the signaling events necessary for the initiation and refinement of Ato expression. Ato is almost certainly a direct TG of the RD network, and likely of Ey itself (Tanaka-Matakatsu and Du, 2008; Zhang *et al.*, 2006). Further, there is evidence that the initiation of Ato expression anterior to the MF is positively regulated by Dpp and negatively regulated by Wg perhaps directly via the Smad and TCF TFs, respectively (Niwa *et al.*, 2004), but these conclusions have been questioned (Tanaka-Matakatsu and Du, 2008; Zhang *et al.*, 2006). Following its initial activation, Ato positively regulates its own expression. In addition, Notch signaling is required for the increase in Ato levels anterior to the MF (Baker and Yu, 1997; Li and Baker, 2001). Perhaps paradoxically, concomitant with the increase in levels of expression comes a spatial restriction in the Ato expression domain from all cells anterior to the MF, to a cluster of about 5–7 cells and then a subcluster of 3 cells within the MF to eventually a single cell posterior to the MF (Baker *et al.*, 1996; Dokucu *et al.*, 1996; Jarman *et al.*, 1995). Interestingly, each of these subsets of cells shows increasing levels of Ato protein. Once again Notch signaling plays a crucial role in this spatial restriction via the classical Notch mediated process of mutual inhibition (Bray, 1998). Briefly, neighboring cells with initially equivalent potential inhibit one another via Notch signaling, which in this context represses Ato expression, itself a direct activator of the Notch ligand Delta. Thus, the slightest imbalance in the levels of Ato, Delta, or Notch is amplified and eventually only a single cell emerges with high levels of Ato while all neighboring cells extinguish its expression. Because Notch signaling requires cell–cell contact, only clusters

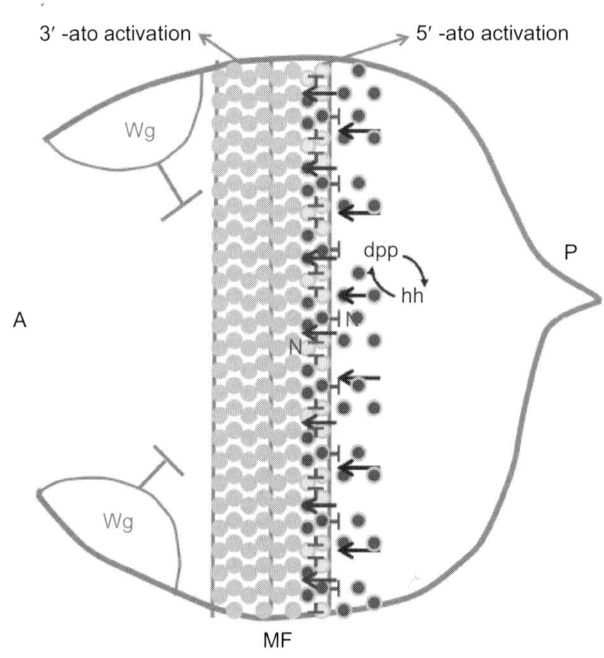


**Figure 10.2** A schematic representation of the regulation of Ato expression in the developing eye by various signaling pathways. Arrows indicate the direction of the progression of the differentiation wave. Color intensity refers to levels of *ato* expression. Ato expression is initiated anterior to the MF by a combination of factors (3′-ato activation) and then maintained through its autoactivation (5′-ato activation) (MF, morphogenetic furrow). (For color version of this figure, the reader is referred to the Web version of this chapter.)

of a few neighboring cells in direct contact participate, resulting in a regularly spaced array of Ato positive cells (reviewed in Frankfort and Mardon, 2002). These are the R8 cells, also known as the founder photoreceptors. The R8 briefly retains Ato expression and then rapidly extinguishes it and expression of Ato is never reactivated during subsequent stages. The specification of the R8 cell is thought to initiate the process of specification and differentiation of all subsequent R cells. It has been amply documented that the R8 is the source of the EGF molecule Spitz and that EGFR signaling is required for the recruitment of all the remaining R cells, with the exception of R7 whose recruitment by R8 requires a distinct Ras-dependent pathway mediated by the Sevenless receptor tyrosine kinase (Freeman, 1994, 1996; Hafen *et al.*, 1987; Kramer *et al.*, 1991; Tio and Moses, 1997; Tio *et al.*, 1994; Tomlinson and Ready, 1986).

Interestingly, the transient expression is a conserved feature of Ato function in various contexts and species where it is required for cell fate specification (Ben-Arie *et al.*, 2000; Bermingham *et al.*, 1999; Yang *et al.*, 2001) suggesting that transient expression is an important feature of this function. However, it is entirely unclear at the molecular level how a positively autoregulatory TF is inactivated with such precision on a short temporal order. It is possible that both posttranscriptional and chromatin silencing mechanisms cooperate to ensure this rapid silencing (Bossuyt *et al.*, 2009). Furthermore, it has not been formally tested to what extent these highly precise expression dynamics (level, space, and time) are necessary for proper retinal differentiation. It would be very interesting to attempt to engineer subtle variations in these dynamics and assay what the consequences for retinal differentiation are.

If the pivotal event in the transition from a multipotent retinal progenitor cell to a committed R8 precursor requires the activity of the Ato transcription factor, then understanding this event requires knowing the transcriptional events downstream of Ato. For nearly a decade since the discovery of the *ato* gene and then its various orthologs in other species, this network has remained more or less a black box. Recently, however, three publications identified tens of genes as confirmed or strong candidate targets of Ato in *Drosophila* and mouse (Aerts *et al.*, 2010; Cachero *et al.*, 2011; Klisch *et al.*, 2011). While the specific functions of most of these genes remain unknown, the majority of them fall within two major categories: intracellular modulators of signaling pathways and other transcription factors. As far as retinal development is concerned, our own work identifying direct TGs for Ato suggests that Ato alters the reception of all the critical signaling pathways that operate earlier to regulate Ato expression, such as Wnt, Dpp, and Notch, as well as pathways that may be involved in later differentiation of R cells such as EGFR and Jun N-terminal Kinase (JNK). Interestingly, the Ato-dependent enhancers of these signal modulators, such as the Nemo kinase (Wnt), Neuralized (Notch), Fasciclin 2 (EGFR), and dTraf1 (JNK), drive expression within the MF and in all or at least most future R cells (Aerts *et al.*, 2010). This suggests that the phase of Ato expression prior to the restriction to 1, 3, or even 5 cells is important for further development of the retina. In other words, Ato may be required for more than just the specification of the founder R8 cell. The identification of Ato TGs in combination with the possibility to generate single cell conditional knockout clones in the postmitotic retina (Choi *et al.*, 2009) should open the door to investigate such a putative function and the molecular mechanisms that may mediate it. The fact that pathways such as Wnt, EGFR, and Notch are involved in various steps of R cell recruitment and differentiation provides a rationale for the regulation of members of these pathways in all future R cells. It is interesting to note that many Ato TGs are also targets of the Supressor of Hairless protein, the key mediator of canonical Notch signaling (Aerts *et al.*, 2010). Whether this reflects coregulation of these gene by Ato and Notch activity, or a Notch-independent function for Su(H) in this particular context, remains to be elucidated.

## 4. Making Sense of Senseless

The second key category of Ato TGs are other transcription factors. Work in the Jarman lab has shown that at least one of these factors, Rfx, is itself a key regulator of ciliogenesis in the embryonic PNS, where Ato regulates the development of the ciliated internal sensory organs, called the chordotonal organs (Cachero *et al.*, 2011; Dubruille *et al.*, 2002). This provides direct evidence for how expression of the master regulator in a progenitor cell translates into specific differentiation events in terminally differentiated cells. In the retina, the zinc-finger transcription factor Senseless (Sens) is a direct target of Ato (Pepple *et al.*, 2008). Several putative targets of Sens have been identified (Aerts *et al.*, 2010), but they remain to be shown to be true targets *in vivo*, and thus precisely how Sens carries out its function in the retina is unclear. Sens was originally identified as a promoter of proneural gene expression (Nolo *et al.*, 2000). Soon afterward it was shown that Sens function was required for the maintenance of R8 identity, as *sens* mutant clones contained no R8 cells, but rather an extra cell that expresses the R2/R5 fate (Frankfort *et al.*, 2001). However, *sens* mutant ommatidia seem to contain the correct number of R cells, suggesting that the mutant "R8" is still able to recruit other R cells. The fact that Sens expression in the R8 is maintained throughout retinal development relates to later functions in R8 maintenance and further differentiation as will be discussed later in this review (Xie *et al.*, 2007).

## 5. The Outside-Rs!

Once the R8 has been specified, recruitment of the other photoreceptors is thought to commence via the release of Spitz from the R8 and its reception by the EGFR in surrounding cells. The R8 itself is thought to be nonresponsive to EGFR signaling through the modulation of transcription factor function. Specifically, there is evidence that Sens may bind to the transcriptional mediator of MAP Kinase signaling called Pointed and prevent it from entering the nucleus (Frankfort and Mardon, 2004). Thus, the R8 is itself immune to the signal emanating from it. One possibility why the R8 changes fate in the absence of Sens is that it becomes susceptible to EGFR signaling. The R1–6 photoreceptors are commonly referred to as the outer photoreceptors, because they surround the inner pair of R8 and R7. They are collectively required for motion detection (Cook and Desplan, 2001), and although they have distinct gene expression profiles and are recruited in sequential pairs and not altogether, it is not clear what the functional differences between them in terms of motion vision are.

It is also not clear what the code for the sequential recruitment is. The first pair of the six outer Rs to be recruited is the R2/5 pair, followed by the R3/4, and finally R1/6. The R2/5 pair expresses the Rough (Ro) transcription factor, which appears to repress Sens expression and thus distinguish R2/5 from R8. Sens in turn represses Ro in the R8 resulting in a dual-repression loop between these two factors (Frankfort *et al.*, 2001). Ro TGs are unknown, and thus the transcriptional cascade of events that specifies R2/5 fate remains to be determined. The next pair of photoreceptor neurons to be recruited from the primary precluster of five cells is R3/4 pair. Although Ro is expressed weakly in R3/4, loss of Ro does not appear to affect R3/4 fate. On the other hand, the zinc-finger transcription factor and nuclear hormone receptor Seven Up (Svp) is expressed by R3/4—as well as R1/6—and is required to repress Sevenless signaling emanating from R8 through the Bride of Sevenless (Boss) protein. In Svp mutant retinal clones, these four outer R cells become R7s (Mlodzik *et al.*, 1990).

In contrast to R8, R2/5 and R3/4, the R1/6 and R7 cells do not derive from the primary precluster expressing Ato and Sens. Rather, the second mitotic wave generates the cells that will eventually give rise to the R1/6, R7, and nonneuronal cells, such as the cone cells (Baker, 2001). In this sense, R1/6 and R7 have a different developmental history than their neighbors. One transcription factor expressed by all future R cells, regardless of their birth date, is Lozenge (Lz). Lz appears to positively regulate the expression of the Bar transcription factor in R1/6 cells, where Bar is required for R1/6 identity (Daga *et al.*, 1996; Higashijima *et al.*, 1992). Another nuclear factor important for R1/6 fate is Phyllopod (Phyl). Phyl appears to be expressed mainly in R1/R6 and R7 precursors and is required to maintain their neuronal fate. In the absence of Phyl, these cells adopt the nonneuron cone cell fate (Chang *et al.*, 1995; Dickson *et al.*, 1995).

## 6. R7: The Special One

The final photoreceptor to be specified by inductive signaling from R8 is the R7. R8 expresses Boss, which activates the Sevenless receptor in R7 whose fate is marked by the absence of Svp—which presumably allows Sevenless signaling to be active—as well as high expression of divergent homeodomain transcription factor Prospero (Pros) (Cagan, 1993; Raabe, 2000; Zipursky and Rubin, 1994). The R7 arises from its equivalence group of five cells, with the other four cells adopting the nonneuronal cone cell fate. Very recent work suggests that this may be due to combinatorial and opposing activities of Pros and dPax2, both targets of Notch and Ras signaling pathways, acting together to specify R7 versus cone cell fate (Charlton-Perkins and Cook, 2010). In addition to specifying R7 neuronal fate, Pros also acts later in development to control R7 differentiation. An intriguing finding is that the role of Pros might be restricted to late differentiation events—such as distal localization of rhabdomere and nucleus on the cell as well as photopigment expression—but not earlier events—localization of R7s on top of R8s at the center of the ommatidia or correct axonal projections in the medulla (Cook *et al.*, 2003; Kauffmann *et al.*, 1996). Therefore, other factors, still unidentified, are needed to achieve proper differentiation of R7. An extensive list of Pros TGs in the CNS has been described (Southall and Brand, 2009). Whether Pros regulates similar targets in R7 remains to be seen.

## 7. Spineless Stochasticity

During differentiation, PRs develop different morphologies, location within the ommatidia and photopigment (Rhodopsin) expression (Cook and Desplan, 2001; Hardie, 1985). The outer PRs, R1–R6, express Rhodopsin 1 (Rh1), which is sensitive to a broad wavelength spectrum, and are involved in motion detection and vision in dim light. The inner PRs, R7 and R8, sit on top of each other at the center of each ommatidium and differentiate into functionally distinct cells involved in color vision. R7s express one of the two UV-sensitive Rhodopsins, Rh3 and Rh4 whereas R8s express either the blue-sensitive (Rh5) or the green-sensitive (Rh6) Rhodopsins. In most ommatidia, expression of Rh4 in the R7 is primarily coupled to Rh6 in the corresponding R8 while Rh3 expressing-R7s are exclusively coupled with R8s that express Rh5. This defines two major functional subtypes of ommatidia: "pale" (Rh3/Rh5) and "yellow" (Rh4/Rh6) that, interestingly, are stochastically located in the retina. Despite their stochastic spatial pattern, the proportion of "pale" versus "yellow" ommatidia follows is roughly 30–70% and this ratio is conserved among flies from *Drosophila* to *Musca* (Franceschini *et al.*, 1981; Jukam and Desplan, 2010; Morante *et al.*, 2007).

The proper expression of the Rhodopsin genes, and more generally, the differentiation of PRs, involves an intricate network of transcriptional activators and repressors. The general logic of this regulatory network is that all PRs have the competency to express all *rhodopsin* genes, due to the combinatorial expression of transcriptional activators, either general to all PRs, like Orthodenticle (Otd), or subtype-specific, like Spalt (Sal) (inner PRs) (Mollereau *et al.*, 2001), Prospero (R7) (Cook *et al.*, 2003) or Senseless (R8) (Xie *et al.*, 2007). Specific expression of Rhodopsins then results from the repression of all but one of them in specific subsets of cells. In some cases, this repression is directly achieved by transcriptional factors controlling PR specific fates. For instance, the R8 determinant Senseless directly represses the expression of the R7-specific Rh3 and Rh4 in R8 cells

(Xie *et al.*, 2007). Similarly, the R7 determinant Prospero directly represses the expression of R8-specific rhodopsins, Rh5 and Rh6 in R7, by binding to their promoters (Cook *et al.*, 2003). A recent beautiful piece of work from the Desplan lab shed light onto a previously unidentified key player in this process, *defective proventriculus* (*dve*), a K50 homeobox protein (Johnston *et al.*, 2011). Johnston and colleagues found that regulation of *dve* expression results from the balance between two opposite signals: activation by Otd and repression by Sal. As a result, Dve is strongly expressed in outer PRs where it represses the expression of inner PRs photopigments Rh3, Rh5, and Rh6. In contrast, it is not expressed in most inner PRs except for the yellow R7 where it represses Rh3 expression.

The expression of "pale" Rh5 versus "yellow" Rh6 in R8 depends on a signal emanating from the committed "pale" R7 (Chou *et al.*, 1999; Papatsenko *et al.*, 1997). The exact nature of this signal remains to be elucidated. In contrast, maintenance of the decision involves a bistable regulatory loop where the Melted pleckstrin-homology (PH) domain protein and the Ser/Thr kinase warts, respectively members of the Tor/Insulin and the Hippo/Salvador signaling pathways, play opposite roles (Mikeladze-Dvali *et al.*, 2005). Therefore, the mosaic of pale versus yellow ommatidia derives from the presence of Dve in a subset of R7 cells (the future yellow R7s) where it escapes the repression by Spalt. Previous work from Desplan's lab had already demonstrated that the stochastic expression of the transcriptional factor Spineless represented a binary switch defining the mosaic of "yellow" versus "pale" PRs in the retina (Wernet *et al.*, 2006). It is thus satisfying to learn from Johnston and coworkers that, Spineless activates the expression of *dve* in yellow R7s. In pale R7s, in contrast, *dve* is repressed by Sal. This, in combination with the activation of Rh3 by Otd and Sal and the repression of Rh5/6 by Pros, results in exclusive Rh3 expression in these cells. Signaling from pale R7s represses Wts activity in their neighboring R8s thus preventing Rh6 expression and leading to pale, Rh5 expressing R8. Figure 10.3 summarizes these findings as an example of the complex combination of gene regulatory networks and signaling networks resulting in terminal PR differentiation. Ultimately, the question of the stochasticity of the formation of "pale" versus "yellow" ommatidia relates to how the stochastic expression of Spineless is achieved during eye development, which remains to be elucidated.

## 8. Hints, Allegations, and Things Left Unsaid

In conclusion, while many transcription factors, their expression and function in photoreceptor cell fate determination have been described, the gene regulatory networks emanating from these factors remain largely

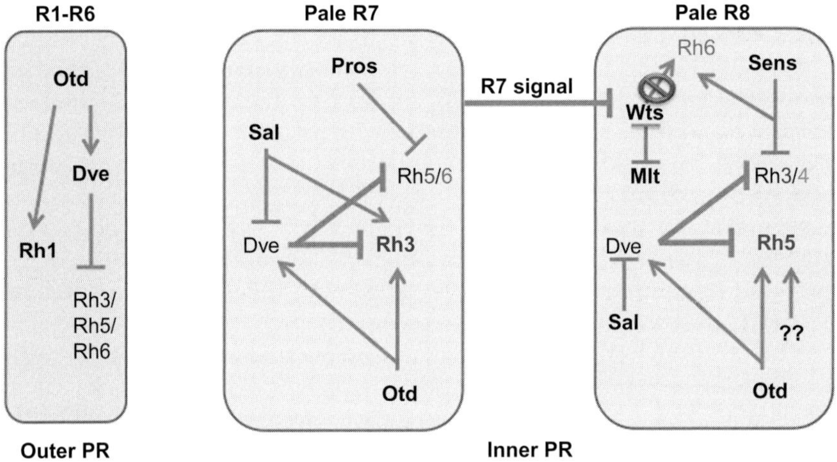


**Figure 10.3** Schematic representation of the gene regulatory network and signaling events thought to specify the terminal differentiation of the "pale" inner PR subtype. For a comprehensive model of how these events control the differentiation of the various PR subtypes across the retina the reader is referred to figure 7 in the original work by Johnston *et al.* (2011). (For color version of this figure, the reader is referred to the Web version of this chapter.)

unresolved. Two exceptions to this may be the gene regulatory network underlying the initial transition from Ey expressing progenitors to Ato expressing precursors, and the regulatory network underlying regulation of Rh expression (see below). We now know enough about the TGs of Ey and Ato to draw a reasonably high confidence—although admittedly only partial—gene regulatory network linking these two transcription factors and their targets (Aerts *et al.*, 2010). In principle, precisely the same combination of expression profiling and bioinformatics analyses that lead to the unraveling of the Ey-Ato network can be applied to all other retinal transcription factors. Because of the relative simplicity of genetically manipulating studying and isolating *Drosophila* retinae and because the fly retina allows the simultaneous visualization of spatial and temporal expression, the fly retina presents a most attractive model for studying the gene regulatory events underlying the development of complex tissues.

One of the very intriguing aspects of photoreceptor specification is its sequential nature. The fact that the sequence of R cell formation appears invariant, rather than stochastic, suggests that it is important for the eventual function of the retina. Particular cells in particular ommatidia need to "look at" specific points in space. Nonetheless, why the UV-sensitive R7 is necessarily the last cell to be recruited, even though it is always next to the R8 is unclear. One of the intriguing concepts to emerge from the analysis of Ato TGs in the retina is that the Ato target enhancers of these genes are in fact expressed in all or most R cell precursors, as of course is Ato. This suggests that the transient expression of Ato in all future R cells is important for their later differentiation. For example, the Nemo Kinase is activated by Ato early on but is required later for ommatidial rotation. Perhaps then the sequential recruitment of R cells reflects in part the length of time during which they expressed Ato, Sens, and other early transcription factors, resulting in a different repertoire of gene expression, or even simply different levels of expression of the same genes. Thus, cells which have expressed Ato and Sens the longest become specified earlier because they express higher levels of "neuronal genes." If, in addition, one imagines that cells inhibit each other from differentiating, until they themselves have differentiated, one can potentially explain how cells with the highest levels of neuronal genes differentiate first, followed by the ones with slightly less, and so on. This model makes a very specific prediction, which is in principle testable: the ablation of R2/5 cells before they adopt their fate, would result in the precautious recruitment of R3/4 for example. Thus, the speed with which differentiation occurs would remain similar, but the resulting ommatidia would be two cells smaller.

Finally, the recruitment of R cells from the first precluster (R8, R2/5, R3/4) depends on the expression of ato, a bHLH proneural factor. Interestingly, however, no proneural factor has been identified for the second precluster (R7, R1/6, cone cells). Instead a complex combinatorial code of Pros, Bar, dPax2, Svp, and Phyl appears to be required in ways that are only partially understood. Thus, the determination of TGs for these factors would allow the integration of the sequence of regulatory control that permits the formation of such an exquisite crystalline pattern as the *Drosophila* eye.

Not only is the formation of the retinal pattern under transcriptional control but so is the differentiation of these cells into specialized photoreceptors through the expression of their unique Rhodopsins. Add to this the stochastic nature of Rhodopsin expression in R7 and R8 cells and one is faced with the fascinating problem of the stochastic regulation of gene expression. The gene regulatory network emerging from the recent work by the Desplan lab (Fig. 10.3) demonstrates how various activators and repressors have their activities coordinated to result in the expression of a unique Rhodopsin in each cell type. How the stochastic expression of Spineless is initiated and superimposed upon this network surely is the key question that remains to be solved in this context.

Essentially all the transcription factors discussed here are highly conserved, and many of them have been studied in the context of mammalian development, homeostasis, and disease. Although it is likely that several of these factors have evolved novel interactions and more complex regulatory networks, there is little doubt that the basic principles as well as the core molecular components of such networks remain operative in mammals. In particular, a better understanding of the mechanisms of transition from progenitor to committed cells and the maintenance of a differentiated state is key to developing clearer ideas about stem cell biology in development, homeostasis, and cancer. In this context, studying the integration of transcriptional regulatory networks with cellular signaling pathways leading to retinal cell fate specification in the fly retina is likely to contribute significantly to these endeavors for the foreseeable future.

## REFERENCES

Aerts, S., Quan, X. J., Claeys, A., Naval Sanchez, M., Tate, P., Yan, J., and Hassan, B. A. (2010). Robust target gene discovery through transcriptome perturbations and genome-wide enhancer predictions in Drosophila uncovers a regulatory basis for sensory specification. *PLoS Biol.* **8,** e1000435.

Baker, N. E. (2001). Cell proliferation, survival, and death in the Drosophila eye. *Semin. Cell Dev. Biol.* **12,** 499–507.

Baker, N. E., and Yu, S. Y. (1997). Proneural function of neurogenic genes in the developing Drosophila eye. *Curr. Biol.* **7,** 122–132.

Baker, N. E., Yu, S., and Han, D. (1996). Evolution of proneural atonal expression during distinct regulatory phases in the developing Drosophila eye. *Curr. Biol.* **6,** 1290–1301.

Ben-Arie, N., Hassan, B. A., Bermingham, N. A., Malicki, D. M., Armstrong, D., Matzuk, M., Bellen, H. J., and Zoghbi, H. Y. (2000). Functional conservation of atonal and Math1 in the CNS and PNS. *Development* **127,** 1039–1048.

Bermingham, N. A., Hassan, B. A., Price, S. D., Vollrath, M. A., Ben-Arie, N., Eatock, R. A., Bellen, H. J., Lysakowski, A., and Zoghbi, H. Y. (1999). Math1: An essential gene for the generation of inner ear hair cells. *Science* **284,** 1837–1841.

Bonini, N. M., Bui, Q. T., Gray-Board, G. L., and Warrick, J. M. (1997). The Drosophila eyes absent gene directs ectopic eye formation in a pathway conserved between flies and vertebrates. *Development* **124,** 4819–4826.

Bossuyt, W., Kazanjian, A., De Geest, N., Van Kelst, S., De Hertogh, G., Geboes, K., Boivin, G. P., Luciani, J., Fuks, F., Chuah, M., VandenDriessche, T., Marynen, P., *et al.* (2009). Atonal homolog 1 is a tumor suppressor gene. *PLoS Biol.* **7,** e39.

Bray, S. (1998). Notch signalling in Drosophila: Three ways to use a pathway. *Semin. Cell Dev. Biol.* **9,** 591–597.

Brennan, C. A., and Moses, K. (2000). Determination of Drosophila photoreceptors: Timing is everything. *Cell. Mol. Life Sci.* **57,** 195–214.

Cachero, S., Simpson, T. I., Zur Lage, P. I., Ma, L., Newton, F. G., Holohan, E. E., Armstrong, J. D., and Jarman, A. P. (2011). The gene regulatory cascade linking proneural specification with differentiation in Drosophila sensory neurons. *PLoS Biol.* **9,** e1000568.

Cagan, R. (1993). Cell fate specification in the developing Drosophila retina. *Dev. Suppl.* 19–28.

Cagan, R. L., and Ready, D. F. (1989). Notch is required for successive cell decisions in the developing Drosophila retina. *Genes Dev.* **3,** 1099–1112.

Chang, H. C., Solomon, N. M., Wassarman, D. A., Karim, F. D., Therrien, M., Rubin, G. M., and Wolff, T. (1995). Phyllopod functions in the fate determination of a subset of photoreceptors in Drosophila. *Cell* **80,** 463–472.

Charlton-Perkins, M., and Cook, T. A. (2010). Building a fly eye: Terminal differentiation events of the retina, corneal lens, and pigmented epithelia. *Curr. Top. Dev. Biol.* **93,** 129–173.

Choi, C. M., Vilain, S., Langen, M., Van Kelst, S., De Geest, N., Yan, J., Verstreken, P., and Hassan, B. A. (2009). Conditional mutagenesis in Drosophila. *Science* **324,** 54.

Chou, W. H., Huber, A., Bentrop, J., Schulz, S., Schwab, K., Chadwell, L. V., Paulsen, R., and Britt, S. G. (1999). Patterning of the R7 and R8 photoreceptor cells of Drosophila: Evidence for induced and default cell-fate specification. *Development* **126,** 607–616.

Cook, T., and Desplan, C. (2001). Photoreceptor subtype specification: From flies to humans. *Semin. Cell Dev. Biol.* **12,** 509–518.

Cook, T., Pichaud, F., Sonneville, R., Papatsenko, D., and Desplan, C. (2003). Distinction between color photoreceptor cell fates is controlled by Prospero in Drosophila. *Dev. Cell* **4,** 853–864.

Daga, A., Karlovich, C. A., Dumstrei, K., and Banerjee, U. (1996). Patterning of cells in the Drosophila eye by Lozenge, which shares homologous domains with AML1. *Genes Dev.* **10,** 1194–1205.

Dickson, B. J., Dominguez, M., van der Straten, A., and Hafen, E. (1995). Control of Drosophila photoreceptor cell fates by phyllopod, a novel nuclear protein acting downstream of the Raf kinase. *Cell* **80,** 453–462.

Dokucu, M. E., Zipursky, S. L., and Cagan, R. L. (1996). Atonal, rough and the resolution of proneural clusters in the developing Drosophila retina. *Development* **122,** 4139–4147.

Dominguez, M. (1999). Dual role for Hedgehog in the regulation of the proneural gene atonal during ommatidia development. *Development* **126,** 2345–2353.

Dominguez, M., and Casares, F. (2005). Organ specification-growth control connection: New in-sights from the Drosophila eye-antennal disc. *Dev. Dyn.* **232,** 673–684.

Dominguez, M., and Hafen, E. (1997). Hedgehog directly controls initiation and propagation of retinal differentiation in the Drosophila eye. *Genes Dev.* **11,** 3254–3264.

Dubruille, R., Laurencon, A., Vandaele, C., Shishido, E., Coulon-Bublex, M., Swoboda, P., Couble, P., Kernan, M., and Durand, B. (2002). Drosophila regulatory factor X is necessary for ciliated sensory neuron differentiation. *Development* **129,** 5487–5498.

Franceschini, N., Kirschfeld, K., and Minke, B. (1981). Fluorescence of photoreceptor cells observed *in vivo*. *Science* **213,** 1264–1267.

Frankfort, B. J., and Mardon, G. (2002). R8 development in the Drosophila eye: A paradigm for neural selection and differentiation. *Development* **129,** 1295–1306.

Frankfort, B. J., and Mardon, G. (2004). Senseless represses nuclear transduction of Egfr pathway activation. *Development* **131,** 563–570.

Frankfort, B. J., Nolo, R., Zhang, Z., Bellen, H., and Mardon, G. (2001). Senseless repression of rough is required for R8 photoreceptor differentiation in the developing Drosophila eye. *Neuron* **32,** 403–414.

Freeman, M. (1994). The spitz gene is required for photoreceptor determination in the Drosophila eye where it interacts with the EGF receptor. *Mech. Dev.* **48,** 25–33.

Freeman, M. (1996). Reiterative use of the EGF receptor triggers differentiation of all cell types in the Drosophila eye. *Cell* **87,** 651–660.

Garcia-Bellido, A., and Merriam, J. R. (1969). Cell lineage of the imaginal discs in Drosophila gynandromorphs. *J. Exp. Zool.* **170,** 61–75.

Hafen, E., Basler, K., Edstroem, J. E., and Rubin, G. M. (1987). Sevenless, a cell-specific homeotic gene of Drosophila, encodes a putative transmembrane receptor with a tyrosine kinase domain. *Science* **236,** 55–63.

Halder, G., Callaerts, P., and Gehring, W. J. (1995). Induction of ectopic eyes by targeted expression of the eyeless gene in Drosophila. *Science* **267,** 1788–1792.

Hardie, R. (1985). Functional organization of the fly retina. *In* "Progress in Sensory Physiology," (D. Ottoson, ed.), pp. 1–79. Springer, New York.

Heberlein, U., Wolff, T., and Rubin, G. M. (1993). The TGF beta homolog dpp and the segment polarity gene hedgehog are required for propagation of a morphogenetic wave in the Drosophila retina. *Cell* **75,** 913–926.

Higashijima, S., Kojima, T., Michiue, T., Ishimaru, S., Emori, Y., and Saigo, K. (1992). Dual Bar homeo box genes of Drosophila required in two photoreceptor cells, R1 and R6, and primary pigment cells for normal eye development. *Genes Dev.* **6,** 50–60.

Jang, C. C., Chao, J. L., Jones, N., Yao, L. C., Bessarab, D. A., Kuo, Y. M., Jun, S., Desplan, C., Beckendorf, S. K., and Sun, Y. H. (2003). Two Pax genes, eye gone and eyeless, act cooperatively in promoting Drosophila eye development. *Development* **130,** 2939–2951.

Jarman, A. P., Sun, Y., Jan, L. Y., and Jan, Y. N. (1995). Role of the proneural gene, atonal, in formation of Drosophila chordotonal organs and photoreceptors. *Development* **121,** 2019–2030.

Johnston, R. J., Jr., Otake, Y., Sood, P., Vogt, N., Behnia, R., Vasiliauskas, D., McDonald, E., Xie, B., Koenig, S., Wolf, R., Cook, T., Gebelein, B., *et al.* (2011). Interlocked feedforward loops control cell-type-specific rhodopsin expression in the Drosophila eye. *Cell* **145,** 956–968.

Jukam, D., and Desplan, C. (2010). Binary fate decisions in differentiating neurons. *Curr. Opin. Neurobiol.* **20,** 6–13.

Kauffmann, R. C., Li, S., Gallagher, P. A., Zhang, J., and Carthew, R. W. (1996). Ras1 signaling and transcriptional competence in the R7 cell of Drosophila. *Genes Dev.* **10,** 2167–2178.

Kenyon, K. L., Ranade, S. S., Curtiss, J., Mlodzik, M., and Pignoni, F. (2003). Coordinating proliferation and tissue specification to promote regional identity in the Drosophila head. *Dev. Cell* **5,** 403–414.

Klisch, T. J., Xi, Y., Flora, A., Wang, L., Li, W., and Zoghbi, H. Y. (2011). *In vivo* Atoh1 targetome reveals how a proneural transcription factor regulates cerebellar development. *Proc. Natl. Acad. Sci. USA* **108,** 3288–3293.

Kramer, H., Cagan, R. L., and Zipursky, S. L. (1991). Interaction of bride of sevenless membrane-bound ligand and the sevenless tyrosine-kinase receptor. *Nature* **352,** 207–212.

Kumar, J. P. (2009). The molecular circuitry governing retinal determination. *Biochim. Biophys. Acta* **1789,** 306–314.

Kumar, J. P., and Moses, K. (2001). The EGF receptor and notch signaling pathways control the initiation of the morphogenetic furrow during Drosophila eye development. *Development* **128,** 2689–2697.

Li, Y., and Baker, N. E. (2001). Proneural enhancement by Notch overcomes Suppressor-of-Hairless repressor function in the developing Drosophila eye. *Curr. Biol.* **11,** 330–338.

Ma, C., and Moses, K. (1995). Wingless and patched are negative regulators of the morphogenetic furrow and can affect tissue polarity in the developing Drosophila compound eye. *Development* **121,** 2279–2289.

Ma, C., Zhou, Y., Beachy, P. A., and Moses, K. (1993). The segment polarity gene hedgehog is required for progression of the morphogenetic furrow in the developing Drosophila eye. *Cell* **75,** 927–938.

Mikeladze-Dvali, T., Wernet, M. F., Pistillo, D., Mazzoni, E. O., Teleman, A. A., Chen, Y. W., Cohen, S., and Desplan, C. (2005). The growth regulators warts/lats and melted interact in a bistable loop to specify opposite fates in Drosophila R8 photoreceptors. *Cell* **122,** 775–787.

Mlodzik, M., Hiromi, Y., Weber, U., Goodman, C. S., and Rubin, G. M. (1990). The Drosophila seven-up gene, a member of the steroid receptor gene superfamily, controls photoreceptor cell fates. *Cell* **60,** 211–224.

Mollereau, B., Dominguez, M., Webel, R., Colley, N. J., Keung, B., de Celis, J. F., and Desplan, C. (2001). Two-step process for photoreceptor formation in Drosophila. *Nature* **412,** 911–913.

Morante, J., Desplan, C., and Celik, A. (2007). Generating patterned arrays of photoreceptors. *Curr. Opin. Genet. Dev.* **17,** 314–319.
Niwa, N., Hiromi, Y., and Okabe, M. (2004). A conserved developmental program for sensory organ formation in Drosophila melanogaster. *Nat. Genet.* **36,** 293–297.
Nolo, R., Abbott, L. A., and Bellen, H. J. (2000). Senseless, a Zn finger transcription factor, is necessary and sufficient for sensory organ development in Drosophila. *Cell* **102,** 349–362.
Ostrin, E. J., Li, Y., Hoffman, K., Liu, J., Wang, K., Zhang, L., Mardon, G., and Chen, R. (2006). Genome-wide identification of direct targets of the Drosophila retinal determination protein Eyeless. *Genome Res.* **16,** 466–476.
Papatsenko, D., Sheng, G., and Desplan, C. (1997). A new rhodopsin in R8 photoreceptors of Drosophila: Evidence for coordinate expression with Rh3 in R7 cells. *Development* **124,** 1665–1673.
Pepple, K. L., Atkins, M., Venken, K., Wellnitz, K., Harding, M., Frankfort, B., and Mardon, G. (2008). Two-step selection of a single R8 photoreceptor: A bistable loop between senseless and rough locks in R8 fate. *Development* **135,** 4071–4079.
Raabe, T. (2000). The sevenless signaling pathway: Variations of a common theme. *Biochim. Biophys. Acta* **1496,** 151–163.
Ready, D. F., Hanson, T. E., and Benzer, S. (1976). Development of the Drosophila retina, a neurocrystalline lattice. *Dev. Biol.* **53,** 217–240.
Shen, W., and Mardon, G. (1997). Ectopic eye development in Drosophila induced by directed dachshund expression. *Development* **124,** 45–52.
Southall, T. D., and Brand, A. H. (2009). Neural stem cell transcriptional networks highlight genes essential for nervous system development. *EMBO J.* **28,** 3799–3807.
Tanaka-Matakatsu, M., and Du, W. (2008). Direct control of the proneural gene atonal by retinal determination factors during Drosophila eye development. *Dev. Biol.* **313,** 787–801.
Tio, M., and Moses, K. (1997). The Drosophila TGF alpha homolog Spitz acts in photoreceptor recruitment in the developing retina. *Development* **124,** 343–351.
Tio, M., Ma, C., and Moses, K. (1994). spitz, a Drosophila homolog of transforming growth factor-alpha, is required in the founding photoreceptor cells of the compound eye facets. *Mech. Dev.* **48,** 13–23.
Tomlinson, A., and Ready, D. F. (1986). Sevenless: A cell-specific homeotic mutation of the Drosophila eye. *Science* **231,** 400–402.
Treisman, J. E., and Rubin, G. M. (1995). Wingless inhibits morphogenetic furrow movement in the Drosophila eye disc. *Development* **121,** 3519–3527.
Weasner, B., Salzer, C., and Kumar, J. P. (2007). Sine oculis, a member of the SIX family of transcription factors, directs eye formation. *Dev. Biol.* **303,** 756–771.
Wernet, M. F., and Desplan, C. (2004). Building a retinal mosaic: Cell-fate decision in the fly eye. *Trends Cell Biol.* **14,** 576–584.
Wernet, M. F., Mazzoni, E. O., Celik, A., Duncan, D. M., Duncan, I., and Desplan, C. (2006). Stochastic spineless expression creates the retinal mosaic for colour vision. *Nature* **440,** 174–180.
Wolff, T., and Ready, D. F. (1991). The beginning of pattern formation in the Drosophila compound eye: The morphogenetic furrow and the second mitotic wave. *Development* **113,** 841–850.
Wolff, T., and Ready, D. F. (1993). Pattern formation in *Drosophila* retina. *In* "The Development of *Drosophila melanogaster*," (M. Bate and A. Martinez-Arias, eds.), pp. 1277–1325. Cold Spring Harbor Laboratory Press, Cold Spring Harbor, NY.
Xie, B., Charlton-Perkins, M., McDonald, E., Gebelein, B., and Cook, T. (2007). Senseless functions as a molecular switch for color photoreceptor differentiation in Drosophila. *Development* **134,** 4243–4253.

Yang, Q., Bermingham, N. A., Finegold, M. J., and Zoghbi, H. Y. (2001). Requirement of Math1 for secretory cell lineage commitment in the mouse intestine. *Science* **294,** 2155–2158.

Zhang, T., Ranade, S., Cai, C. Q., Clouser, C., and Pignoni, F. (2006). Direct control of neurogenesis by selector factors in the fly eye: Regulation of atonal by Ey and So. *Development* **133,** 4881–4889.

Zipursky, S. L., and Rubin, G. M. (1994). Determination of neuronal cell fate: Lessons from the R7 neuron of Drosophila. *Annu. Rev. Neurosci.* **17,** 373–397.

CHAPTER ELEVEN

# Diversification of Muscle Types in *Drosophila*: Upstream and Downstream of Identity Genes

Cyrille de Joussineau, Laetitia Bataillé,[1] Teresa Jagla, *and* Krzysztof Jagla

## Contents



## Abstract

Understanding gene regulatory pathways underlying diversification of cell types during development is one of the major challenges in developmental

GReD INSERM UMR1103, CNRS UMR6293, University of Clermont-Ferrand, Clermont-Ferrand, France
[1] Present address: Developmental Biology Centre, CNRS UMR5547/Université Paul Sabatier, Toulouse, France

*Current Topics in Developmental Biology*, Volume 98
ISSN 0070-2153, DOI: 10.1016/B978-0-12-386499-4.00011-2

biology. Progressive specification of mesodermal lineages that are at the origin of body wall muscles in *Drosophila* embryos has been extensively studied during past years, providing an attractive framework for dissecting cell type diversification processes. In particular, it has been found that muscle founder cells that are at the origin of individual muscles display specific expression of transcription factors that control diversification of muscle types. These factors, encoded by genes collectively called muscle identity genes, are activated in discrete subsets of muscle founders. As a result, each founder cell is thought to carry a unique combinatorial code of identity gene expression. Considering this, to define temporally and spatially restricted expression of identity genes, a set of coordinated upstream regulatory inputs is required. But also, to realize the identity program and to form specific muscle types with distinct properties, an efficient battery of downstream identity gene targets needs to be activated. Here we review how the specificity of expression and action of muscle identity genes is acquired.

## 1. Introduction

Somatic muscle system of the fruit fly embryo, composed of a set of easy to distinguish muscle fibers, represents a model of choice for studying diversification of cell types within a tissue. Following this assumption, we present here an overview of cellular and gene regulatory events underlying diversification of initially equipotent mesodermal cells into a set of somatic muscles displaying distinct properties. The initial steps of diversification of mesodermal cells have been analyzed at both individual gene and genome-wide levels (for review, see Beckett and Baylies, 2006; Ciglar and Furlong, 2009; Maqbool and Jagla, 2007), a set of muscle cell fate-specifying genes has been identified and characterized (for review, see Tixier *et al.*, 2010), and recently, first genes operating downstream of these specifiers in acquisition of distinct muscle properties have been described (Bataillé *et al.*, 2010). Below we present a current state of knowledge for all these diversification steps.

## 2. Development of Fruit Fly Muscles: An Overview

During *Drosophila* life cycle, two waves of myogenesis take place (Fig. 11.1). The initial embryonic wave leads to the formation of a simple array of mono-fiber body wall muscles ensuring mobility of the larva. In each abdominal hemisegment, 30 different muscles are formed, each of them displaying specific properties such as number of nuclei composing the

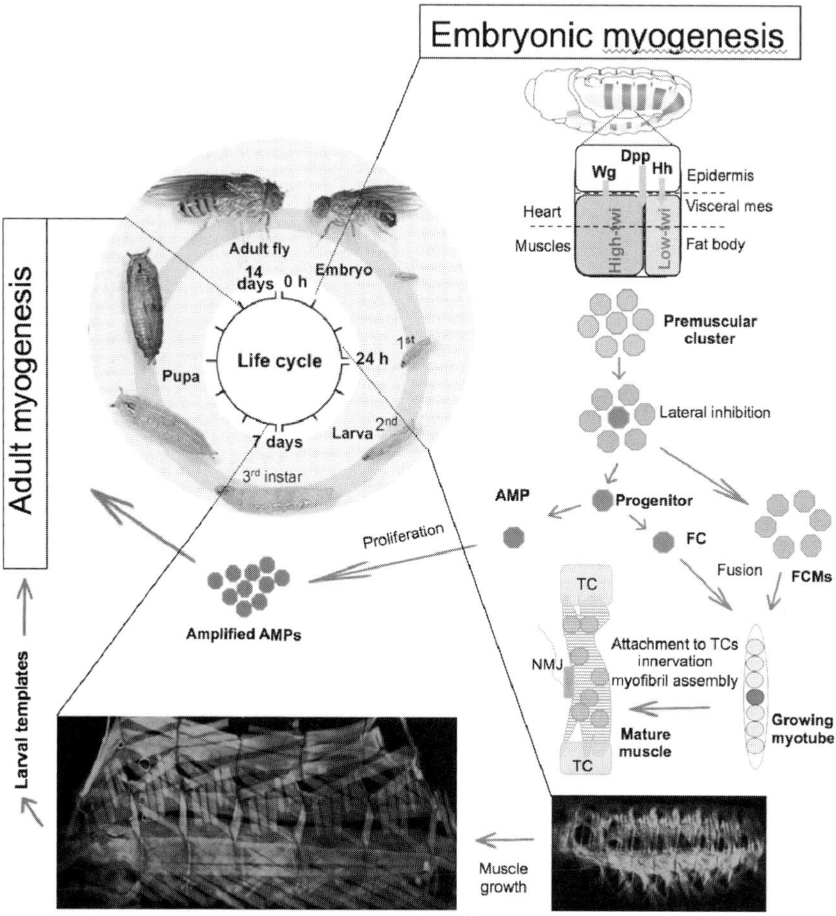


**Figure 11.1** Two waves of myogenesis during *Drosophila* life cycle. The main developmental stages of Drosophila life cycle including embryo, larva, pupa, and adult are depicted in the center of the scheme. Notice that the entire life cycle takes about 2weeks. The embryonic muscles (right side of the scheme) form between 5 and 18 h after egg laying (AEL) and then increase in size (about 50 times) during larval life. Muscles of the adult flies (left side of the scheme) form during pupal stages from the amplified AMPs. The flight and leg muscles develop from the AMPs associated with wing and leg imaginal discs, respectively, whereas abdominal body wall muscles originate from the embryonic abdominal AMPs. During embryonic myogenesis, mesoderm is segmentally patterned and subdivided into promuscular clusters expressing the proneural gene, *lethal of scute* (*l'sc*). A process of lateral inhibition restricts *l'sc* expression to one cell per cluster. These cells become the muscle progenitors and after asymmetric division gives rise to founder cells (FCs), whereas nonselected cells give rise to fusion competent myoblasts (FCMs). Each FC will seed the formation of a specific muscle fiber by fusion with FCMs leading to establishment of a stereotyped pattern of 30 muscles per hemisegment. During differentiation, growing myotubes will interact with tendon cells (TCs) and with motor neurons leading to establishment of neuromuscular

syncytial fiber, position, orientation along the anterior–posterior and dorsal–ventral axes, attachment point to the exoskeleton and innervation pattern (Fig. 11.1). Further, the embryonic myogenesis leads to the specification of myogenic stem-like cells, called adult muscle precursors (AMPs). The AMPs keep quiescent and nondifferentiated during the embryonic and the majority of larval life and then proliferate to enter differentiation process and form musculature of the adult fly (Figeac *et al.*, 2010). In contrast to larval, the adult muscles are composed of several myofibers and are thus structurally more similar to vertebrate skeletal muscles. However, because of its simplicity (invariant pattern of single-fiber muscles) and its temporal continuity in the development, embryonic/larval muscle system represents an attractive model in which to dissect mechanisms of diversification of muscle types. Considering this, here, we focus on genetic pathways and transcriptional switches controlling embryonic/larval myogenesis.

In *Drosophila,* all larval muscles including cardiac, visceral, and somatic muscles arise from the mesoderm layer determined in the ventral part of the blastoderm embryo by a high concentration of Dorsal morphogen, which induces expression of transcription factors Twist (Twi) and Snail (Sna) (Leptin, 1991). As revealed by extensive genetic analyses (Baylies and Bate, 1996) and recent genome-wide study (Sandmann *et al.*, 2007), Twi activates a large range of genes involved in specification of different mesodermal derivatives. In parallel, the ectodermal cells send instructive signals for mesoderm development such as Decapentaplegic (Dpp), a TGF-β orthologue, and Wingless (Wg), a Wnt orthologue (Baker and Schubiger, 1995; Maggert *et al.*, 1995). The secreted molecules act across the germ layers and activate mesodermal genes in cell type-specific manner (Staehling-Hampton *et al.*, 1994). In such a way, Dpp signaling maintains dorsally the expression of homeobox gene *tinman* (*tin*) and represses ventrally expressed *Pox meso* (*Poxm*) thus dividing the mesoderm into dorsal and ventral (D/V) domains (Frasch, 1995; Staehling-Hampton *et al.*, 1994). The mesoderm also undergoes partitioning along the anterior–posterior (A/P) axis by the combined action of Wg signals and the segmentation genes *even-skipped* (*eve*) and *sloppy paired* (*slp*)

junctions (NMJs). Finally, a process of myofibrillogenesis will take place to end by the formation of a fully functional larval muscle. A myosin heavy chain-GFP revealed pattern of embryonic muscles at the end of embryogenesis is shown in right bottom corner of the scheme. The pattern of twist-positive AMPs in the late stage embryo is also shown. Notice that six AMPs are present in each abdominal hemisegment and that a larger amount of AMPs is detected in thoracic segments from which originate the wing and leg imaginal discs. During larval stages, muscles undergo important growth but their pattern (revealed by actin-bound phalloidin staining—middle bottom panel) remains unchanged. Finally, at the beginning of metamorphosis in pupa stages, larval muscles are hydrolyzed and only some of them persist as templates for adult myogenesis (reviewed in Maqbool and Jagla, 2007). (For interpretation of the references to color in this figure legend, the reader is referred to the Web version of this chapter.)

(Azpiazu *et al.*, 1996; Riechmann *et al.*, 1997). This leads to the refinement of Twi expression from uniform levels to a modulated pattern (Fig. 11.1; Baylies and Bate, 1996). In the "*high Twi*" domain, dorsal cells develop into the heart, and the remainder forms the body wall muscles of the embryo, whereas in the "*low Twi*" domain, the most dorsal cells become precursors of the visceral muscles and those more ventrally, the fat body (Azpiazu *et al.*, 1996; Borkowski *et al.*, 1995; Riechmann *et al.*, 1997).

The anterior part of each segment patterned by the extrinsic Wg signals and, dorsally, the combination of Wg and Dpp signals make mesodermal cells competent to enter myogenic program. The first event underlying specification of the myogenic competence domain is the activation of proneural gene *lethal of scute* (*l'sc*) (Carmena *et al.*, 1995) followed by its subdivision in promuscular clusters (Figs. 11.1 and 11.2). This is a highly dynamic process involving combinatorial action of intrinsic fibroblast growth factor (FGF) and epidermal growth factor (EGF) signals, which progressively restrain *l'sc* expression to spatially defined group of cells, the promuscular clusters (Buff *et al.*, 1998; Michelson *et al.*, 1998). Within promuscular clusters, cells start to express muscle identity genes, thus initiating diversification process (Carmena *et al.*, 1998) that will lead to the formation of distinct types of muscles.

Promuscular clusters are also at the origin of two types of myoblasts, the founder cells (FCs) and the fusion competent myoblasts (FCMs). Establishment of these two distinct subsets of myoblasts requires a process of lateral inhibition involving Notch and the activity of Argos (Aos), a diffusible inhibitor of epidermal growth factor receptor (EGFR; Carmena *et al.*, 2002). In a context of competition, each promuscular cell expresses *l'sc* and displays MAPK pathway activity, this MAPK activity leading in turn to the induction of the Notch pathway in the surrounding cells, decreasing their own *l'sc* expression and MAPK pathway activity. As a result, *l'sc* expression and MAPK activity remain in only one cell in each promuscular cluster. The *l'sc* positive cells acquire the identity of muscle progenitors that are at the origin of all FCs, whereas Notch pathway activity in nonsegregating surrounding cells induces expression of the FCM-specific gene *lame duck* (*lmd*) (Carmena *et al.*, 2002; Duan *et al.*, 2001). Upregulation of *l'sc* in segregating progenitors suggests its important role in myogenesis. However, as deduced from minor muscle defects in *l'sc* mutant embryos (Carmena *et al.*, 1995), other genes with redundant functions such as *Pox meso* (Duan *et al.*, 2007) contribute to progenitor's specification. Each progenitor undergoes then Numb/Notch-dependent asymmetric cell division so that cell inheriting Numb becomes the FC of a larval muscle, while the sibling cell that retains Notch activity becomes another FC, a pericardial cell, or an AMP (Fig. 11.2; Baylies *et al.*, 1998; Ruiz Gomez and Bate, 1997). Thus, as a result of progressive cell fate specification in the somatic mesoderm, three distinct classes of myoblasts are specified: the FCMs, the FCs, and the AMPs.

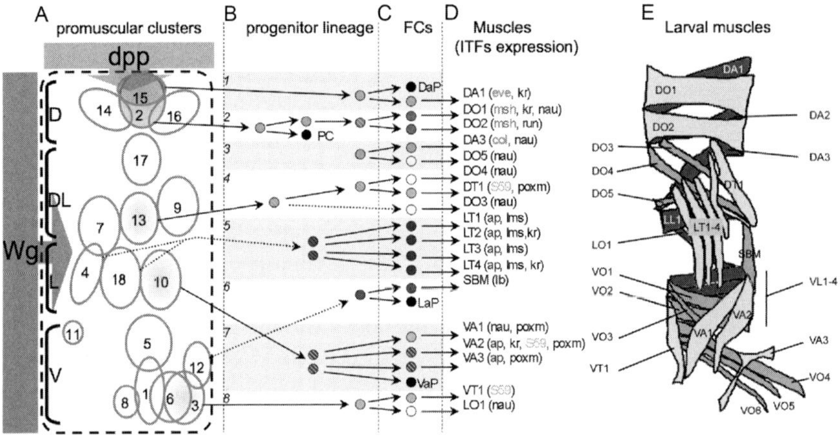


**Figure 11.2** Selection of the FCs in an abdominal segment of the mesoderm. Most representative iTFs to determine the origin of the muscle founder cells are shown in the figure. (A) Early patterning by wg/dpp and others induce definition of four domains in the presumptive somatic mesoderm: D (dorsal), DL (dorsolateral), L (lateral), and V (ventral). To note (not shown), some iTFs could be implicated in this patterning before later role in FCs and muscles: Msh (red) is expressed in the ectoderm above D and L domains, ap (blue) in the lateral domain, Poxm in V and posterior L domains. Lateral inhibition occurs in the 18 promuscular clusters of l'sc expressing cells (blue–green circles) and selects progenitors for muscle founder cells (figure according to Carmena *et al.*, 1995). Cells of clusters 2 and 15 express eve (pink) as 1–4 cells expressing S59 can be seen in clusters 3, 10, and 13 (green). (B, C) Probable (plain arrows) or supposed (dotted) connections between promuscular clusters, progenitors (P), and founders cells (FCs): (1) clearly established lineage by expression of eve (Carmena *et al.*, 1998); (2) lineage established by early expression of eve (Carmena *et al.*, 1998) and then later expression of msh in P9/10 (progenitor for muscles 9 and 10, respectively, DO1 and DO2; Nose *et al.*, 1998); (3) col expression in P3/20 (Crozatier and Vincent, 1999); (4) conflicting data indicated a common origin for DO3 and DT1 based on S59 expression (Carmena *et al.*, 1995; Knirr *et al.*, 1999), but col is expressed in P18/19 (progenitor for DT1/DO4; Crozatier and Vincent, 1999); (5) transient msh expression in P21/22 and P23/24 (Nose *et al.*, 1998); constant expression of ap during lineage of all LTs suggests a lateral domain origin (Bourgouin *et al.*, 1992); (6) specific expression of lb in SBM, its founder and its progenitor (Jagla *et al.*, 1998). Experiments with misexpression of S59 suggest that the progenitor emerges from cluster 12 (Knirr *et al.*, 1999); (7) lineage determined by l'sc/S59 expression (Carmena *et al.*, 1995) and apterous expression (Bourgouin *et al.*, 1992); (8) deduced by S59 expression (Dohrmann *et al.*, 1990). (D) iTFs expressed in muscle fibers. Additional muscles (and their iTF expression, when described): DA2, LL1 (kr, nau), VL1, VL2, VL3 (kr), VL4, VO1, VO2 (kr), VO3 (run), VO4 (run), VO5 (kr), VO6. (E) Schematic representation of the larval somatic muscles (figure from Tixier *et al.* according to Ruiz Gomez and Bate, 1997). PC: pericardial cell; D, L, or Vamp: dorsal, lateral, or ventral adult muscle precursor. (See Color Insert.)

The FCs are at the origin of all embryonic muscles and are thought to carry all information required to define the shape, position, and all properties of individual muscle (Bate *et al.*, 1991). Thus, a general muscle differentiation program common to all FCs and muscle precursors appears to be coordinated with a muscle-specific program ensuring diversification of muscle types. Over the years, several transcription factors such as S59/slouch (slou), ladybird (lb), Kruppel (Kr), msh, Pox meso (Poxm), Collier (Col) expressed in subsets of FCs/muscles have been identified and functionally analyzed (reviewed in Tixier *et al.*, 2010) revealing their roles in specification of FC identities. Due to their capacities to specify different subpopulations of FCs and thus the properties of resulting muscles, they are called muscle identity factors (iTFs). There is growing experimental evidence that a combinatorial action of iTFs is required to determine individual FC fates. However, as shown in Fig. 11.2, several gaps remain in assigning iTF code to all embryonic muscle lineages. Also, how a given FC executes identity program transmitted by iTFs and how this program is coordinated with differentiation program common to all muscles is not well understood.

## 3. Intrinsic and Extrinsic Factors Controlling Diversification of Embryonic Muscles

From 18 promuscular clusters present in each abdominal hemisegment, 6 AMPs and 30 different FCs will emerge to give rise to the somatic muscle pattern of the embryo (see scheme in Fig. 11.2). This diversification implies integration of multiple signals including those already implicated in the earlier patterning of the embryo.

### 3.1. Role of signaling pathways

#### 3.1.1. Wingless

In addition to its major role in the patterning of the embryo, segment-polarity gene *wingless* (*wg*) has been implicated in the selection of muscle progenitor cells. Particularly, singling out of dorsal progenitors, defined by either *even-skipped* (*eve*) or *S59* (*slouch*) expression was found to require permissive clue from ectodermally expressed Wg (Baylies *et al.*, 1995; Brennan *et al.*, 1999; Carmena *et al.*, 1998; Lawrence *et al.*, 1995). In more ventral groups, a similar phenomenon occurs for *nautilus* (*nau*) expressing muscle progenitors (Ranganayakulu *et al.*, 1996), Wg signal coming from the neuroectoderm first promoting expression of myogenic transcription regulator Twist and then positively regulating *S59* expression in a subset of FCs (Cox and Baylies, 2005; Cox *et al.*, 2005). Interestingly, FCs in both lateral and medial promuscular clusters that are expressing *nau* do not require Wg signaling to be defined, showing that this permissive action of Wg is restricted to some specific regions of the mesoderm by still unknown mechanisms (Baylies *et al.*, 1995).

### 3.1.2. Notch

Asymmetric division of a progenitor generally produces FCs for two different muscles. For example, FCs for VA1 and VA2 originate from the same progenitor. They differ by Kr/S59 expression, which is repressed by Notch in VA1. In VA2 FC, numb negatively regulates Notch pathway activity, allowing the expression of Kr and S59 in this FC (Ruiz Gomez and Bate, 1997). The repression of Notch pathway in one single daughter cell is induced by the segregation of numb protein in this cell, this segregation being itself dependent on *inscuteable* gene expression (Kraut and Campos-Ortega, 1996). Numb asymmetrical distribution has been implicated in repression of Notch activity in many FCs, contributing to diversification of muscle fates.

### 3.1.3. RTKs: The receptor tyrosine kinases

Two RTKs have been implicated in embryonic muscle patterning: Heartless (Htl), an orthologue of fibroblast growth factor receptor (Shishido *et al.*, 1993); and DER, the *Drosophila* EGFR (Shilo *et al.*, 1986). Downstream of these two receptors lies the Ras1/MAPK pathway (van der Geer *et al.*, 1994), which is itself implicated in the selection of muscle progenitors in parallel with the Notch pathway (see above and Carmena *et al.*, 2002). First, MAPK pathway, which is downstream of Htl, was shown to be important for the mesoderm differentiation (Shishido *et al.*, 1997), and Htl was then directly implicated in the emergence of dorsal muscle precursors (Carmena *et al.*, 1998; Michelson *et al.*, 1998). In the same way, the selection of ventral muscle precursors has also been shown to be influenced by signaling dependent of Htl, this phenomenon implicating their neighboring neuroblasts (Schulz and Gajewski, 1999). Convincingly, mutants for Htl ligands, *thisbe* (*ths*) and *pyramus* (*pyr*), induced a phenotype similar to *htl* mutations (Gryzik and Muller, 2004; Stathopoulos *et al.*, 2004). Recent works implicated both ligands in the selection of a subset of muscle progenitors, even though they showed not completely redundant action, pyr being, for example, the only ligand necessary for differentiation of *eve*-expressing precursors (Kadam *et al.*, 2009; Klingseisen *et al.*, 2009). Similar analyses implicated both DER and DER ligands spitz and vein in the singling out of DA1, VA2, and LL1 FCs (Buff *et al.*, 1998; Yarnitzky *et al.*, 1998). Interestingly, DER and Htl were shown to cooperate in the specification of the cluster where DA1 FC is selected (Carmena *et al.*, 1998), implicating them further in the phenomenon of mesoderm fate diversification.

## 3.2. *Hox* genes and profile diversification of muscle fibers

Homeotic genes are clustered in the *Antennapedia* complex (*ANT-C*) and the *Bithorax* complex (*BX-C*). They are expressed along the anterior–posterior axis of the embryo and participate to the definition of the segmented pattern of both the embryo and the adult fly (McGinnis and Krumlauf, 1992). In these segments, embryonic musculature varies along the A/P axis of the embryo. *Hox* genes have further been directly implicated in the late steps of FC selection. Early works indicated that presence of founder for VT1 depended on Ubx for abdominal segments A1–3 (Hooper, 1986; Michelson, 1994). Later, muscle identity gene *apterous* (*ap*) was demonstrated to be transcriptionally activated by *Antennapedia* (*Antp*), showing for the first time the direct implication of *Hox* genes in muscle progenitor selection (Capovilla *et al.*, 2001), as Hox proteins not only promote *collier* (*col*) and *nautilus* (*nau*) expression in DA3 FC *via* specific *cis*-regulatory modules but also partly regulate fusion of this FC to its surrounding FCMs (Enriquez *et al.*, 2010). Recently, it was also shown that anterior abdominal muscles of the embryo were specifically affected when some Ubx splicing forms were not expressed (Reed *et al.*, 2010). Knowing that these different spliced variants present different patterns of expression (Lopez and Hogness, 1991), this might explain how Hox proteins could exert different activities at different times of the *Drosophila* embryonic development. On the whole, Hox proteins seem strongly implicated in specification of some FCs, and, given the latest works, this role may be still largely underestimated.

## 3.3. Identity factors and specification of muscle cell fates

Genes encoding identity factors (iTFs) are regarded as being expressed in one or a few FCs and responsible for inducing in these FCs their specific program of development to give one specific muscle fiber (Baylies *et al.*, 1998; Frasch, 1999) (see Fig. 11.2).

iTFs include Even-skipped (Eve; Carmena *et al.*, 1998), S59 (slouch; Dohrmann *et al.*, 1990), Krüppel (Kr; Ruiz-Gomez *et al.*, 1997), Muscle segment homeobox (Msh; Lord *et al.*, 1995; Nose *et al.*, 1998), Collier (Col; Crozatier and Vincent, 1999), Apterous (Ap; Bourgouin *et al.*, 1992), Runt (Carmena *et al.*, 2002), Ladybird (Lb; Jagla *et al.*, 1998), Lateral muscle scarcer (Lms; Muller *et al.*, 2010), Nautilus (Nau; Keller *et al.*, 1998), Pox meso (Poxm; Duan *et al.*, 2007), and to a certain extent transcription regulators Vestigial (Vg) and Scalloped (Sd; Deng *et al.*, 2010), transcription factor D-Ptx1 (Vorbruggen *et al.*, 1997), and transcription inhibitor teyrha-meyhra (tey; Inaki *et al.*, 2010). Some of them are specifically expressed in one single fiber, as Lb (SBM), Col (DA3) or Eve (DA1) (see Fig. 11.2). Others are expressed in subset of fibers (for review, see Tixier *et al.*, 2010).

Most of the FCs express a combination of iTFs, all of them being important for proper differentiation of the muscle fiber and defining a unique combinatorial iTF code for individual FCs and developing muscle precursors (Fig. 11.2). The coexpression of iTFs can be temporary, as in VA2 FC, where Kr is expressed to promote later maintenance of *S59* (Ruiz-Gomez *et al.*, 1997). In other cases, several iTFs can truly cooperate to promote the right development of a muscle fiber: in VA2 still, Poxm acts in concert with S59 and control proper development of this muscle (Duan *et al.*, 2007) (Fig. 11.2). iTFs also participate to the patterning, in repressing each other and thus refining their expression domains to small subsets of FCs and resulting muscles. For example, S59 is transiently expressed in LO1 and VT1 FCs, both cells originating from the same progenitor. When *S59* expression is missing, there is a derepression of *lb* gene in these cells and transformation of the resulting muscles in ectopic segment border muscles (SBMs; Knirr *et al.*, 1999). On the contrary, *lb* expression in SBM FC represses *S59* expression in this cell, and absence of *lb* induces loss or transformation of the SBM into one of its neighboring muscle precursors (Jagla *et al.*, 1998; Knirr *et al.*, 1999). Finally, when cross-repression between three iTFs, Eve, Lb, and Msh was experimentally disturbed, it could induce changes in dorsal muscles identities and switches in both directions from muscle precursor to heart precursor identity (Jagla *et al.*, 1999, 2002). Thus, expression of one single iTF can induce a whole developmental program, and ectopic or loss of expression of one of these factors is sufficient to modify FC identity and acquisition of unique properties of resulting muscle by inducing a heterogeneity of expression of large sets of downstream genes (Estrada *et al.*, 2006; Junion *et al.*, 2007).

## 4. Muscle Differentiation: Common Rules for All Muscles

All larval muscles are composed of a single multinucleate fiber seeded by individual FC myoblasts. They share several muscle-specific characteristics including the capacity to fuse and form syncytial cells, to connect to both tendon cells and motor neurons, and to build highly complex contractile machinery (Figs. 11.1 and 11.3). As revealed by genome-wide screens (Elgar *et al.*, 2008; Junion *et al.*, 2005; Sandmann *et al.*, 2006), many genes involved in these processes are positively regulated by Mef2, a MADS-box transcription factor that is expressed in all muscle cells and is known to promote muscle differentiation. Below we provide a short overview of genetic control of these common steps of muscle differentiation before discussing how each muscle will acquire its specificity in size, attachment, and innervation.

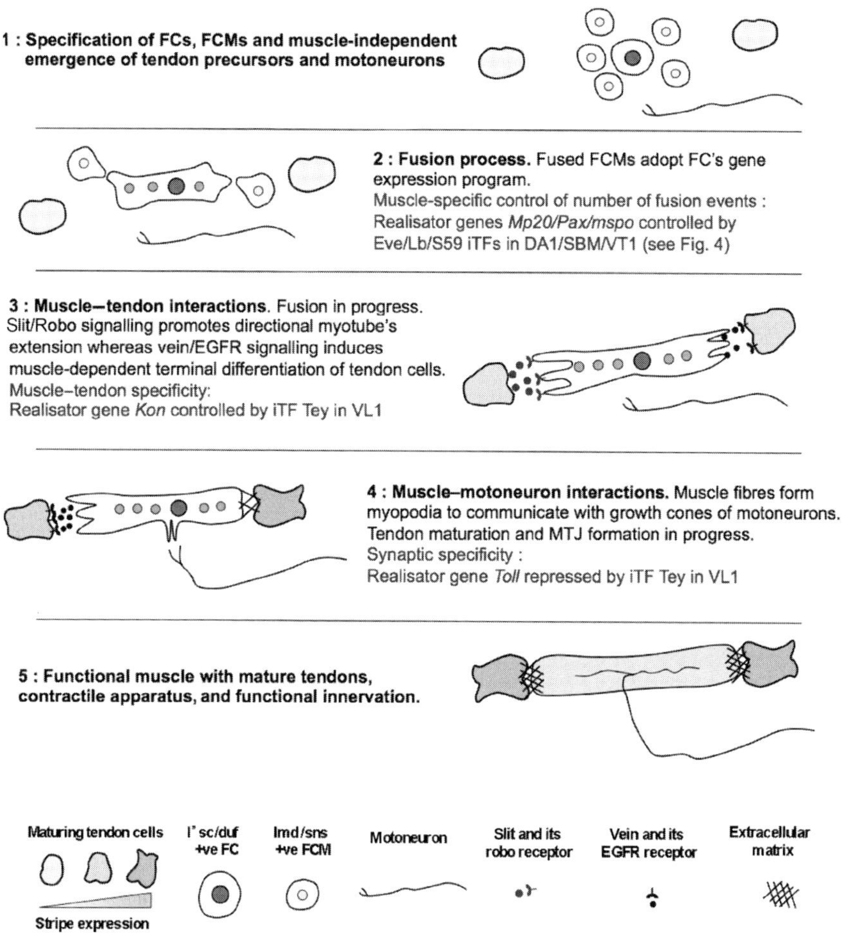


**Figure 11.3** Common rules of muscle formation. Formation of functional muscle in embryo requires regulated fusion process and high degree of communication with tendon cells and motoneurons. When known, transcriptional switches ensuring acquisition of muscle type-specific properties such as number of fusion events, tendon choice, and specificity of innervation are indicated in blue. (For interpretation of the references to color in this figure legend, the reader is referred to the Web version of this chapter.)

## 4.1. Myoblast fusion and muscle–tendon interaction

The final size of each muscle fiber is attained through multiple rounds of fusion of the FCs and then developing myotubes with the FCMs. Studies in *Drosophila* embryos have led to identification of a large number of genes that control different steps of myoblast fusion (for review, see Schejter and Baylies, 2010). These include *dumfounded/Kin of Irre C* (*duf/kirre*;

Ruiz-Gomez *et al.*, 2000) *and irregular-chiasm-C/roughest* (*irreC/rst*; Strunkelnberg *et al.*, 2001) expressed in FCs, but not in FCMs and *sticks and stone* (*sns*; Bour *et al.*, 2000) expressed in FCMs only. It has been proposed that Sns acts as a receptor for its ligands Duf and Rst, thus ensuring directionality of the fusion process between FCMs and FCs (Fig. 11.3; Baylies *et al.*, 1998).

In parallel with fusion events, the developing myotubes extend growth cone-like filopodia at both their extremities to contact tendon cells and to form myotendinous junctions (MTJs; Schweitzer *et al.*, 2010; Volk, 1999). Genetic analyses in *Drosophila* embryo suggest that an initial induction of tendon progenitors is muscle independent, but their differentiation into mature tendons is triggered by interaction with growing myotubes (Fig. 11.3). The initial pattern of tendon cells within the embryonic ectoderm is achieved through activation of *stripe* (*sr*) gene, encoding an early growth response (EGR)-like nuclear protein. Stripe was found to positively regulate several genes involved in muscle targeting to tendon cells, including *slit*, *Thrombospondin* (*Tsp*), *Leucine-rich tendon-specific protein* (*Lrt*), and *slowdown* (*slow*) (Chanana *et al.*, 2009; Gilsohn and Volk, 2010; Kramer *et al.*, 2001; Subramanian *et al.*, 2007; Wayburn and Volk, 2009). At the same time, the leading edges of migrating muscles accumulate a spectrin-related molecule Msp-300, which promotes the myotube extension toward attachment sites (Volk, 1992). Growing myotubes also secrete the Neuregulin homolog Vein, which acts as a muscle-derived signal to activate EGFR signaling in tendon cells and thus to promote the final differentiation of tendon cells (Yarnitzky *et al.*, 1997) (Fig 11.3). Finally, the stabilization of muscle attachment to tendons is achieved by the expression of position-specific integrins, αPS1βPS and αPS2βPS, which function as anchors to strengthen the muscle–tendon cell interactions and to build MTJs able to support the forces of contraction.

## 4.2. Establishment of motor neuron–muscle connections and building contractile apparatus

Fully differentiated muscles have the capacity to contract in answer to neural stimuli. To achieve this, they have to build sarcomeric contractile apparatus and have to connect to the motor neuron network. The developmental rules and gene pathways that control these processes in *Drosophila* embryos only begin to be characterized. As described by Landgraf *et al.* (1999), motor axons leave the central nervous system (CNS) as two separate bundles, the segmental nerve (SN) and intersegmental nerve (ISN). To innervate muscles, axons defasciculate progressively from the SN and the ISN, initially as nerve branches and then as individual axons. The early steps of development of motor neuron network appear to be muscle independent. However, in the absence of developing muscles, motor axons fail to defasciculate from the nerve bundles. Moreover, since a single FC can trigger the defasciculation of an entire nerve branch, it has been proposed that the muscle field is structured into sets of muscles, each set expressing a common defasciculation cue for a particular nerve branch (Landgraf *et al.*, 1999). Thus, muscles are expected to synthesize specific target-recognition molecules that allow the establishment of synaptic contacts. Interestingly, the discovery of cellular extensions, called myopodia, at the surface of embryonic muscle cells revealed that neuromuscular recognition is a reciprocal process (Ritzenthaler and Chiba, 2001; Ritzenthaler *et al.*, 2000) in which muscle cells by sending myopodia play active role in the guidance of motoneuronal growth cones (Kohsaka and Nose, 2009).

In parallel to establishing connections with motor neurons, muscle cells undergo the myofibrillogenesis. This process consists in replacement of the "nonmuscle" cytoskeletal architecture present in myoblasts by the near crystalline arrays of muscle sarcomeres, the smallest functional contractile units present in muscle. Myofibrillogenesis requires the coordination of multiple structural events. However, if it has been extensively studied in cardiomyocyte cell culture and during flight muscle formation in *Drosophila* (reviewed by Sparrow and Schock, 2009), virtually nothing is known about the myofibril assembly in the *Drosophila* embryonic muscles, whose efficient contraction is critical for hatching and entering the larval life.

## 5. Muscle Type-Specific Tuning of Differentiation and Acquisition of Individual Muscle Properties

In addition to common differentiation, each muscle has to acquire its specific size, shape, and connection pattern. Particularly, combinatorial codes of iTF expression play key role in this process. To identify genes implicated in the acquisition of specific muscle properties and acting downstream of iTFs, a promising strategy based on large-scale screens and *in silico* approaches has emerged in the past few years.

### 5.1. Genome-wide approaches to identify genes expressed in founders and operating downstream of iTFs

The first attempt to identify new determinants of muscle diversification process (Artero *et al.*, 2003) was based on the observation that during specification of muscle progenitors from clusters of equivalent myoblasts, the Ras pathway induces the muscle progenitor fate, whereas the Notch pathway promotes the FCM fate. By using a cDNA microarray approach on embryos enriched in Ras-dependent FCs, or in Notch-dependent FCMs, Artero *et al.*

identified a set of new genes with potential roles in the acquisition of muscle identities. Among identified candidates expressed in FCs and involved in diversification of muscle types are *phyllopod* and *polychaetoid* genes encoding signaling-related molecules as well as encoding an adhesion protein *tartan*. However, how these new regulators of muscle identity are connected to gene networks specifying distinct muscle fates remains unknown. Further, an *in silico* approach called CodeFinder (Philippakis *et al.*, 2006) and a combined genomic approach (Estrada *et al.*, 2006) have led to identification of a set of candidate genes potentially involved in diversification of muscle types. Finally, the first and, to date, the only genome-wide approach attempting to identify genes operating downstream of a muscle iTF has been performed in our laboratory. It was designed to identify downstream targets of *lb* (Junion *et al.*, 2007). To approach this issue, we developed a targeted transcriptional profiling strategy and used it in combination with the ChIP-enriched *in silico* target (ChEST) technique (Junion *et al.*, 2005). The phenotypes observed in *lb* mutant embryos (Jagla *et al.*, 1998) suggested a role of *lb* genes in axial positioning of muscle precursors and in the acquisition of muscle-specific shape and attachment to tendon cells. Hence, the unveiling of the global gene expression program operating downstream of *lb* was expected to provide a general view of how the acquisition of cell identity is executed at the genome level. Indeed, the genome-wide analyses (Junion *et al.*, 2007) showed that *lb* acts at multiple levels and over a long time period. Initially, it represses a subset of identity genes and contributes to the cell fate-specifying code of transcription factors. Later, it contributes to the acquisition of individual properties of muscle cells by acting as a regulator of late acting transcriptional cofactors, and genes involved in setting muscle cell motility, myoblast fusion, muscle shapes and attachment, such as *Mp20* (described in next chapter), *vkg*, and *inflated* but also in muscle function, such as *CG8698*, which is required for muscle contraction. Thus, identification of this first set of targets of muscle iTF opened an attractive perspective to make a link between the identity gene function and the acquisition of distinct muscle properties.

## 5.2. Muscle type-specific regulation of myoblast fusion

A syncytial myotube is obtained by one FC fusing with a determined number of FCMs to generate a muscle displaying a specific number of nuclei. The fusion process occurs between the embryonic stages 12 and 15. Initially from stage 12 to stage 13, all FCs fuse with only one or two FCMs to generate a muscular precursor. Then, additional fusions occur until stage 15 (Beckett and Baylies, 2007). Our recent work (Bataillé *et al.*, 2010) provides evidence that identity genes specifically control the second phase of fusion by determining a muscle-specific rate of nuclei recruitment, thus leading to the formation of muscles that differ by number of nuclei.

As stated before, many genes that are required for the general regulation of myoblast fusion have been previously identified (Kesper *et al.*, 2007; Schejter and Baylies, 2010). However, all these fusion genes act in all muscles and do not provide cues into muscle type-specific tuning of the fusion process. Only very recently, by analyzing myoblast fusion in muscles that differ by number of nuclei, the first genes involved in muscles-specific regulation of the fusion process have been identified (Bataillé *et al.*, 2010). They encode regulators of actin cytoskeleton Muscle protein 20 (Mp20) and Paxillin (Pax) and an extracellular matrix component m-spondin (mspo), known to influence cell adhesion. Consistent with their function as modulators of the fusion process, *Mp20, Pax*, and *mspo* are expressed in the growing myotubes during the second phase of fusion that is controlled by identity genes (Bataillé *et al.*, 2010). They are expressed in a muscle type-specific manner in relatively large subsets of muscles and levels of their transcripts appear to vary from a myotube to the other. When analyzing five different muscles (DA1, DT1, VA2, VT1, and SBM) whose identities depend on the three iTFs, Eve, S59, and Lb, we first observed that each of these muscles can be characterized by unique combinatorial code of expression levels of *Mp20, Pax*, and *mspo* (Fig. 11.4) and second that this code critically depends on Eve, S59, and Lb activity. Moreover, the given combination of expression levels of *Mp20, Pax*, and *mspo* can be correlated with the number of fusion events so that, for example, increasing levels of *Mp20* and/or *Pax* expression in DT1 muscle lead to additional fusion events (Bataillé *et al.*, 2010) (Fig. 11.4). Thus, the identity genes determine a specific combinatorial realizator code of their targets allowing the muscle-specific modulation of the fusion process. In agreement with this view, *Mp20, Pax*, and *mspo* regulate the number of fusion events in a dose-dependent manner acting either as positive (*Mp20* and *Pax*) or as negative modulators (*mspo)* of these events (Bataillé *et al.*, 2010).

Thus, these data reveal first components of gene regulatory cascade acting downstream of iTFs to regulate number of fusion events and *in fine* the muscle size. However, how the two levels of regulation of the fusion process, common to all muscles and the muscle type-specific, are coordinated remains to be elucidated. Previous studies have enlightened the importance of remodeling the actin cytoskeleton during fusion process (for review, see Guerin and Kramer, 2009; Schejter and Baylies, 2010). The three genes described here, that control the muscle-specific number of fusion events, are coding for proteins involved in the regulation of the actin cytoskeleton dynamics, suggesting that they could modulate the general mechanism at this level.

## 5.3. Genes regulating muscle-specific attachment

As they grow in size through cell fusion, myotubes also elongate and select proper attachment sites to the epidermis. The dynamics of the elongation and attachment process have been precisely analyzed in the cases of VT1,

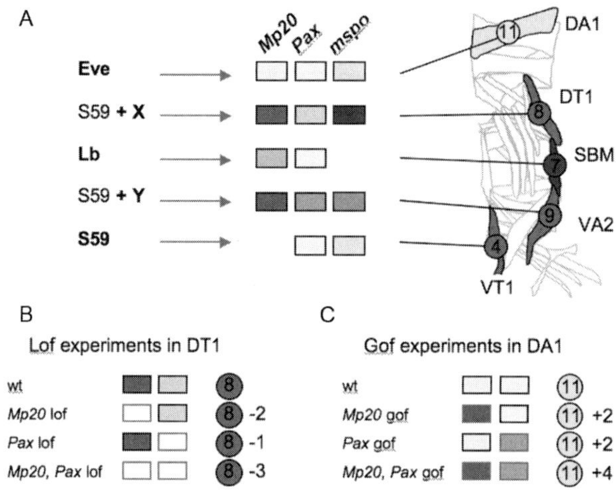


**Figure 11.4** iTFs regulate the final number of fusion events by determining muscle-specific combinatorial code of target gene expression. (A) The identity genes *eve*, *lb*, and *S59* control muscle-specific fusion programs of DA1, DT1, SBM, VA2, and VT1 muscles. Corresponding number of nuclei is indicated on the scheme. Notice that *S59* alone is not sufficient to explain the DT1 and VA2 fusion programs and is proposed to act in combination with unidentified factors (X and Y). In each growing muscle, iTFs establish a specific combinatorial code of target's expression. High, median, and low expression levels of *Mp20*, *Pax*, and *mspo* are represented by the color intensity. (B, C) Decreased (B) or increased (C) expression levels of realizator genes induce changes in final number of nuclei. Modification of the number of recruited nuclei is noted above each context. Because muscle expressing a low endogenous level of *Mp20* and *Pax* are not affected in loss of function (lof) contexts, results of lof experiments are illustrated for DT1 muscle in which the endogenous *Mp20* and *Pax* expression levels are, respectively, high and median. In lof contexts, muscles expressing a high level and a median level display a proportional loss of nuclei (i.e., −2 nuclei for high-level muscles, −1 nuclei for median-level muscles). The double loss of function shows an additive effect of realizator genes. Opposite effects are observed in gain of function contexts (C), as illustrated in DA1 muscle, which display low endogenous expression levels of *Pax* and *Mp20*. In this muscle, changes in expression levels of realizator genes induce an additive increase of nuclei number. (See Color Insert.)

LO1, and VL1 ventral muscles (Schnorrer and Dickson, 2004; Schnorrer *et al.*, 2007). Although the regulation of elongation and attachment is still poorly understood, few genes involved in each of these steps have been described (Schweitzer *et al.*, 2010). Among them are signaling molecules mediating muscle guidance toward tendon cells such as Slit-Robo and the receptor tyrosine kinase Derailed (Callahan *et al.*, 1996; Kramer *et al.*, 2001; Ypsilanti *et al.*, 2010). Robo, which acts as a receptor, is expressed in the elongating myotubes while its ligand Slit acting as attractant is secreted by tendon cells (Kramer *et al.*, 2001) (Fig. 11.3). Based on observed mutant phenotypes, Slit-Robo signaling plays rather general than muscle type-specific guidance role. In contrast, Derailed, which is expressed in LT muscle subset (Callahan *et al.*, 1996), and *kon-tiki* (*kon*)/*perdido* that encodes a transmembrane protein involved in the targeting ventral–longitudinal muscles (Estrada *et al.*, 2007; Schnorrer *et al.*, 2007) appear to have muscle type-specific functions in guiding myotubes to their proper attachment sites. Thus, these first examples of identified genes required for targeting of muscles to specific attachment sites (Schnorrer and Dickson, 2004; reviewed in Schweitzer *et al.*, 2010; Tixier *et al.*, 2010) indicate that, similarly to what is described for the fusion process, a muscle attachment realizator code may exist and operate in a muscle type-specific manner. Whether muscle identity genes are upstream regulators of this code remains to be tested.

## 5.4. Genes regulating muscle-specific innervation

During embryonic development, muscle fibers are innervated in a stereotyped manner by more than 30 motor neurons (Bate and Broadie, 1995; Landgraf and Thor, 2006). Some muscles are innervated by a single motor neuron, whereas others are polyinnervated by different classes of motor neurons (Bate and Broadie, 1995). Interestingly, when a muscle is duplicated, the appropriate neuron expands and connects both muscles to form neuromuscular junctions (NMJs; Chiba *et al.*, 1993), suggesting that motor neurons select their specific muscle fibers through an active recognition process. Consistent with this observation, nerve branching appears to be regulated by individual muscle FCs and the muscles that they seed (Landgraf *et al.*, 1999). Although the cellular mechanisms governing the specificity of motor neuron–muscle recognition remain poorly understood, classical and molecular genetic approaches begin to reveal how motor axons find and recognize their appropriate muscles targets. Among the genes encoding "recognition" molecules expressed on the surface of specific muscle are Netrin A (NetA), Netrin B (NetB), knockout (ko), Fasciclin III (Fas III), capricious (caps), Toll (Tl), Wnt4, and hattifattener (haf) (reviewed in Tixier *et al.*, 2010). Two types of recognition mechanisms appear to contribute to specificity of NMJ formation. First, the homophilic adhesion to occur requires that both muscle and motor neuron express a same molecule at their cell surface. Connectin and Fas III that are expressed on a subset of muscles and the motor neurons innervating them (Halpern *et al.*, 1991; Nose *et al.*, 1992; Patel *et al.*, 1987) are known to promote NMJ formation by homophilic interactions. Although *fasciclin III* or *connectin* loss of function do not affect the neuromuscular system (Chiba *et al.*, 1995; Nose *et al.*, 1994), the pan-muscular overexpression of Fas III is sufficient to establish novel connections, thus suggesting a redundancy in Connectin and Fas III functions. An alternative recognition model proposes that each motor neuron and its corresponding target muscle express a complementary code of interacting proteins. For example, NetA and NetB are expressed in a subset of dorsal muscles (Winberg *et al.*, 1998) and act as axons attractants through DCC family receptor Frazzled (Fra; Kolodziej *et al.*, 1996). It is also possible that these two mechanisms are not exclusive and that target recognition results from both homo- and heterotypic interactions between neurons and muscles, which could be modulated by both attractive and repulsive cues. However, the link between the muscular identity information conferred by identity genes and the expression of muscle-specific "recognition" proteins allowing a targeted innervation remains largely unknown. Recently, Inaki *et al.* (2010) have shown that a novel muscle identity gene *teyrha-meyhra* (*tey*) that is expressed in the ventral longitudinal muscle 1 (VL1) acts as a negative regulator of a repulsive cue, Toll, thus providing a first link between iTFs and the acquisition of specific muscle innervation (Fig. 11.3).

## 6. Final Remarks

Analyses of complex developmental processes such as progressive diversification of cell fates require adapted model systems. Embryonic somatic muscles of the fruit fly represent one of the well-adapted models for studying cell diversification, documented by a large number of original and review publications referenced here. The attractiveness of the muscular system resides in its highly stereotyped developmental program and large set of available markers to visualize individual muscle precursors and resulting differentiated muscles. Another important advantage is its simplicity based on restricted number (30 per hemisegment) of easily identifiable muscles, each of them being composed of one single fiber, but also evolutionarily conservation of genetic pathways governing myogenesis.

Moreover, the logic of transcriptional switches governing cell diversification appears common to cells in different developing tissues, a view that is supported by finding that some of muscle iTFs such as lb and slou are also acting in specification of neural lineages (Jagla *et al.*, 2001).

As stated in Chapters 3 and 5 the timing of iTF expression is finely regulated so that some iTFs are expressed over the entire muscle diversification process whereas others are only transiently ON in specific time windows. The role of different timing of iTFs' activity and their internal hierarchy remains to be investigated. However, the timely precise iTF transcription is likely to involve paused RNA Pol II checkpoint, recently found to contribute to developmental pattern formation in the *Drosophila* mesoderm (reviewed in Levine, 2011). Once activated, the iTFs have to exert their functions coordinately with common myogenic factors such as Twi and Mef2.

One way to achieve this, reported by Junion *et al.* (2007) for *inflated* (*if*), the direct target of Mef2 and Lb, is *via* binding to coregulated enhancers. However, a more complex coordination mechanism certainly exists that allows propagation of transcriptional identity program from FC to FCMs nuclei that integrate syncytial muscle cell. This remains entirely unknown.

Finally, the main challenge in understanding how different properties of individual muscles are acquired during development appears not only at conceptual but also at technological level. Analyses of recently identified muscle-specific modulators of myoblast fusion (Bataillé *et al.*, 2010) indicated that they act cooperatively by setting a combinatorial code of transcription levels specific to a given muscle. However, many gaps remain to be filled in genetic pathways operating downstream of muscle iTFs. To fill these gaps, genome-wide muscle type-specific screens for downstream targets of iTFs have to be performed. New technological approaches that allow cell type-specific transcriptional (Heiman *et al.*, 2008; Miller *et al.*, 2009) and chromatin-profiling (Deal and Henikoff, 2011) analyses have been recently described and need to be tested for rare populations of muscle cell subsets. The large ChIP and genetic screens performed nowadays in Furlong (reviewed in Furlong, 2011) and Schnorrer (Schnorrer *et al.*, 2011) labs have also shed light on gene networks operating in developing muscles and logic of transcriptional switches underlying developmental decisions providing a framework for studying muscle diversification process at whole genome scale.

## ACKNOWLEDGMENTS

Research in the Jagla lab is supported by the ANR, the FRM, the AFM, and the EraSysBio ModHeart grants.

## REFERENCES

Artero, R., Furlong, E. E., Beckett, K., Scott, M. P., and Baylies, M. (2003). Notch and Ras signaling pathway effector genes expressed in fusion competent and founder cells during Drosophila myogenesis. *Development* **130,** 6257–6272.

Azpiazu, N., Lawrence, P. A., Vincent, J. P., and Frasch, M. (1996). Segmentation and specification of the Drosophila mesoderm. *Genes Dev.* **10,** 3183–3194.

Baker, R., and Schubiger, G. (1995). Ectoderm induces muscle-specific gene expression in Drosophila embryos. *Development* **121,** 1387–1398.

Bataillé, L., Delon, I., Da Ponte, J. P., Brown, N. H., and Jagla, K. (2010). Downstream of identity genes: Muscle-type-specific regulation of the fusion process. *Dev. Cell* **19,** 317–328.

Bate, M., and Broadie, K. (1995). Wiring by fly: The neuromuscular system of the Drosophila embryo. *Neuron* **15,** 513–525.

Bate, M., Rushton, E., and Currie, D. A. (1991). Cells with persistent twist expression are the embryonic precursors of adult muscles in Drosophila. *Development* **113,** 79–89.

Baylies, M. K., and Bate, M. (1996). Twist: A myogenic switch in Drosophila. *Science* **272,** 1481–1484.

Baylies, M. K., Martinez Arias, A., and Bate, M. (1995). Wingless is required for the formation of a subset of muscle founder cells during Drosophila embryogenesis. *Development* **121,** 3829–3837.

Baylies, M. K., Bate, M., and Ruiz Gomez, M. (1998). Myogenesis: A view from Drosophila. *Cell* **93,** 921–927.

Beckett, K., and Baylies, M. K. (2006). The development of the Drosophila larval body wall muscles. *Int. Rev. Neurobiol.* **75,** 55–70.

Beckett, K., and Baylies, M. K. (2007). 3D analysis of founder cell and fusion competent myoblast arrangements outlines a new model of myoblast fusion. *Dev. Biol.* **309,** 113–125.

Borkowski, O. M., Brown, N. H., and Bate, M. (1995). Anterior–posterior subdivision and the diversification of the mesoderm in Drosophila. *Development* **121,** 4183–4193.

Bour, B. A., Chakravarti, M., West, J. M., and Abmayr, S. M. (2000). Drosophila SNS, a member of the immunoglobulin superfamily that is essential for myoblast fusion. *Genes Dev.* **14,** 1498–1511.

Bourgouin, C., Lundgren, S. E., and Thomas, J. B. (1992). Apterous is a Drosophila LIM domain gene required for the development of a subset of embryonic muscles. *Neuron* **9,** 549–561.

Brennan, K., Baylies, M., and Arias, A. M. (1999). Repression by Notch is required before Wingless signalling during muscle progenitor cell development in Drosophila. *Curr. Biol.* **9,** 707–710.

Buff, E., Carmena, A., Gisselbrecht, S., Jimenez, F., and Michelson, A. M. (1998). Signalling by the Drosophila epidermal growth factor receptor is required for the specification and diversification of embryonic muscle progenitors. *Development* **125,** 2075–2086.

Callahan, C. A., Bonkovsky, J. L., Scully, A. L., and Thomas, J. B. (1996). Derailed is required for muscle attachment site selection in Drosophila. *Development* **122,** 2761–2767.

Capovilla, M., Kambris, Z., and Botas, J. (2001). Direct regulation of the muscle-identity gene apterous by a Hox protein in the somatic mesoderm. *Development* **128,** 1221–1230.

Carmena, A., Bate, M., and Jimenez, F. (1995). Lethal of scute, a proneural gene, participates in the specification of muscle progenitors during Drosophila embryogenesis. *Genes Dev.* **9,** 2373–2383.

Carmena, A., Gisselbrecht, S., Harrison, J., Jimenez, F., and Michelson, A. M. (1998). Combinatorial signaling codes for the progressive determination of cell fates in the Drosophila embryonic mesoderm. *Genes Dev.* **12,** 3910–3922.

Carmena, A., Buff, E., Halfon, M. S., Gisselbrecht, S., Jimenez, F., Baylies, M. K., and Michelson, A. M. (2002). Reciprocal regulatory interactions between the Notch and Ras signaling pathways in the Drosophila embryonic mesoderm. *Dev. Biol.* **244,** 226–242.

Chanana, B., Steigemann, P., Jackle, H., and Vorbruggen, G. (2009). Reception of Slit requires only the chondroitin-sulphate-modified extracellular domain of Syndecan at the target cell surface. *Proc. Natl. Acad. Sci. USA* **106,** 11984–11988.

Chiba, A., Hing, H., Cash, S., and Keshishian, H. (1993). Growth cone choices of Drosophila motoneurons in response to muscle fiber mismatch. *J. Neurosci.* **13,** 714–732.

Chiba, A., Snow, P., Keshishian, H., and Hotta, Y. (1995). Fasciclin III as a synaptic target recognition molecule in Drosophila. *Nature* **374,** 166–168.

Ciglar, L., and Furlong, E. E. (2009). Conservation and divergence in developmental networks: A view from Drosophila myogenesis. *Curr. Opin. Cell Biol.* **21,** 754–760.

Cox, V. T., and Baylies, M. K. (2005). Specification of individual Slouch muscle progenitors in Drosophila requires sequential Wingless signaling. *Development* **132,** 713–724.

Cox, V. T., Beckett, K., and Baylies, M. K. (2005). Delivery of wingless to the ventral mesoderm by the developing central nervous system ensures proper patterning of individual slouch-positive muscle progenitors. *Dev. Biol.* **287,** 403–415.

Crozatier, M., and Vincent, A. (1999). Requirement for the Drosophila COE transcription factor Collier in formation of an embryonic muscle: Transcriptional response to Notch signalling. *Development* **126,** 1495–1504.

Deal, R. B., and Henikoff, S. (2011). The INTACT method for cell type-specific gene expression and chromatin profiling in Arabidopsis thaliana. *Nat. Protoc.* **6,** 56–68.

Deng, H., Bell, J. B., and Simmonds, A. J. (2010). Vestigial is required during late-stage muscle differentiation in Drosophila melanogaster embryos. *Mol. Biol. Cell* **21,** 3304–3316.

Dohrmann, C., Azpiazu, N., and Frasch, M. (1990). A new Drosophila homeo box gene is expressed in mesodermal precursor cells of distinct muscles during embryogenesis. *Genes Dev.* **4,** 2098–2111.

Duan, H., Skeath, J. B., and Nguyen, H. T. (2001). Drosophila Lame duck, a novel member of the Gli superfamily, acts as a key regulator of myogenesis by controlling fusion-competent myoblast development. *Development* **128,** 4489–4500.

Duan, H., Zhang, C., Chen, J., Sink, H., Frei, E., and Noll, M. (2007). A key role of Pox meso in somatic myogenesis of Drosophila. *Development* **134,** 3985–3997.

Elgar, S. J., Han, J., and Taylor, M. V. (2008). mef2 activity levels differentially affect gene expression during Drosophila muscle development. *Proc. Natl. Acad. Sci. USA* **105,** 918–923.

Enriquez, J., Boukhatmi, H., Dubois, L., Philippakis, A. A., Bulyk, M. L., Michelson, A. M., Crozatier, M., and Vincent, A. (2010). Multi-step control of muscle diversity by Hox proteins in the Drosophila embryo. *Development* **137,** 457–466.

Estrada, B., Choe, S. E., Gisselbrecht, S. S., Michaud, S., Raj, L., Busser, B. W., Halfon, M. S., Church, G. M., and Michelson, A. M. (2006). An integrated strategy for analyzing the unique developmental programs of different myoblast subtypes. *PLoS Genet.* **2,** e16.

Estrada, B., Gisselbrecht, S. S., and Michelson, A. M. (2007). The transmembrane protein Perdido interacts with Grip and integrins to mediate myotube projection and attachment in the Drosophila embryo. *Development* **134,** 4469–4478.

Figeac, N., Jagla, T., Aradhya, R., Da Ponte, J. P., and Jagla, K. (2010). Drosophila adult muscle precursors form a network of interconnected cells and are specified by the rhomboid-triggered EGF pathway. *Development* **137,** 1965–1973.

Frasch, M. (1995). Induction of visceral and cardiac mesoderm by ectodermal Dpp in the early Drosophila embryo. *Nature* **374,** 464–467.

Frasch, M. (1999). Controls in patterning and diversification of somatic muscles during Drosophila embryogenesis. *Curr. Opin. Genet. Dev.* **9,** 522–529.

Furlong, E. E. (2011). Molecular biology: A fly in the face of genomics. *Nature* **471,** 458–459.

Gilsohn, E., and Volk, T. (2010). Slowdown promotes muscle integrity by modulating integrin-mediated adhesion at the myotendinous junction. *Development* **137,** 785–794.

Gryzik, T., and Muller, H. A. (2004). FGF8-like1 and FGF8-like2 encode putative ligands of the FGF receptor Htl and are required for mesoderm migration in the Drosophila gastrula. *Curr. Biol.* **14,** 659–667.

Guerin, C. M., and Kramer, S. G. (2009). Cytoskeletal remodeling during myotube assembly and guidance: Coordinating the actin and microtubule networks. *Commun. Integr. Biol.* **2,** 452–457.

Halpern, M. E., Chiba, A., Johansen, J., and Keshishian, H. (1991). Growth cone behavior underlying the development of stereotypic synaptic connections in Drosophila embryos. *J. Neurosci.* **11,** 3227–3238.

Heiman, M., Schaefer, A., Gong, S., Peterson, J. D., Day, M., Ramsey, K. E., Suarez-Farinas, M., Schwarz, C., Stephan, D. A., Surmeier, D. J., Greengard, P., and Heintz, N. (2008). A translational profiling approach for the molecular characterization of CNS cell types. *Cell* **135,** 738–748.

Hooper, J. E. (1986). Homeotic gene function in the muscles of Drosophila larvae. *EMBO J.* **5,** 2321–2329.

Inaki, M., Shinza-Kameda, M., Ismat, A., Frasch, M., and Nose, A. (2010). Drosophila Tey represses transcription of the repulsive cue Toll and generates neuromuscular target specificity. *Development* **137,** 2139–2146.

Jagla, T., Bellard, F., Lutz, Y., Dretzen, G., Bellard, M., and Jagla, K. (1998). Ladybird determines cell fate decisions during diversification of Drosophila somatic muscles. *Development* **125,** 3699–3708.

Jagla, T., Bellard, F., Vonesch, J. L., Bellard, M., Dastugue, B., and Jagla, K. (1999). Plasticity within the lateral somatic mesoderm of Drosophila embryos. *Int. J. Dev. Biol.* **43,** 571–573.

Jagla, K., Bellard, M., and Frasch, M. (2001). A cluster of Drosophila homeobox genes involved in mesoderm differentiation programs. *Bioessays* **23,** 125–133.

Jagla, T., Bidet, Y., Da Ponte, J. P., Dastugue, B., and Jagla, K. (2002). Cross-repressive interactions of identity genes are essential for proper specification of cardiac and muscular fates in Drosophila. *Development* **129,** 1037–1047.

Junion, G., Jagla, T., Duplant, S., Tapin, R., Da Ponte, J. P., and Jagla, K. (2005). Mapping Dmef2-binding regulatory modules by using a ChIP-enriched in silico targets approach. *Proc. Natl. Acad. Sci. USA* **102,** 18479–18484.

Junion, G., Bataille, L., Jagla, T., Da Ponte, J. P., Tapin, R., and Jagla, K. (2007). Genome-wide view of cell fate specification: Ladybird acts at multiple levels during diversification of muscle and heart precursors. *Genes Dev.* **21,** 3163–3180.

Kadam, S., McMahon, A., Tzou, P., and Stathopoulos, A. (2009). FGF ligands in Drosophila have distinct activities required to support cell migration and differentiation. *Development* **136,** 739–747.

Keller, C. A., Grill, M. A., and Abmayr, S. M. (1998). A role for nautilus in the differentiation of muscle precursors. *Dev. Biol.* **202,** 157–171.

Kesper, D. A., Stute, C., Buttgereit, D., Kreiskother, N., Vishnu, S., Fischbach, K. F., and Renkawitz-Pohl, R. (2007). Myoblast fusion in Drosophila melanogaster is mediated through a fusion-restricted myogenic-adhesive structure (FuRMAS). *Dev. Dyn.* **236,** 404–415.

Klingseisen, A., Clark, I. B., Gryzik, T., and Muller, H. A. (2009). Differential and overlapping functions of two closely related Drosophila FGF8-like growth factors in mesoderm development. *Development* **136,** 2393–2402.

Knirr, S., Azpiazu, N., and Frasch, M. (1999). The role of the NK-homeobox gene slouch (S59) in somatic muscle patterning. *Development* **126,** 4525–4535.

Kohsaka, H., and Nose, A. (2009). Target recognition at the tips of postsynaptic filopodia: Accumulation and function of Capricious. *Development* **136,** 1127–1135.

Kolodziej, P. A., Timpe, L. C., Mitchell, K. J., Fried, S. R., Goodman, C. S., Jan, L. Y., and Jan, Y. N. (1996). Frazzled encodes a Drosophila member of the DCC immunoglobulin subfamily and is required for CNS and motor axon guidance. *Cell* **87,** 197–204.

Kramer, S. G., Kidd, T., Simpson, J. H., and Goodman, C. S. (2001). Switching repulsion to attraction: Changing responses to slit during transition in mesoderm migration. *Science* **292,** 737–740.

Kraut, R., and Campos-Ortega, J. A. (1996). Inscuteable, a neural precursor gene of Drosophila, encodes a candidate for a cytoskeleton adaptor protein. *Dev. Biol.* **174,** 65–81.

Landgraf, M., and Thor, S. (2006). Development of Drosophila motoneurons: Specification and morphology. *Semin. Cell Dev. Biol.* **17,** 3–11.

Landgraf, M., Baylies, M., and Bate, M. (1999). Muscle founder cells regulate defasciculation and targeting of motor axons in the Drosophila embryo. *Curr. Biol.* **9,** 589–592.

Lawrence, P. A., Bodmer, R., and Vincent, J. P. (1995). Segmental patterning of heart precursors in Drosophila. *Development* **121,** 4303–4308.

Leptin, M. (1991). Twist and snail as positive and negative regulators during Drosophila mesoderm development. *Genes Dev.* **5,** 1568–1576.

Levine, M. (2011). Paused RNA polymerase II as a developmental checkpoint. *Cell* **145,** 502–511.

Lopez, A. J., and Hogness, D. S. (1991). Immunochemical dissection of the Ultrabithorax homeoprotein family in Drosophila melanogaster. *Proc. Natl. Acad. Sci. USA* **88,** 9924–9928.

Lord, P. C., Lin, M. H., Hales, K. H., and Storti, R. V. (1995). Normal expression and the effects of ectopic expression of the Drosophila muscle segment homeobox (msh) gene suggest a role in differentiation and patterning of embryonic muscles. *Dev. Biol.* **171,** 627–640.

Maggert, K., Levine, M., and Frasch, M. (1995). The somatic-visceral subdivision of the embryonic mesoderm is initiated by dorsal gradient thresholds in Drosophila. *Development* **121,** 2107–2116.

Maqbool, T., and Jagla, K. (2007). Genetic control of muscle development: Learning from Drosophila. *J. Muscle Res. Cell Motil.* **28,** 397–407.

McGinnis, W., and Krumlauf, R. (1992). Homeobox genes and axial patterning. *Cell* **68,** 283–302.

Michelson, A. M. (1994). Muscle pattern diversification in Drosophila is determined by the autonomous function of homeotic genes in the embryonic mesoderm. *Development* **120,** 755–768.

Michelson, A. M., Gisselbrecht, S., Zhou, Y., Baek, K. H., and Buff, E. M. (1998). Dual functions of the heartless fibroblast growth factor receptor in development of the Drosophila embryonic mesoderm. *Dev. Genet.* **22,** 212–229.

Miller, M. R., Robinson, K. J., Cleary, M. D., and Doe, C. Q. (2009). TU-tagging: Cell type-specific RNA isolation from intact complex tissues. *Nat. Methods* **6,** 439–441.

Muller, D., Jagla, T., Bodart, L. M., Jahrling, N., Dodt, H. U., Jagla, K., and Frasch, M. (2010). Regulation and functions of the lms homeobox gene during development of embryonic lateral transverse muscles and direct flight muscles in Drosophila. *PLoS One* **5,** e14323.

Nose, A., Van Vactor, D., Auld, V., and Goodman, C. S. (1992). Development of neuromuscular specificity in Drosophila. *Cold Spring Harb. Symp. Quant. Biol.* **57,** 441–449.

Nose, A., Takeichi, M., and Goodman, C. S. (1994). Ectopic expression of connectin reveals a repulsive function during growth cone guidance and synapse formation. *Neuron* **13,** 525–539.

Nose, A., Isshiki, T., and Takeichi, M. (1998). Regional specification of muscle progenitors in Drosophila: The role of the msh homeobox gene. *Development* **125,** 215–223.

Patel, N. H., Snow, P. M., and Goodman, C. S. (1987). Characterization and cloning of fasciclin III: A glycoprotein expressed on a subset of neurons and axon pathways in Drosophila. *Cell* **48,** 975–988.

Philippakis, A. A., Busser, B. W., Gisselbrecht, S. S., He, F. S., Estrada, B., Michelson, A. M., and Bulyk, M. L. (2006). Expression-guided in silico evaluation of candidate cis regulatory codes for Drosophila muscle founder cells. *PLoS Comput. Biol.* **2,** e53.

Ranganayakulu, G., Schulz, R. A., and Olson, E. N. (1996). Wingless signaling induces nautilus expression in the ventral mesoderm of the Drosophila embryo. *Dev. Biol.* **176,** 143–148.

Reed, H. C., Hoare, T., Thomsen, S., Weaver, T. A., White, R. A., Akam, M., and Alonso, C. R. (2010). Alternative splicing modulates Ubx protein function in Drosophila melanogaster. *Genetics* **184,** 745–758.

Riechmann, V., Irion, U., Wilson, R., Grosskortenhaus, R., and Leptin, M. (1997). Control of cell fates and segmentation in the Drosophila mesoderm. *Development* **124,** 2915–2922.

Ritzenthaler, S., and Chiba, A. (2001). Multiple personalities: Synaptic target cells as introverts and extroverts. *Dev. Growth Differ.* **43,** 503–508.

Ritzenthaler, S., Suzuki, E., and Chiba, A. (2000). Postsynaptic filopodia in muscle cells interact with innervating motoneuron axons. *Nat. Neurosci.* **3,** 1012–1017.

Ruiz Gomez, M., and Bate, M. (1997). Segregation of myogenic lineages in Drosophila requires numb. *Development* **124,** 4857–4866.

Ruiz-Gomez, M., Romani, S., Hartmann, C., Jackle, H., and Bate, M. (1997). Specific muscle identities are regulated by Kruppel during Drosophila embryogenesis. *Development* **124,** 3407–3414.

Ruiz-Gomez, M., Coutts, N., Price, A., Taylor, M. V., and Bate, M. (2000). Drosophila dumbfounded: A myoblast attractant essential for fusion. *Cell* **102,** 189–198.

Sandmann, T., Jakobsen, J. S., and Furlong, E. E. (2006). ChIP-on-chip protocol for genome-wide analysis of transcription factor binding in Drosophila melanogaster embryos. *Nat. Protoc.* **1,** 2839–2855.

Sandmann, T., Girardot, C., Brehme, M., Tongprasit, W., Stolc, V., and Furlong, E. E. (2007). A core transcriptional network for early mesoderm development in Drosophila melanogaster. *Genes Dev.* **21,** 436–449.

Schejter, E. D., and Baylies, M. K. (2010). Born to run: Creating the muscle fiber. *Curr. Opin. Cell Biol.* **22,** 566–574.

Schnorrer, F., and Dickson, B. J. (2004). Muscle building: Mechanisms of myotube guidance and attachment site selection. *Dev. Cell* **7,** 9–20.

Schnorrer, F., Kalchhauser, I., and Dickson, B. J. (2007). The transmembrane protein Kon-tiki couples to Dgrip to mediate myotube targeting in Drosophila. *Dev. Cell* **12,** 751–766.

Schnorrer, F., Schonbauer, C., Langer, C. C., Dietzl, G., Novatchkova, M., Schernhuber, K., Fellner, M., Azaryan, A., Radolf, M., Stark, A., Keleman, K., and Dickson, B. J. (2011). Systematic genetic analysis of muscle morphogenesis and function in Drosophila. *Nature* **464,** 287–291.

Schulz, R. A., and Gajewski, K. (1999). Ventral neuroblasts and the heartless FGF receptor are required for muscle founder cell specification in Drosophila. *Oncogene* **18,** 6818–6823.

Schweitzer, R., Zelzer, E., and Volk, T. (2010). Connecting muscles to tendons: Tendons and musculoskeletal development in flies and vertebrates. *Development* **137,** 2807–2817.

Shilo, B. Z., Schejter, E. D., Segal, D., Ginsberg, D. S., and Glazer, L. (1986). The Drosophila epidermal growth factor receptor homolog: Structure, evolution, and possible functions. *Symp. Fundam. Cancer Res.* **39,** 87–97.

Shishido, E., Higashijima, S., Emori, Y., and Saigo, K. (1993). Two FGF-receptor homologues of Drosophila: One is expressed in mesodermal primordium in early embryos. *Development* **117,** 751–761.

Shishido, E., Ono, N., Kojima, T., and Saigo, K. (1997). Requirements of DFR1/Heartless, a mesoderm-specific Drosophila FGF-receptor, for the formation of heart, visceral and somatic muscles, and ensheathing of longitudinal axon tracts in CNS. *Development* **124,** 2119–2128.

Sparrow, J. C., and Schock, F. (2009). The initial steps of myofibril assembly: Integrins pave the way. *Nat. Rev. Mol. Cell Biol.* **10,** 293–298.

Staehling-Hampton, K., Hoffmann, F. M., Baylies, M. K., Rushton, E., and Bate, M. (1994). dpp induces mesodermal gene expression in Drosophila. *Nature* **372,** 783–786.

Stathopoulos, A., Tam, B., Ronshaugen, M., Frasch, M., and Levine, M. (2004). pyramus and thisbe: FGF genes that pattern the mesoderm of Drosophila embryos. *Genes Dev.* **18,** 687–699.

Strunkelnberg, M., Bonengel, B., Moda, L. M., Hertenstein, A., de Couet, H. G., Ramos, R. G., and Fischbach, K. F. (2001). rst and its paralogue kirre act redundantly during embryonic muscle development in Drosophila. *Development* **128,** 4229–4239.

Subramanian, A., Wayburn, B., Bunch, T., and Volk, T. (2007). Thrombospondin-mediated adhesion is essential for the formation of the myotendinous junction in Drosophila. *Development* **134,** 1269–1278.

Tixier, V., Bataille, L., and Jagla, K. (2010). Diversification of muscle types: Recent insights from Drosophila. *Exp. Cell Res.* **316,** 3019–3027.

van der Geer, P., Hunter, T., and Lindberg, R. A. (1994). Receptor protein-tyrosine kinases and their signal transduction pathways. *Annu. Rev. Cell Biol.* **10,** 251–337.

Volk, T. (1992). A new member of the spectrin superfamily may participate in the formation of embryonic muscle attachments in Drosophila. *Development* **116,** 721–730.

Volk, T. (1999). Singling out Drosophila tendon cells: A dialogue between two distinct cell types. *Trends Genet.* **15,** 448–453.

Vorbruggen, G., Constien, R., Zilian, O., Wimmer, E. A., Dowe, G., Taubert, H., Noll, M., and Jackle, H. (1997). Embryonic expression and characterization of a Ptx1 homolog in Drosophila. *Mech. Dev.* **68,** 139–147.

Wayburn, B., and Volk, T. (2009). LRT, a tendon-specific leucine-rich repeat protein, promotes muscle–tendon targeting through its interaction with Robo. *Development* **136,** 3607–3615.

Winberg, M. L., Mitchell, K. J., and Goodman, C. S. (1998). Genetic analysis of the mechanisms controlling target selection: Complementary and combinatorial functions of netrins, semaphorins, and IgCAMs. *Cell* **93,** 581–591.

Yarnitzky, T., Min, L., and Volk, T. (1997). The Drosophila neuregulin homolog Vein mediates inductive interactions between myotubes and their epidermal attachment cells. *Genes Dev.* **11,** 2691–2700.

Yarnitzky, T., Min, L., and Volk, T. (1998). An interplay between two EGF-receptor ligands, Vein and Spitz, is required for the formation of a subset of muscle precursors in Drosophila. *Mech. Dev.* **79,** 73–82.

Ypsilanti, A. R., Zagar, Y., and Chedotal, A. (2010). Moving away from the midline: New developments for Slit and Robo. *Development* **137,** 1939–1952.

# Index

Note: Page numbers followed by "*f*" indicate figures, and "*t*" indicate tables

**H**



**I**



**L**



**M**

# Contents of Previous Volumes

## Volume 47



## Volume 48

## Volume 49

## Volume 50



## Volume 51

## Volume 54

## Volume 57

## Volume 60



## Volume 61

## Volume 62

## Volume 63



## Volume 64

## Volume 65

## Volume 66



## Volume 67

## Volume 68

## Volume 69

## Volume 70

## Volume 71



## Volume 72

## Volume 73

## Volume 74



## Volume 75

## Volume 76

## Volume 77

## Volume 78



## Volume 79

## Volume 80

## Volume 81

## Volume 82

## Volume 83



## Volume 84

## Volume 85

## Volume 86

## Volume 87



## Volume 88

## Volume 89

## Volume 90

## Volume 91

## Volume 92

## Volume 93



## Volume 94

## Volume 95

## Volume 96

## Volume 97

1. **New Insights into Signaling During Myelination in Zebrafish**
*Alya R. Raphael and William S. Talbot*

2. **Wnt Signaling: Signaling at and Above the Receptor Level**
*Tina Buechling and Michael Boutros*

3. **Notch-Independent Functions of CSL**
*Jane E. Johnson and Raymond J. MacDonald*

4. **Sonic Hedgehog Signaling in the Developing CNS: Where It Has Been and Where It Is Going**
*Michael P. Matise and Hui Wang*

5. **Activity-Mediated Synapse Formation: A Role for Wnt-Fz Signaling**
*Macarena Sahores and Patricia C. Salinas*

6. **Signaling Pathways and Axis Formation in the Lower Metazoa**
*Thomas W. Holstein, Hiroshi Watanabe, and Suat Özbek*

7. **FGF and ROR2 Receptor Tyrosine Kinase Signaling in Human Skeletal Development**
*Sigmar Stricker and Stefan Mundlos*

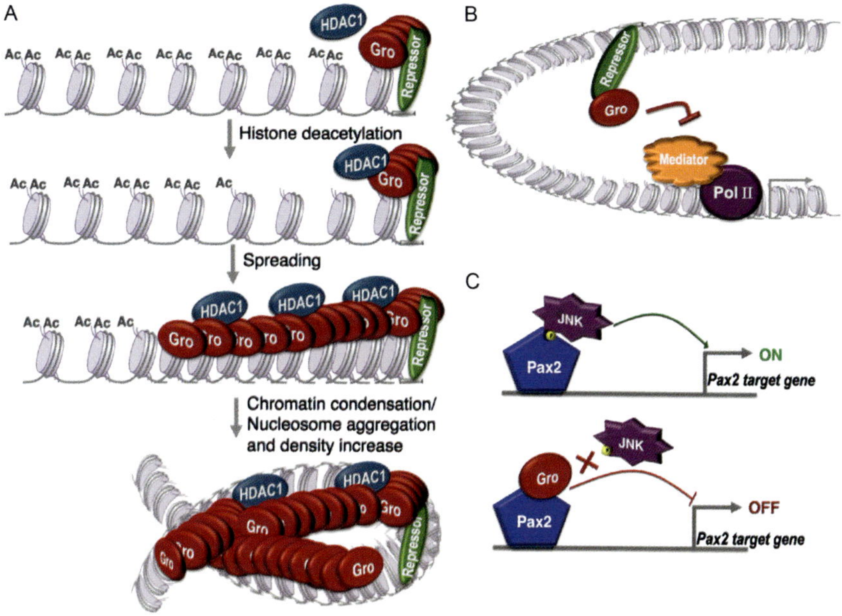


**Wiam Turki-Judeh and Albert J. Courey, Figure 3.2** Models for Gro/TLE-mediated repression. Gro/TLE appears to direct repression by multiple mechanisms. (A) A variety of studies indicate that the ability of Gro/TLE proteins to self-associate, recruit histone deacetylase, bind to deacetylated histones, and direct chromatin condensation (or increased nucleosome density) are required for Gro-mediated repression. This multistep model accounts for all of these biochemical functions. (B) Gro/TLE may interact with the Mediator, a component of the basal transcriptional machinery, thereby interfering with the formation or function of the preinitiation complex. (C) Gro/TLE may mask the activation domains of transcriptional activators. In the example shown here, Gro binding prevents phosphorylation of the activation domain in Pax2, thereby preventing Pax2 activation. Additional details regarding these three models and references are given in the text.

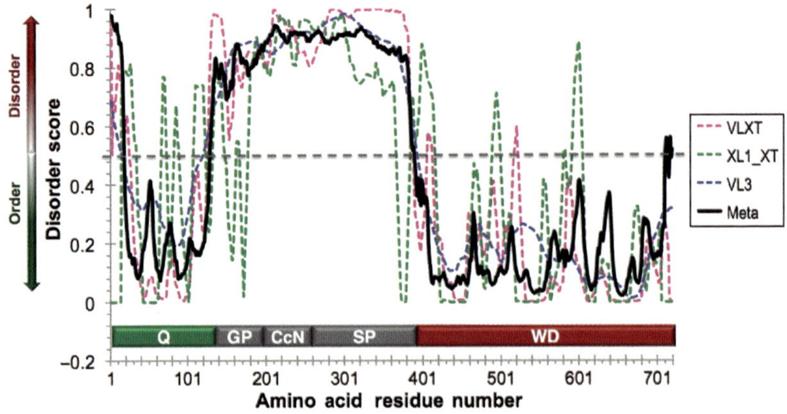


**Wiam Turki-Judeh and Albert J. Courey, Figure 3.3** The Gro/TLE central region appears to be intrinsically disordered. Prediction of disorder in *Drosophila* Gro using four disorder prediction algorithms: PONDR-FIT™ (Meta, black solid line) (Xue *et al.*, 2010); PONDR®VLXT (pink dashed line) (Li *et al.*, 1999; Romero *et al.*, 2001); PONDR®XL1_XT (green dashed line) (Romero *et al.*, 1997); and PONDR®VL3 (blue dashed line) (Radivojac *et al.*, 2003). Amino acid residues with scores exceeding 0.5 are likely to be disordered. All four prediction tools strongly suggest that the central region of Gro is disordered, while the Q and WD-repeat domains are ordered.

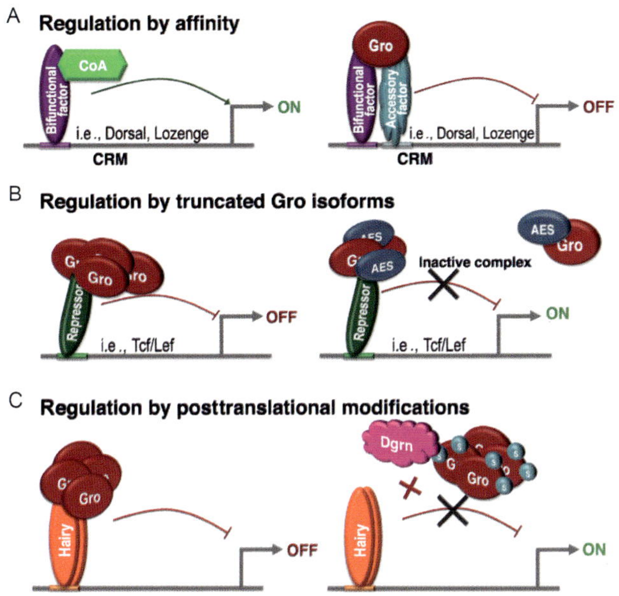


**Wiam Turki-Judeh and Albert J. Courey, Figure 3.4** The regulation of Gro/TLE-mediated repression. Gro/TLE function is regulated in a variety of ways. (A) Gro/TLE function depends on the affinity of the Gro/TLE-interaction motif in the repressor protein for Gro. The dual functionality of Lozenge and Dorsal relies on the fact that these two factors have an intrinsically low affinity for Gro. They therefore require assistance from accessory proteins to create a high-affinity platform for Gro/TLE recruitment. (B) The AES proteins, which consist only of the Q and GP domains, inhibit Gro/TLE-mediated transcriptional repression in a dominant negative manner perhaps by forming mixed oligomers with full-length Gro/TLE proteins through the Q domain, resulting in an inactive complex. (C) Gro/TLE is also regulated by posttranslational modification. For example, sumoylated Gro/TLE is targeted by the ubiquitin ligase Degringloade (Dgrn) for inactivation, thereby antagonizing Gro/TLE function. See text for references and details.

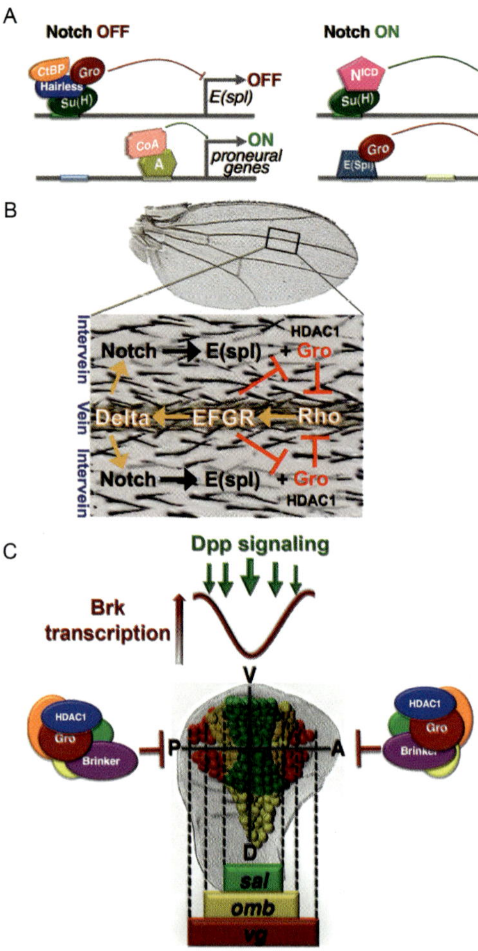


**Wiam Turki-Judeh and Albert J. Courey, Figure 3.5** Roles for Gro/TLE family proteins in signaling. (A) Gro plays multiple roles in Notch signaling, working both cooperatively and antagonistically with Notch. In the absence of Notch signaling, Gro is recruited by Suppressor of Hairless (Su(H)) to a corepressor complex at Notch target genes such as the genes of the E(spl) complex. Upon activation of Notch signaling, the Notch intracellular domain translocates to the nucleus displacing the Gro-containing complex from Su(H) and resulting in the activation of E(spl) complex genes. bHLH factors encoded by E(spl) then employ Gro as a corepressor to silence proneural genes. (B) Gro/Notch antagonism is required for wing veination. Wing vein specification is initiated by the activation of *rhomboid* (*rho*) in the wing disc in territories where the veins will develop (the provein region). This leads to activation of EGFR and therefore of the RTK/Ras/MAPK signaling pathway triggering expression of genes that lead to vein formation. (These so-called provein genes have been omitted from the diagram for simplicity). RTK/Ras/MAPK signaling also leads to expression of the Notch ligand Delta in the provein region, which signals to adjacent intervein cells. Notch signaling in the intervein region antagonizes wing veination via the

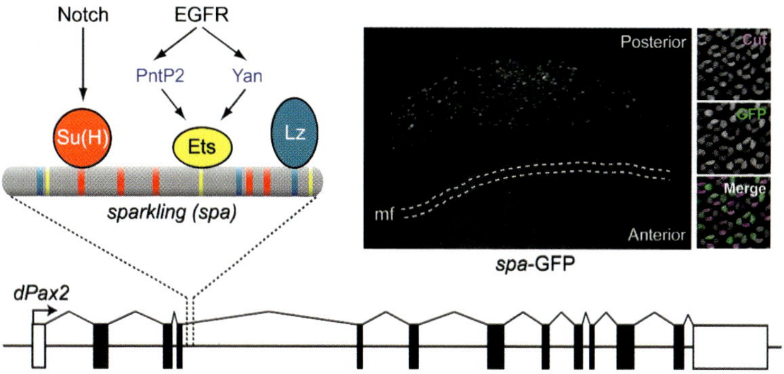


**Nicole C. Evans *et al.*, Figure 4.1** The *sparkling* (*spa*) enhancer of the *Drosophila dPax2 gene* is sufficient to drive cone-cell-specific gene expression *in vivo*. Top left: diagram of the minimal 362-bp *spa* element, with Su(H), Ets, and Lz sites, as determined by Flores *et al.* (2000), indicated with colored stripes. Top right: fluorescence micrograph showing GFP expression, driven by the wild-type *spa* enhancer, in a transgenic third-instar eye imaginal disc. (mf, morphogenic furrow) Smaller images show coexpression of GFP with the cone cell marker Cut in pupae. Bottom: diagram of the structure of the *dPax2* gene, showing the location of *spa* in the fourth intron.

Gro-dependent repression of *rho*, thereby preventing inappropriate expansion of veins into the intervein region. Gro phosphorylation by RTK/Ras/MAPK signaling in the provein region leads to Gro inactivation. Thus, the RTK/Ras/MAPK signal in the provein region prevents the Notch signal from spreading into the provein region and blocking normal wing vein formation. (C) Anteroposterior patterning in the developing wing results in large part from the repression of *brk* in response to the Dpp signal in the region surrounding the anteroposterior midline. Brk protein acts, in turn, as a Gro-dependent repressor to turn off the expression of genes such as *vg*, *omb*, and *sal* at the anterior and posterior edges of the developing wing. These Brk targets are responsive to different concentrations of Brk, thus leading to different size expression domains.

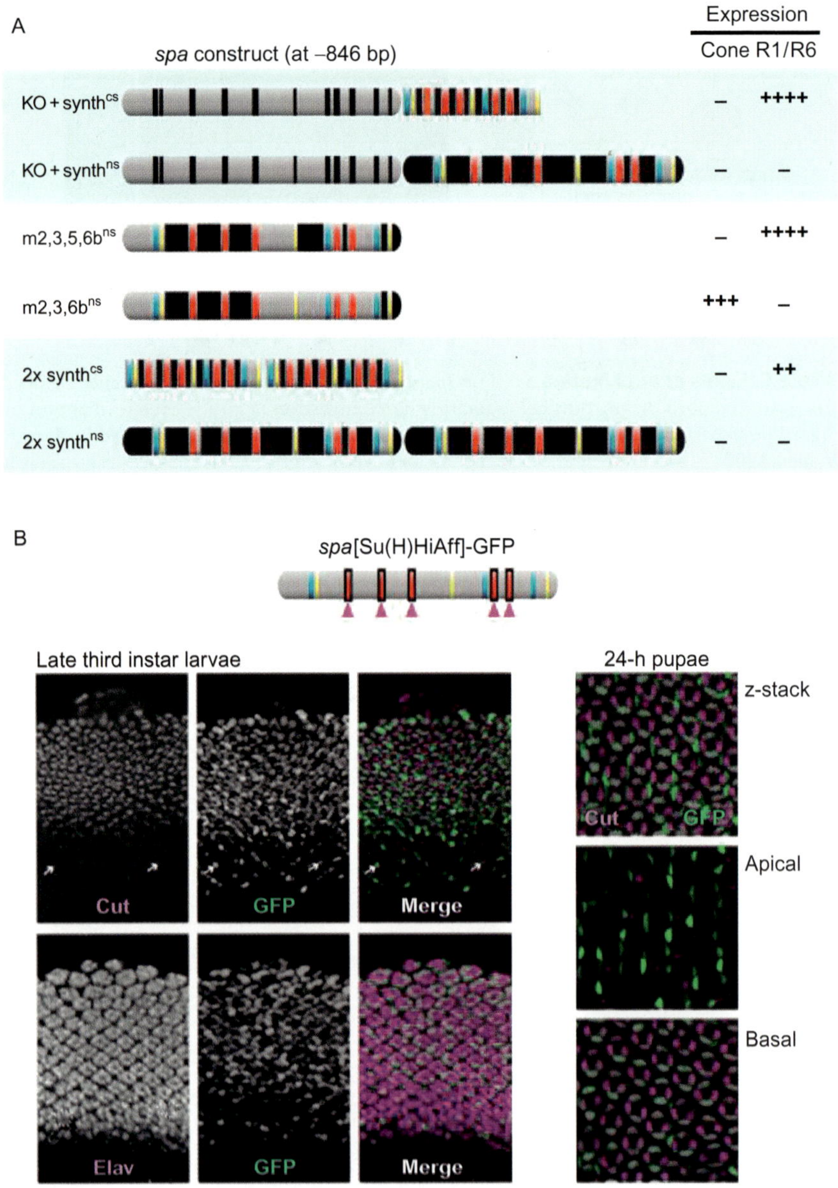


**Nicole C. Evans *et al.*, Figure 4.5** The linear arrangement, spacing, and affinity of binding sites profoundly affects the patterning of gene expression driven by *spa*. (A) Summary of results of experiments testing the effects of binding site organization on cell-type specificity of gene expression. (B) Increasing the affinity of Su(H) binding sites within *spa* creates an enhancer that is oversensitive to Notch signaling, resulting in ectopic gene expression in multiple Notch-responsive cell types. Left: *spa*[Su(H)HiAff]-GFP is expressed ectopically in Cut-negative photoreceptors (arrows) in the larval eye imaginal

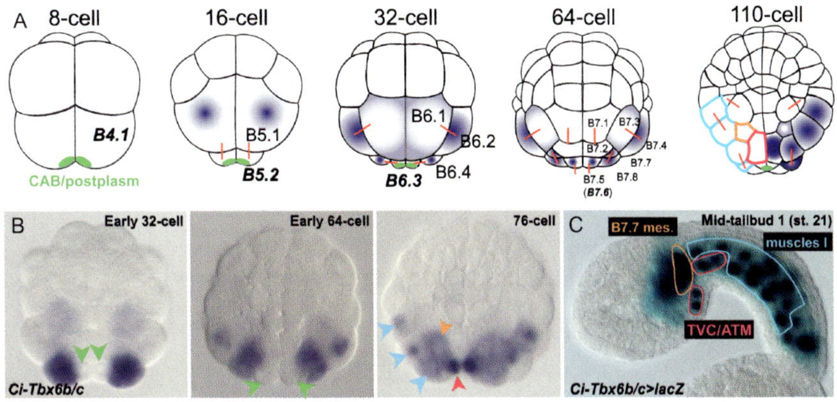


**Wei Wang and Lionel Christiaen, Figure 6.2** Expression of *Tbx6b/c* and reporter construct in *Ciona intestinalis* embryos. (A) Schematic representations of vegetal views of early stage embryos showing the B-line blastomeres. Short red lines connect sister cells following cleavages. The blue staining represents *Tbx6b/c* transcripts as detected by *in situ* hybridization (panel B, see also Takatori *et al.*, 2004; Yagi *et al.*, 2005). The green ellipses represent the posterior vegetal cortical cytoplasm or postplasm and centrosome attracting body (CAB), where the postplasmic mRNAs localize, including macho-1 and PEM-1 (see text for details). Note that postplasmic mRNAs are segregated into the B7.6 germline precursors by the 64-cell stage (not represented on that drawing). Colored outlines on the left side of 110-cell embryo represent the prospective fate of the blastomeres [orange: mesenchyme, blue: primary muscles, red: trunk ventral cells (TVC: heart and atrial siphon muscles precursors) and anterior tail muscles (ATM)]. (B) Chromogenic *in situ* hybridizations showing *Tbx6b/c* expression at indicated stages. Note the absence of expression in cells carrying the postplasmic RNA PEM-1 (green arrowheads). Colored arrowheads indicate the blastomere outlines in A with the same colors. (C) Lateral view of a mid-tailbud 1 (stage 21) embryo electroporated with a *Tbx6b>lacZ* construct (Christiaen *et al.*, 2009a) and stained with X-gal (blue staining). Beta-galactosidase activity is detected in all descendants of the early blastomeres that express *Tbx6b* (outlined with the same colors as in A and B).

---

disc, as shown by coexpression with the neuronal marker Elav. Right: *spa*[Su(H)HiAff]-GFP is also expressed ectopically in primary pigment cells in the 24-h pupal retina, identifiable by lack of Cut staining, apical position within the retina, and characteristic elongated cell shape.

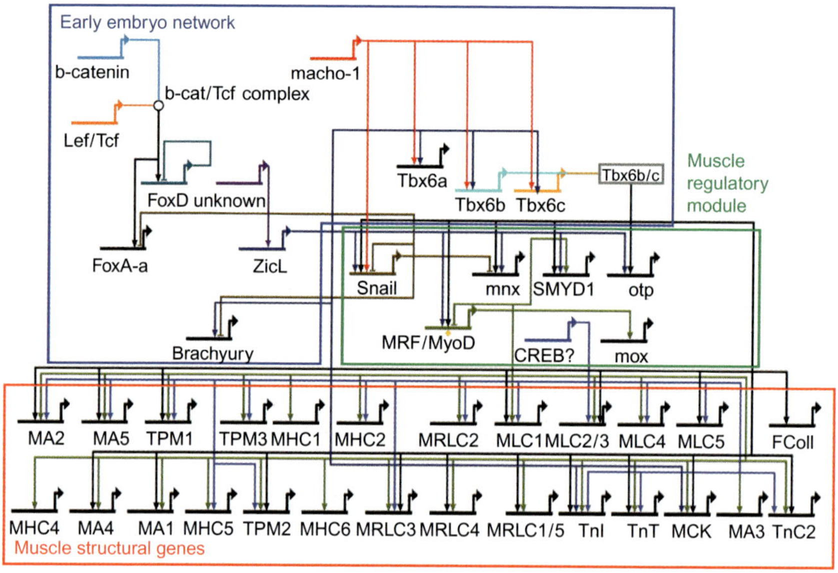


**Wei Wang and Lionel Christiaen, Figure 6.3** Tentative genomic regulatory network model for the early specification of primary muscle precursors in the B-line blastomeres. This model was built using Biotapestry (Longabaugh *et al.*, 2005) and the regulatory interactions inferred from data presented in the literature including Imai *et al.* (2006), Johnson *et al.* (2005), Kubo *et al.* (2010), Kusakabe *et al.* (2004) and Yagi *et al.* (2004a,b, 2005)). Note that no regulatory connection was predicted between the unknown CREB factor and *myosin heavy chain-1 (MHC-1)* because the published analysis focused on the upstream genomic DNA, but the regulatory element may be located in the fourth intron (Fig. 6.1).

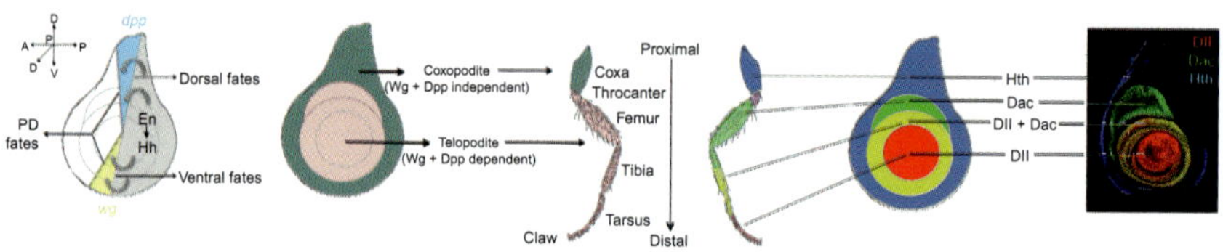


**Carlos Estella *et al.*, Figure 7.1** Overview of fly leg development. On the left shows the relationship between En, Hh, *wg*, and *dpp* and the definition of the telopodite (Hh, Wg, and Dpp-dependent domain) and the coxopodite (Hh, Wg, and Dpp-independent domain). On the right shows the relationship between the three primary PD gene expression domains established by Hth, Dac, and Dll.

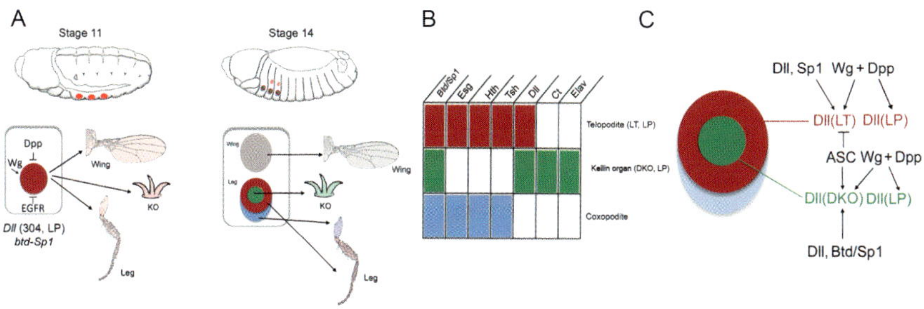


**Carlos Estella *et al.*, Figure 7.4** Embryonic appendage fate map. (A) Cells expressing *Dll* at stage 11 via the 304 CRM can give rise to the entire adult thorax, while those expressing *Dll* at stage 14 give rise to either the KO or telopodite, depending whether DKO or LT is driving expression, respectively. (B) Genes expressed in the progenitors to the telopodite, KO, and coxopodite. (C) Fate map and regulatory network defining the activity of *Dll* CRMs. The blue cells, expressing *esg* but not *Dll*, are fated to become coxopodite.

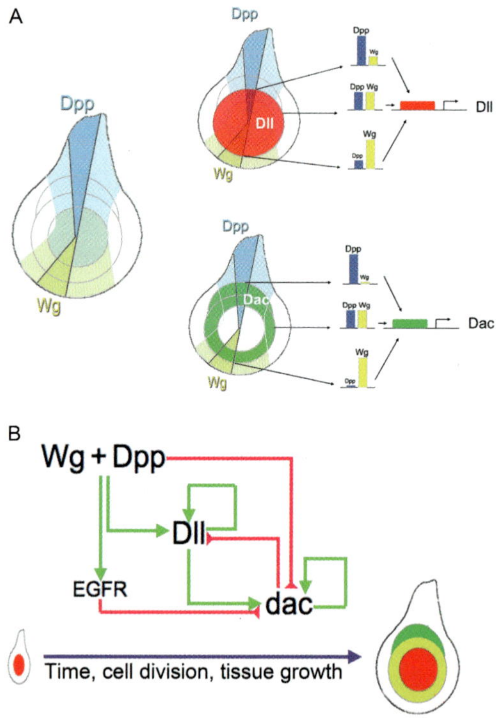


**Carlos Estella *et al.*, Figure 7.5** Gradient versus cascade models. (A) The gradient model, highlighting that, depending on a cell's position in the disc, *Dll* and *dac* CRMs must interpret very different ratios of Dpp:Wg signaling. (B) The cascade model, in which Wg+Dpp are only required to initiate PD axis formation by activating *Dll* and ligands for the EGFR pathway. Dll in turn activates *dac*, and both *Dll* and *dac* maintain their expression in a Wg+Dpp-independent manner. EGFR activity maintains *dac* repression, while Wg+Dpp repress *dac* in the center of the leg disc early in leg development.

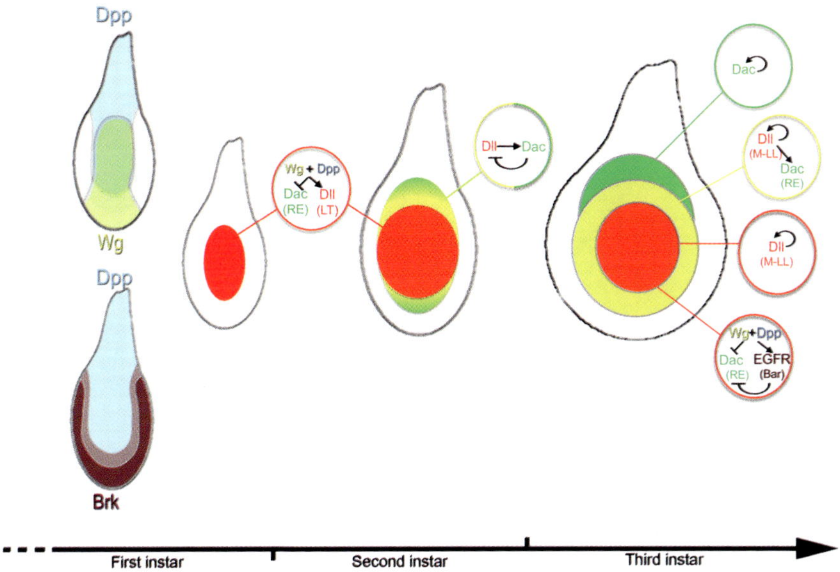


**Carlos Estella *et al.*, Figure 7.6** Summary of the dynamic network establishing the PD axis. The relationship between Dpp, Wg, and Brk triggers the expression of *Dll* and represses *dac* in the center of the disc. As the disc grows, Dll activates *dac* in cells that escape repression by Wg+Dpp. These initial domains are likely maintained by a combination of autoregulation, cross-regulation, and transcriptional memory systems.

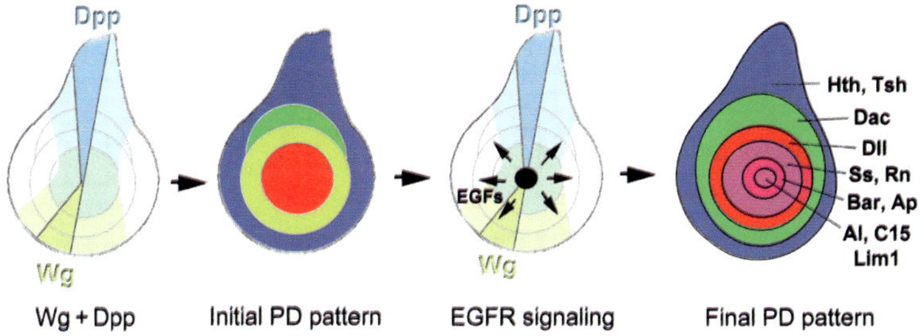


**Carlos Estella *et al.*, Figure 7.7** EGFR signaling patterns the tarsal segments. After the initial PD domains are established, EGFR ligands are produced at the center of the disc and activate a series of secondary PD targets in the progenitors to the tarsal segments.

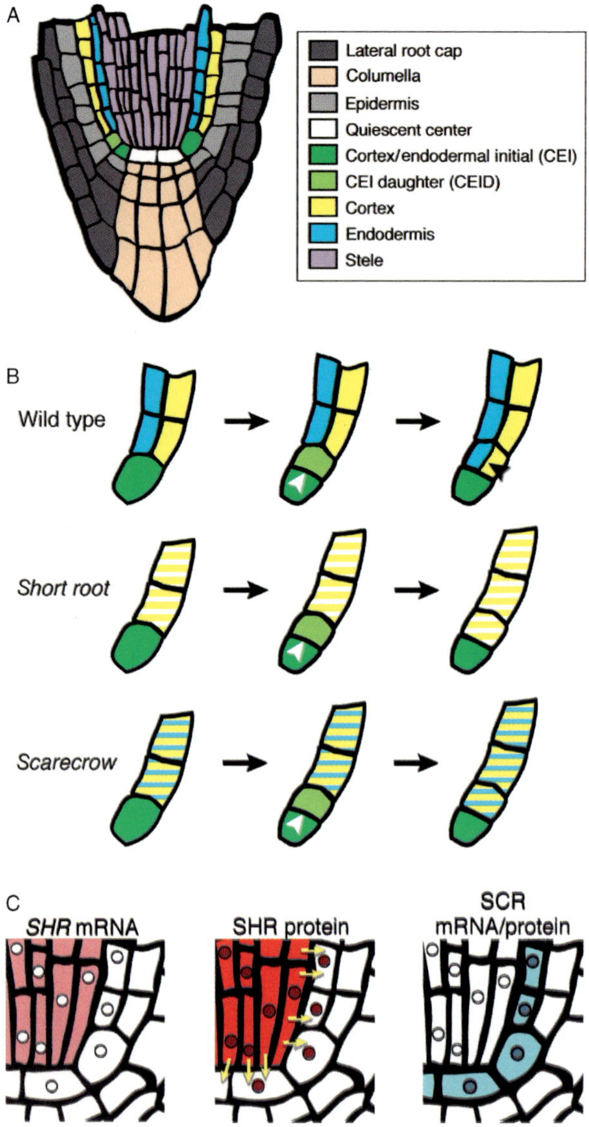


**Miguel A. Moreno-Risueno *et al.*, Figure 9.1** Ground tissue formation in the root. (A) Schematic of a longitudinal section of the Arabidopsis root tip. Individual or groups of cell types are depicted in different colors. (B) Cut away of the ground tissue cell types from (A) with an emphasis on the asymmetric divisions and cellular defects in *short root* (*shr*) and *scarecrow* (*scr*) mutants. (B, upper panel) Ground tissue formation in wild type. (B upper panel, left to right) The CEI (dark green) divides transversely (white arrowheads) to regenerate itself and produce the CEID (light green). The CEID then divides longitudinally (black arrowhead) to generate the cells of the endodermis (blue) and cortex (yellow). (B, center panel) Ground tissue formation in *short root*. In the absence of the longitudinal asymmetric cell division, a single layer of ground tissue with cortical cell features (yellow with white stripes) forms. (B, lower panel) Ground tissue formation in *scarecrow*. The longitudinal asymmetric cell division also does not occur; however, the single layer of ground tissue exhibits both endodermal and cortical cell features (yellow with blue stripes). (C) Schematic of a portion of the Arabidopsis root tip, focusing on the localization of *SHR* mRNA, SHR protein and *SCR* mRNA and protein. Yellow arrows depict SHR protein movement from the stele into the adjacent cell layer. Note that SHR and SCR proteins are colocalized in the nuclei (small circles within the cells) of the adjacent cell layer. This figure was adapted from Petricka *et al.* (2009).

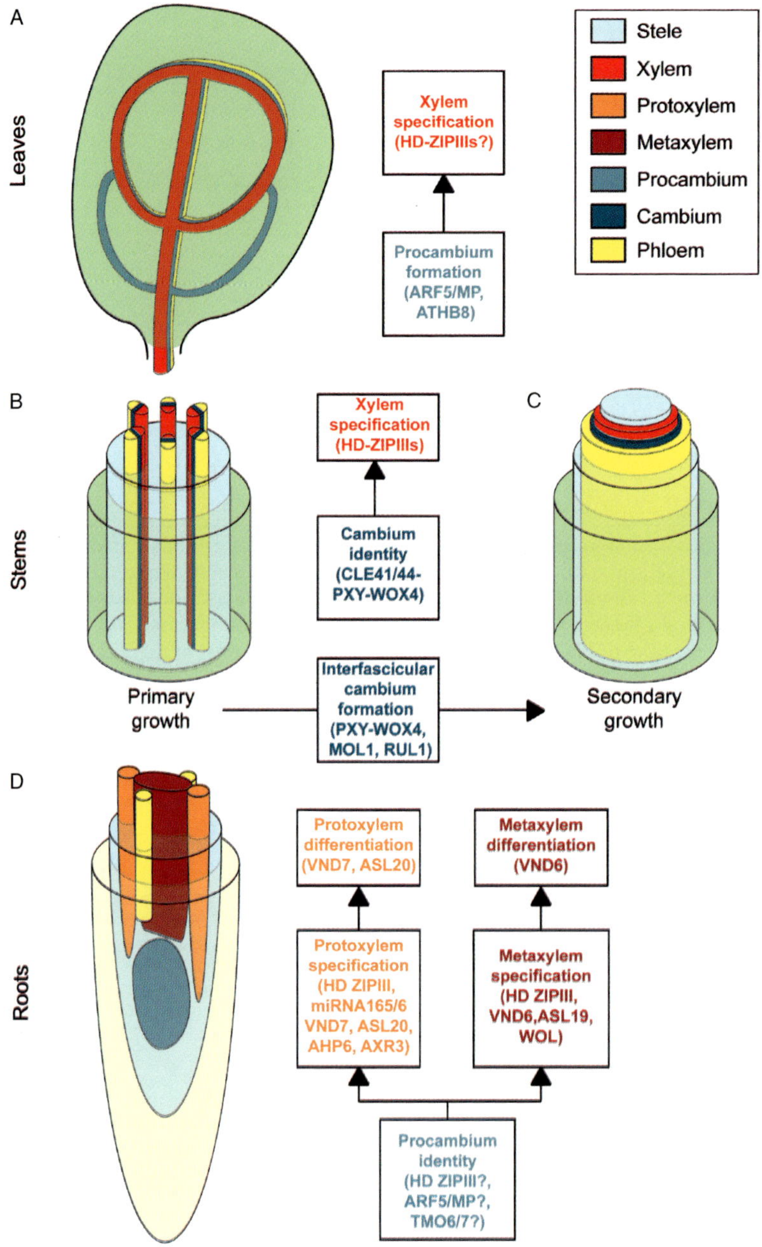


**Miguel A. Moreno-Risueno *et al.*, Figure 9.2** Schematic of vascular patterning in Arabidopsis. (A) Patterning of leaf vasculature occurs through establishment of procambial cells and subsequent specification of vascular tissues from these cells. Regulators of procambium formation during venation are indicated. (B) Vascular tissues in stems are organized into vascular bundles. These bundles are comprised of xylem toward the inside and phloem toward the outside separated by cambial cells. Xylem is specified from cambium by specific regulators that switch cell fate. (C) During secondary growth, a ring of vascular tissue is generated through formation of cambium that closes the spaces between bundles. Common regulators during primary and secondary growth are indicated. (D) Vascular tissues in the root are organized in a central cylinder. Xylem constitutes a symmetry arch with metaxylem being specified toward the center of the arch and two protoxylem poles toward the outside. Regulators of xylem fate act as multistep switch specifying metaxylem and protoxylem.

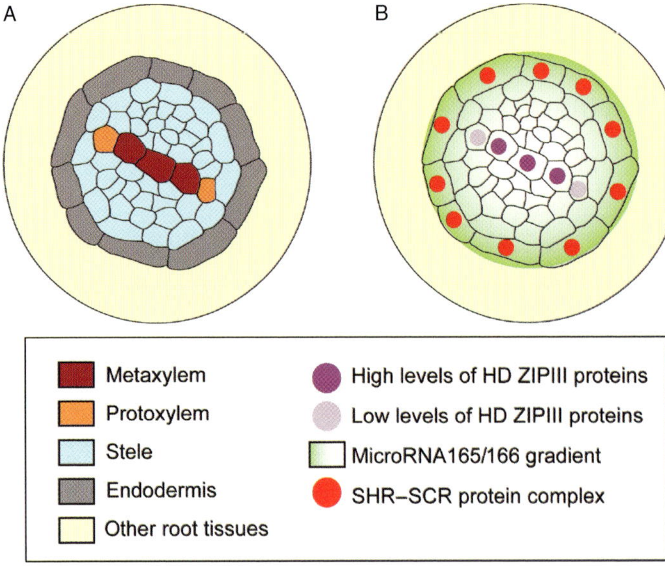


**Miguel A. Moreno-Risueno *et al.*, Figure 9.3** Non-cell autonomous specification of xylem cell types by SHORT ROOT and SCARECROW. (A) Schematic of metaxylem and protoxylem tissues in a transverse section of the Arabidopsis root. Note that the endodermis forms a concentric layer of cells, whereas protoxylem and metaxylem constitute a single row of internal cells surrounded by other cells of the stele. (B) Specification of protoxylem versus metaxylem involves movement of SHORT ROOT to the endodermis where, together with SCARECROW, activates the microRNA165 and 166. The microRNA165/166 then travels back to the stele forming a gradient that targets the transcripts of CLASS III HOMEODOMAIN LEUCINE ZIPPER (HD-ZIPIII) transcription factors. This generates different levels of HD-ZIPIII proteins that determine cell fate in a dose-dependent manner. Low levels of HD-ZIPIII specify protoxylem and high levels specify metaxylem.

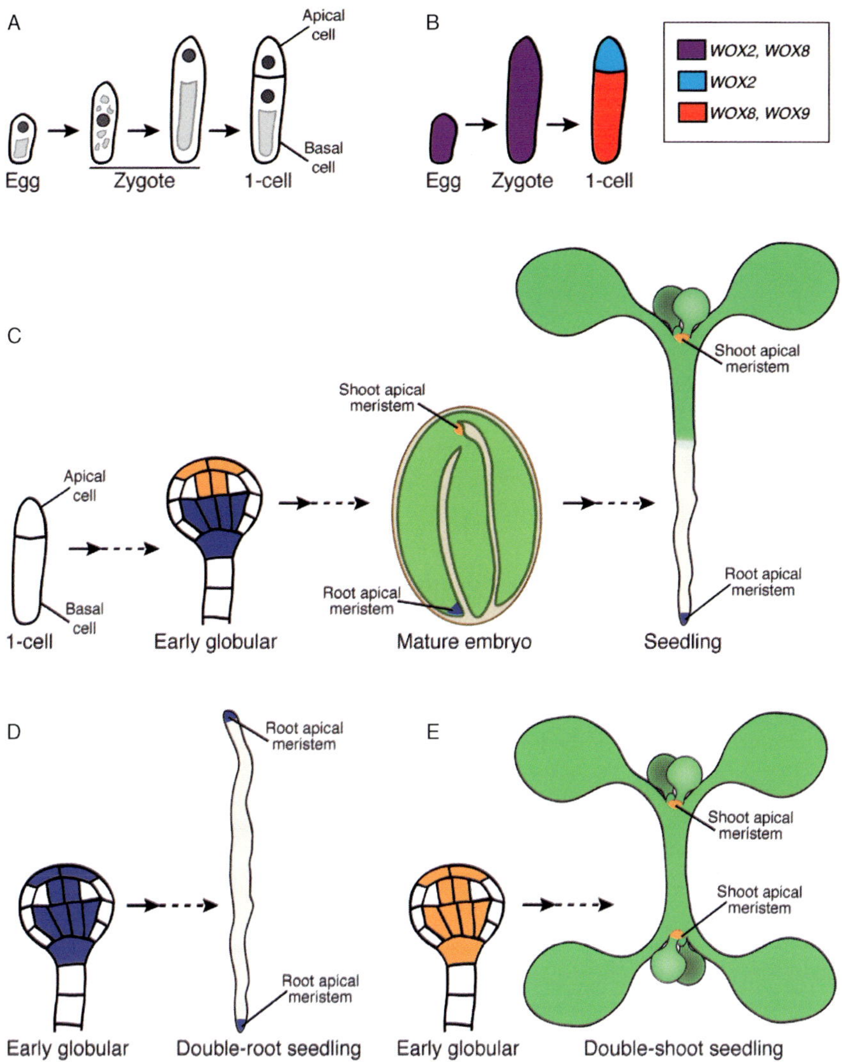


**Miguel A. Moreno-Risueno *et al.*, Figure 9.4** Apical–basal polarity and specification of meristem fate in the Arabidopsis embryo through establishment of distinct transcriptional domains. (A) Schematic of early embryo development from the unfertilized egg cell through the first zygotic division. (A, left to right) The egg cell is polarized with the nucleus (dark gray) at the apical end and a vacuole at the basal end (light gray). After fertilization, the zygote is transiently symmetrical, then it elongates and is repolarized. The first zygotic division is asymmetric and producing an apical and basal cell with distinct fates. Note that embryo stages are based on the cell number in the apical domain, thus the first zygotic division results in one-cell stage embryo. (B) Schematic of the *WOX2/8/9* gene expression patterns in the egg cell through the first zygotic division. (C) Schematic of developmental snapshots from the one-celled embryo to the seedling focused on specification of the root and shoot apical meristems. (C, left to right) Apical–basal polarity is established after the first zygotic division. By the early globular embryo stage, *PLETHORA*s (*PLT*s; blue) and the *HD-ZIPIIIs* (orange) expression is restricted to the domains that will give rise to the root (blue) and shoot (orange) apical meristems, respectively. In the mature embryo, these meristems are quiescent; however, become active after germination. The root meristem (blue) will give rise to all the cells of the primary root and the shoot meristem (orange) will give rise to all the cells of the aerial organs. (D) In the *topless-1* (*tpl-1*) mutant, the *PLT* expression domain has expanded into more apical regions and *HD-ZIPIII* expression is absent leading to a double-root phenotype. (E) Expansion of *HD-ZIPIII* expression into more basal embryo regions represses *PLT* expression leading to a double-shoot phenotype.

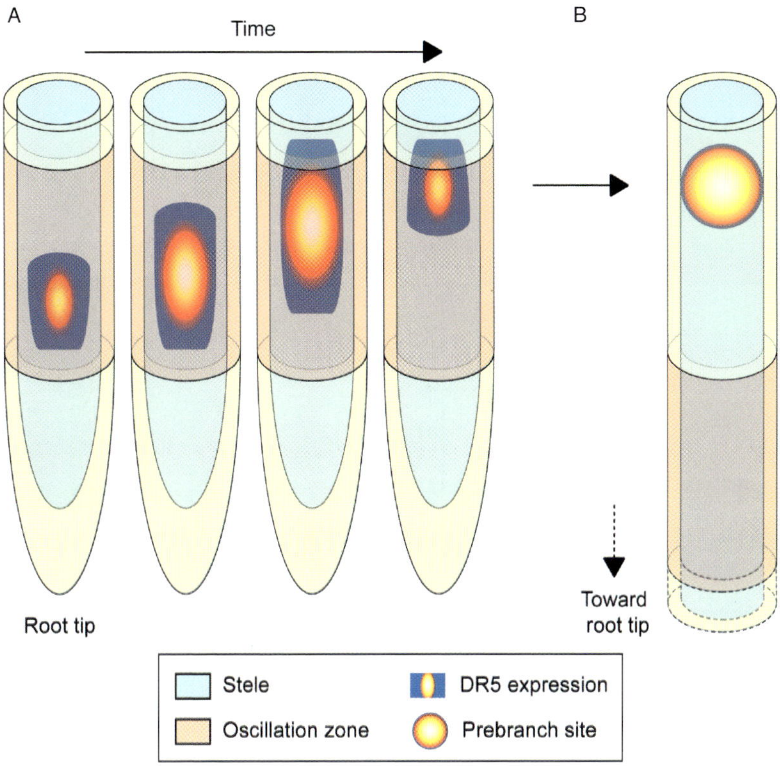


**Miguel A. Moreno-Risueno *et al.*, Figure 9.5** The DR5 oscillation marks position of future lateral roots through establishment of prebranch sites. (A) DR5 expression in the oscillation zone (OZ) increases and moves shootward over time. At the beginning of one oscillation cycle, DR5 is expressed at the proximal part of the OZ (rootward). Over time DR5 expression increases and moves toward the distal or shootward part of the OZ. At the end of the cycle, DR5 expression is turned off. A new oscillation then begins in the OZ. Similar expression patterns have been observed for genes oscillating in phase with DR5. (B) Following a DR5 oscillation in the OZ, a prebranch site is established. Prebranch sites are observed outside the OZ but their locations coincide with the region of the root where an oscillation was observed. Prebranch sites mark the position of future lateral roots.

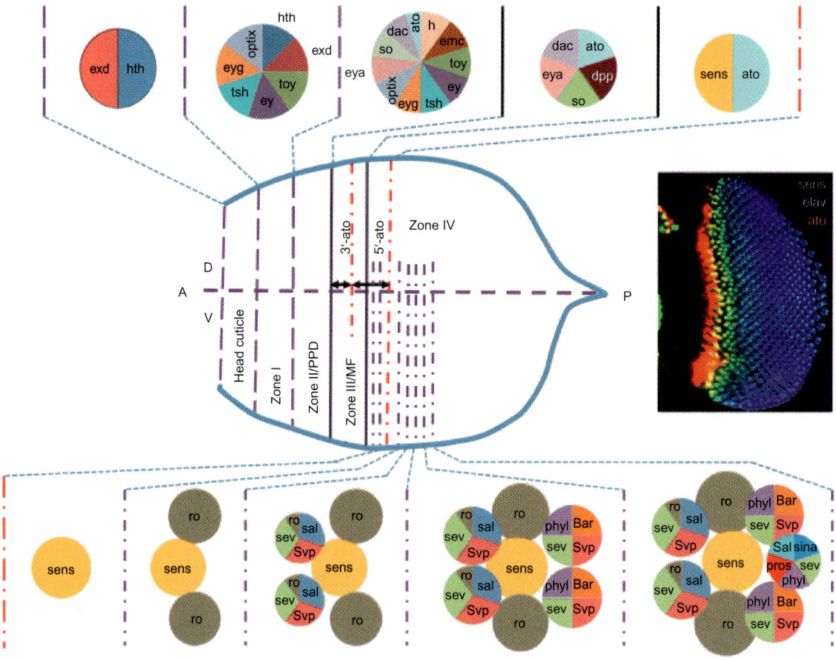


**Xiao-jiang Quan *et al.*, Figure 10.1** A generalized schematic representing the expression code of factors discussed in this review that have been found to be necessary for photoreceptor specification of *Drosophila*. The different zones corresponding to the head cuticle and successive developmental stages of the eye field are represented. Pie charts indicate the specific combinations of factors expressed at each stage. Subdivision of Zone IV corresponds to successive steps of photoreceptor R1–R8 formation (PPD, pre-proneural domain; MF, morphogenetic furrow).

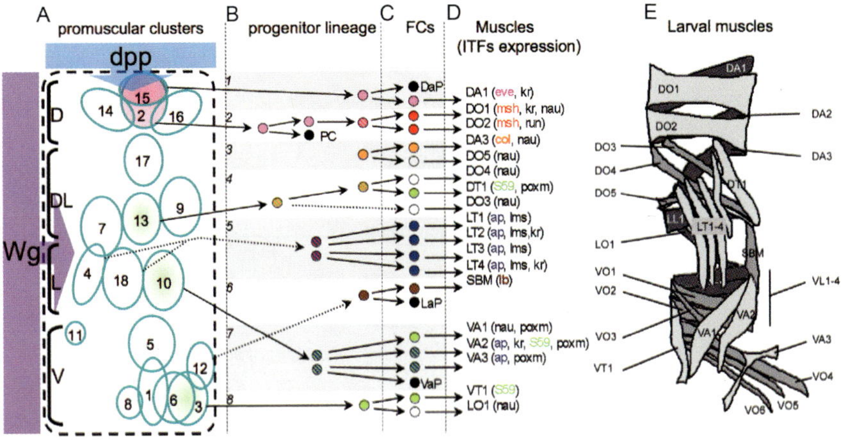


**Cyrille de Joussineau *et al.*, Figure 11.2** Selection of the FCs in an abdominal segment of the mesoderm. Most representative iTFs to determine the origin of the muscle founder cells are shown in the figure. (A) Early patterning by wg/dpp and others induce definition of four domains in the presumptive somatic mesoderm: D (dorsal), DL (dorsolateral), L (lateral), and V (ventral). To note (not shown), some iTFs could be implicated in this patterning before later role in FCs and muscles: Msh (red) is expressed in the ectoderm above D and L domains, ap (blue) in the lateral domain, Poxm in V and posterior L domains. Lateral inhibition occurs in the 18 promuscular clusters of l'sc expressing cells (blue–green circles) and selects progenitors for muscle founder cells (figure according to Carmena *et al.*, 1995). Cells of clusters 2 and 15 express eve (pink) as 1–4 cells expressing S59 can be seen in clusters 3, 10, and 13 (green). (B, C) Probable (plain arrows) or supposed (dotted) connections between promuscular clusters, progenitors (P), and founders cells (FCs): (1) clearly established lineage by expression of eve (Carmena *et al.*, 1998); (2) lineage established by early expression of eve (Carmena *et al.*, 1998) and then later expression of msh in P9/10 (progenitor for muscles 9 and 10, respectively, DO1 and DO2; Nose *et al.*, 1998); (3) col expression in P3/20 (Crozatier and Vincent, 1999); (4) conflicting data indicated a common origin for DO3 and DT1 based on S59 expression (Carmena *et al.*, 1995; Knirr *et al.*, 1999), but col is expressed in P18/19 (progenitor for DT1/DO4; Crozatier and Vincent, 1999); (5) transient msh expression in P21/22 and P23/24 (Nose *et al.*, 1998); constant expression of ap during lineage of all LTs suggests a lateral domain origin (Bourgouin *et al.*, 1992); (6) specific expression of lb in SBM, its founder and its progenitor (Jagla *et al.*, 1998). Experiments with misexpression of S59 suggest that the progenitor emerges from cluster 12 (Knirr *et al.*, 1999); (7) lineage determined by l'sc/S59 expression (Carmena *et al.*, 1995) and apterous expression (Bourgouin *et al.*, 1992); (8) deduced by S59 expression (Dohrmann *et al.*, 1990). (D) iTFs expressed in muscle fibers. Additional muscles (and their iTF expression, when described): DA2, LL1 (kr, nau), VL1, VL2, VL3 (kr), VL4, VO1, VO2 (kr), VO3 (run), VO4 (run), VO5 (kr), VO6. (E) Schematic representation of the larval somatic muscles (figure from Tixier *et al.* according to Ruiz Gomez and Bate, 1997). PC: pericardial cell; D, L, or Vamp: dorsal, lateral, or ventral adult muscle precursor.

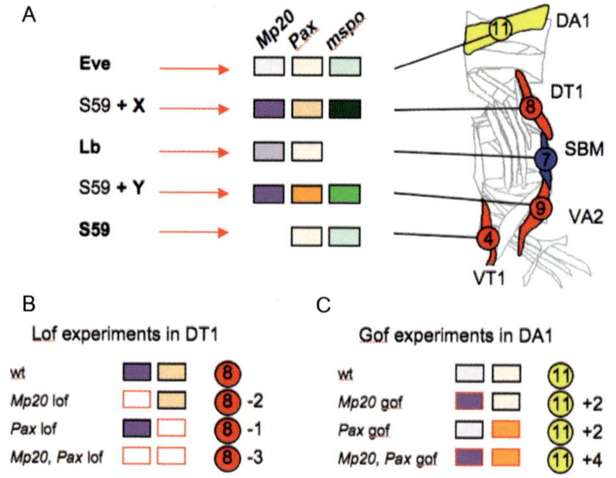


**Cyrille de Joussineau *et al.*, Figure 11.4** iTFs regulate the final number of fusion events by determining muscle-specific combinatorial code of target gene expression. (A) The identity genes *eve*, *lb*, and *S59* control muscle-specific fusion programs of DA1, DT1, SBM, VA2, and VT1 muscles. Corresponding number of nuclei is indicated on the scheme. Notice that *S59* alone is not sufficient to explain the DT1 and VA2 fusion programs and is proposed to act in combination with unidentified factors (X and Y). In each growing muscle, iTFs establish a specific combinatorial code of target's expressions. High, median, and low expression levels of *Mp20*, *Pax*, and *mspo* are represented by the color intensity. (B, C) Decreased (B) or increased (C) expression levels of realizator genes induce changes in final number of nuclei. Modification of the number of recruited nuclei is noted above each context. Because muscle expressing a low endogenous level of *Mp20* and *Pax* are not affected in loss of function (lof) contexts, results of lof experiments are illustrated for DT1 muscle in which the endogenous *Mp20* and *Pax* expression levels are, respectively, high and median. In lof contexts, muscles expressing a high level and a median level display a proportional loss of nuclei (i.e., −2 nuclei for high-level muscles, −1 nuclei for median-level muscles). The double loss of function shows an additive effect of realizator genes. Opposite effects are observed in gain of function contexts (C), as illustrated in DA1 muscle, which display low endogenous expression levels of *Pax* and *Mp20*. In this muscle, changes in expression levels of realizator genes induce an additive increase of nuclei number.